World Of Construction

McKNIGHT

World Of Construction

Donald G. Lux, *Ph.D.*

Willis E. Ray, *Ed.D.*

E. Keith Blankenbaker, *Ed.D.*

William Umstattd, *Ed.D.*

Industrial Technology Education
The Ohio State University
Columbus, Ohio

McKNIGHT
Publishing Company
Bloomington, Illinois

FIFTH EDITION

Lithographed in U.S.A.

SBN: 87345-406-5

Library of Congress
Card Catalog No. 81-83071

Editor: Carole Fletcher

Copy Editing: Joan Dunbar
Production: Robert West
Layout: Elizabeth Purcell
Design: Coventry Creative Graphics
Elizabeth Purcell
Art: Stan Crum
Howard Davis
Richard Draney
Photography: Earl Dunbar
Lawrence Funk
Daryl Littlefield
Barry Moline

Foreword

For a short but important period of your life you will become part of the construction industry — when you use this book, *The World of Construction*. You will learn *about* the industry. You will actually perform many of the same jobs that are done by the skilled men and women who build our office buildings, homes, highways, power plants, bridges, churches, playgrounds, missile sites, warehouses, water and sewage treatment plants, and industrial projects. You will quickly understand why more than four million Americans are proud of the work they do in construction. And you will see why this two hundred and thirty billion dollar industry is the envy of the world.

The construction industry thrives in America's free enterprise system. This economic system allows us to start almost any kind of business we wish and to compete on an even footing with anyone else. Thus we have construction firms that build everything from one-car garages to thousand-family apartment houses, from a simple garden patio to a massive airfield. There are 30,000 firms who are members of the Associated General Contractors of America. These firms compete with each other for construction contracts: to be successful they must do their work well and continually try to improve their construction capabilities. These firms are managed by contractors, who invest their time, their abilities, and their money in America's growth. This great industry depends on the talents, skills, and hard work of many people. Millions of skilled workers — carpenters, electricians, ironworkers, and others from many different trades — work together to build on empty spaces. Very important to this labor force are the engineers and architects, who work with contractors and skilled workers, designing and helping build complex and inspiring structures.

Everyone must work together. The construction industry is characterized by cooperation and trust. *Cooperation* is necessary between skilled workers who must complete their work in order for others to start theirs, as carpenters follow surveyors and painters follow carpenters. Cooperation is also necessary among those who are working side by side on a project — often under different contracts, for different companies, and with different bosses.

The *trust* on which this industry is based is evident in many places. Owners trust the contractors to follow the building codes and to use appropriate and safe materials. Both materials and work may be hidden from view as the job progresses, so trust must go beyond inspection. Contractors trust employees to do good work and not waste materials. The safety and comfort of many depend on the honest efforts of all workers in construction. And pride in the completed project is shared by every planner and "hard hat" who had anything to do with its construction.

Yes, you are about to take part in America's pride, the construction industry. As a representative of this industry, I wish you success in learning about construction work — the skills, the cooperation, the trust, and the pride in doing a difficult job well.

Hubert Beatty

Hubert Beatty
Executive Vice President
Associated General Contractors of America
1957 E Street, N.W.
Washington, D.C. 20006

Preface

Imagine yourself on a remote mountaintop. All that you see below you is trackless wilderness — trees, rocks, perhaps a raging stream of water. This is the natural world.

Imagine yourself on top of the tallest skyscraper. Below you are buildings made of the lumber from trees, stone from the earth, or materials formed from other natural resources. Streets and highways thread paved paths through the city and across the countryside. A stream of water, perhaps controlled by some distant dam, is spanned by bridges. What you see is the constructed world. People have changed nature to serve their needs better. One of the major means of making the changes was construction technology (constructing). This program, *The World of Construction*, is about constructing.

You don't need to be atop a skyscraper to see the importance of construction. Look around you right now. Every day you and everyone in your community uses constructed products. These products very much affect the quality of your life. Safe streets, efficient schools, convenient playgrounds, and low-cost utilities — these don't just happen. Building them or having them built requires knowledge of construction technology. Using this knowledge helps to maintain and improve the quality of life.

You will gain knowledge of construction technology from the words and illustrations in this book. But complete learning requires practical experience as well. Laboratory practice will help you know construction technology more fully. In this program, you will read about activities in the construction industry, and you will actually perform many of these activities in class.

Some of you will use what you learn to find a life's work in the construction industry. Many more of you will use it when servicing your property. All of you can use this knowledge to make better personal and community decisions about what should be built.

The study of construction technology is important. What you learn will be useful to you. And learning it in *The World of Construction* will be exciting and interesting as well.

—The Authors

Table of Contents

Reading

Reading

Reading 1

People and Technology

Technology

People have lived on Earth for millions of years. The first people used only the natural things around them. As time passed, people learned to change these natural resources to serve their needs better. For example, early people changed stones into *tools*. Some stone tools were used to kill wild game for food. Others were used for cutting and shaping.

People found more ways to improve their lives. Fire was discovered. People could cook food and stay warm in winter. They learned to make shelters. See Fig. 1-1. They developed languages and began to live in groups. Knowledge of how to do things better and more easily grew. It was the beginning of civilization (a higher level of living). It was the beginning of technology.

Fig. 1-1

Over the centuries, people have used technology to improve their homes.

Fred E. Mang, Jr. — U.S. Department of the Interior

A. *These ancient Indian cliff dwellings were built between the 11th and 14th centuries.*

C. *A Victorian house, built in the late 1800s.*

B. *Log cabins such as this were common around the time of the Civil War.*

D. *A pleasant modern home.*

Thousands of tools have been developed since early people used stone tools. As new kinds of tools were made, people could do more kinds of work. As tools were improved, work could be done with less effort. This freed time and energy. People had more time to think of new ways of doing things. Through tools, people developed technology.

Technology is thoughtful and skillful doing. It is understanding how to use tools well. It is knowing how to get work done. Technology includes welding metal, sawing wood, and mixing concrete. But it is more. Actions such as reading drawings and hiring workers are parts of technology, too. They don't involve using tools. But they require thoughtful and skillful doing.

Technology, in itself, is neither good nor bad. It is how we choose to use technology that makes it good or bad. For example, technology has improved the way we live and work. But it has also caused problems. Most of the problems resulted from ignorance (not knowing). See Fig. 1-2. We must learn about technology in order to use it well. Proper use of technology is caring how it affects others.

Technology is often thought to be a part of science or science put into practice. Actually, it is a knowledge of its own. Technology and science serve each other. The theory (scientific idea) of the laser (a beam of intense light) is an example. It led to improved surveying instruments. These are used for measuring land. On the other hand, building early flying machines led to the development of several sciences that relate to flight. Aerodynamics is one of them. Technology, once used, tends to lead to further science and technology. People continue to search for better ways of doing work.

Words that describe technology end in "ing." There are many of these words. Sanding, drawing, and nailing are examples. These words identify ways of doing things. They describe human work. They describe technology.

Technology can be subdivided (put in classes) many ways. In this reading, we will examine three levels or classes of technology.

- Economic technology
- Industrial technology
- Construction technology

Economic Technology

Years ago, people worked all day just to meet their basic needs. As technology was developed, they could get work done faster. They could make products more quickly. Some people made more products than they could use. They needed a way of sharing. They needed an *economic system*.

People began to trade their surplus (extra) products. But it was difficult to place a value on products when trading for other products. An economic system using money was developed.

Goods and services are produced and traded within an economic system. See Fig. 1-3.

Goods are material things such as houses, radios, and milk shakes. Some goods can be found in nature. These are called *raw materials*. Some raw materials can be extracted (taken) from nature once, such as by mining. Gravel and diamonds are examples of these. Other raw materials can be grown. They can be reproduced (the same kind made) again and again, such as by farming. Corn and trees are examples of these.

Fig. 1-2
Smoke and dust from industrial plants may pollute the air. Industry is continuing to develop better ways to control air pollution.

Phillips Petroleum Company

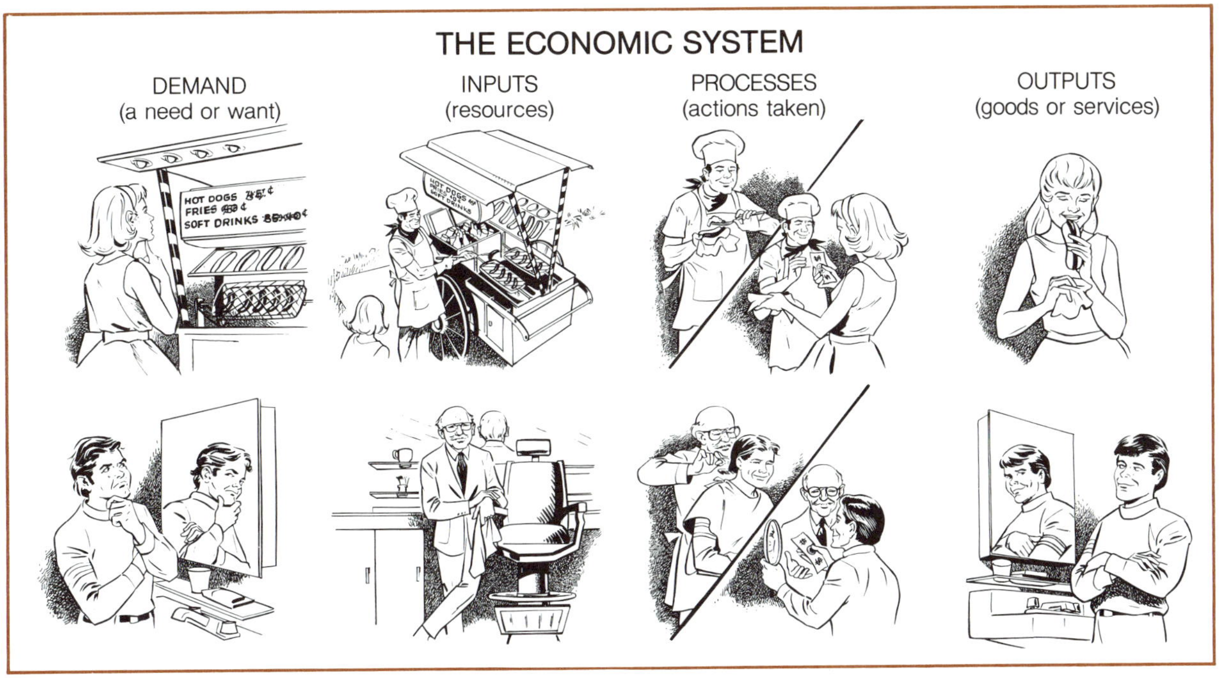

Fig. 1-3
Our economic system provides people with the goods and services they want.

Services are such things as haircuts, bus rides, and bike repairs. Usually, goods are needed to provide services. For example, tools are used to fix bicycles.

Economic technology is the special knowledge of the workers in the economic system. It makes goods and services available to us. Economic technology involves knowing

- what products to make and how many.
- what kinds of work to do, and the best ways to do the work.
- how much money to spend to produce the products.
- what resources to use.

Economic technology makes it possible for people to be more productive. This means that people can get more work done in less time. For example, one person can add just so many numbers in one hour. With a calculator (a result of technology), more numbers can be added more quickly. And a computer can be used to add many more numbers in a flash.

Technology makes tasks easier. Work is done more quickly and with fewer errors. Because of economic technology, we have more goods and services, and they are better.

Industrial Technology

The form of raw materials or goods can be changed to make them more useful. This also makes them more valuable. Materials can be made into manufactured goods or constructed goods. *Manufactured goods* are movable products. Cars, radios, and TV dinners are examples. *Constructed goods* cannot be moved. These products are fixed upon a *site* (location). These include homes, airports, and bridges. Most of the goods we use are manufactured or constructed.

Industry is the part of the economic system that changes the form of materials and goods. *Manufacturing* and *construction* make up industry. About one-third of the labor force (people working for pay) in the United States is employed by industry.

Products made by manufacturing and constructing are known as *industrial goods*. These goods satisfy (provide for) our many needs and wants. The knowledge and skills that make this possible are called *industrial technology*.

In both manufacturing and construction, tools and machines are used to perform *processes*. Processes are actions taken to change materials into products. For example, gold is made into jewelry, a

Betty Medsger

Fig. 1-4
Millions of people work in the construction industry and in related fields. Each worker uses special skills and knowledge to perform an important job.

manufactured product. Trees are changed into lumber, also a manufactured product. Lumber is used to build a house, a constructed product.

Construction Technology

You are surrounded by constructed products. Stores, office buildings, factories, and roads are some of these products. Knowing how to build them requires a vast body of knowledge called *construction technology*.

Construction technology involves *managing*. People, time, work, and materials are managed. Construction technology involves using the knowledge and skills of people. Workers are hired, trained, assigned to jobs, advanced, and retired (*personnel practices*). Construction technology also involves *producing*. This includes all the processes needed in building and servicing constructed products.

Construction technology affects where and how you live. It is important to know about it. Construction knowledge is made up of many smaller parts. Each part can be studied. But a full understanding of construction technology requires doing. For example, knowing how to draw plans for a building is part of construction technology. Another is knowing how to dig a basement.

You need thoughtful practice in addition to reading about technology. In this course, you will learn what construction is all about. You will also develop skills through experience. Through this knowledge and experience, you will learn to build and fix things for yourself. You will understand how things around you were constructed. What you learn may even help you choose a career. See Fig. 1-4.

Fig. 1-5
The many constructed works around you all have common parts.

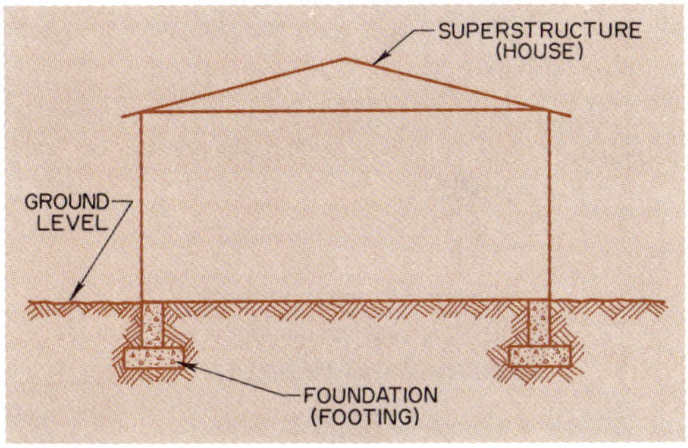

Soil Testing Services

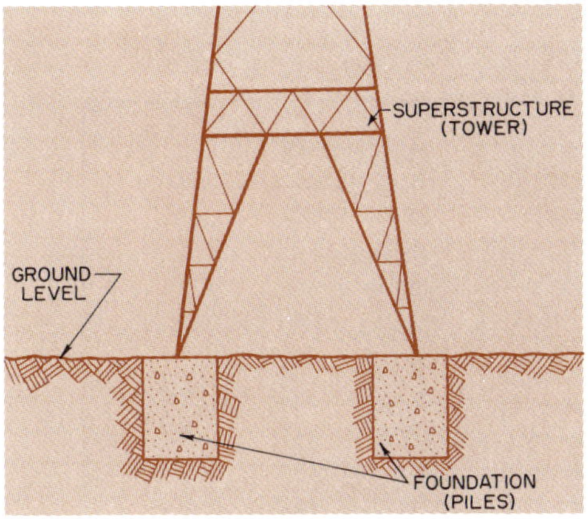

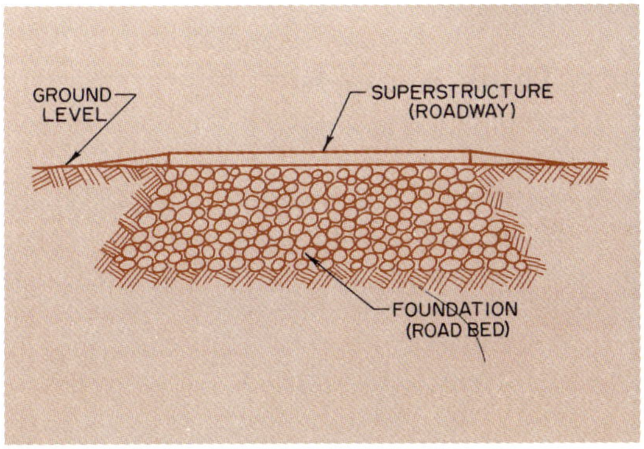

Products of Construction

All constructed products are *structures*. Each has a part that rests upon the earth. This is the *foundation*. Each structure also has another part that rests upon the foundation. It is called the *superstructure*. Together, they make a structure. See Fig. 1-5.

Many kinds of structures are built. Dams, skyscrapers, and tunnels are examples. All structures are built to satisfy human wants and needs. They make up the constructed world. Constructing can make living safer and easier. But it could make life unpleasant and more difficult. Thoughtful and skillful constructing (construction technology) makes the difference.

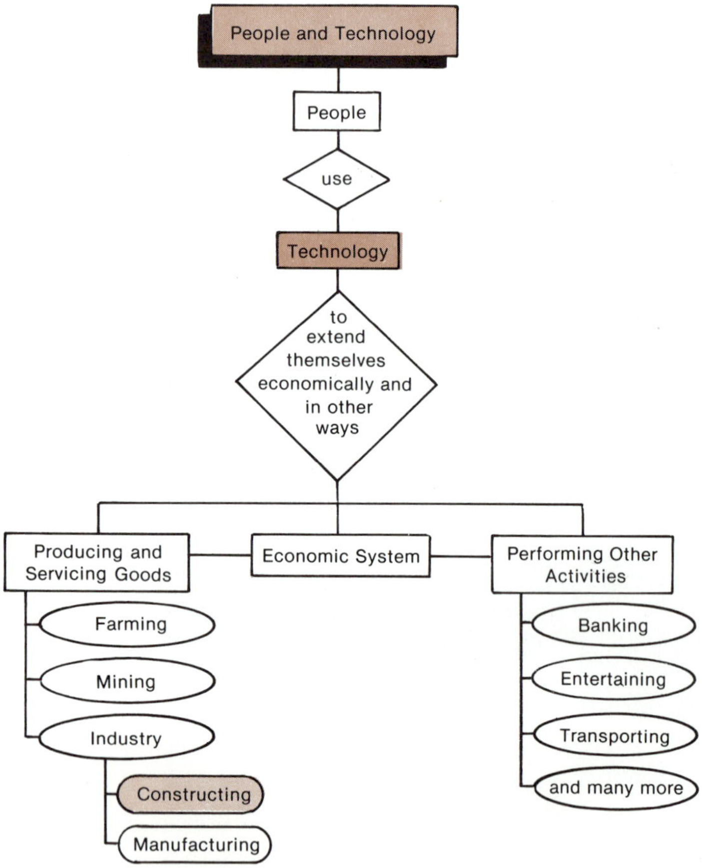

Summary

People have spent centuries developing the technology that frees them from hard labor. Technology is thoughtful and skillful doing.

Technology has many parts. Important among these are economic technology, industrial technology, and construction technology.

Economic technology provides us with the goods and services we need and want. Industrial technology involves the many processes by which products are made. Products may be manufactured or constructed. Construction technology is managing, affecting people through personnel practices, and producing and servicing structures.

Human-made structures that rest upon the earth are constructed products. They make up the constructed world in which we live. Learning about this world helps us make it a more pleasant place in which to live. That makes this study of construction important.

Terms to Know

construction
constructed goods
construction technology
economic system
economic technology
foundation
goods
industrial goods
industrial technology
industry
managing
manufacturing
manufactured goods
personnel practices
processes
producing
raw materials
services
site
structures
superstructure
technology
tools

Think About It!

1. What is technology?
2. How does industry serve people?
3. Identify the foundations and superstructures of each of the following:
   a. Bridge
   b. Dam
   c. Highway
   d. School building
   e. Utility pole

Reading 2

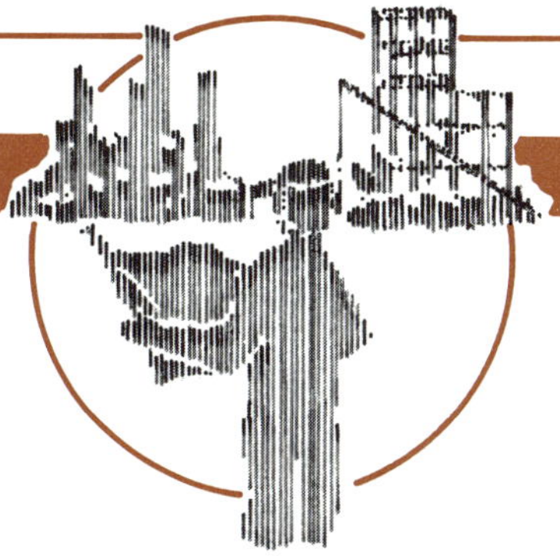

Managing Construction

Managing

Managing is often defined as "getting work done through other people." Managing helps people to work toward a common goal. When people and jobs are managed properly, tasks are done efficiently. Workers know what to do. They know how, when, and where work must be done. Time and money are not wasted. Directing the work of others requires knowledge and skill. It is part of technology.

Managing involves

- planning,
- organizing, and
- controlling.

See Fig. 2-1. These tasks do not have to take place in this order. They often occur at the same time. Managing is done at all levels of work, by many types of *managers*.

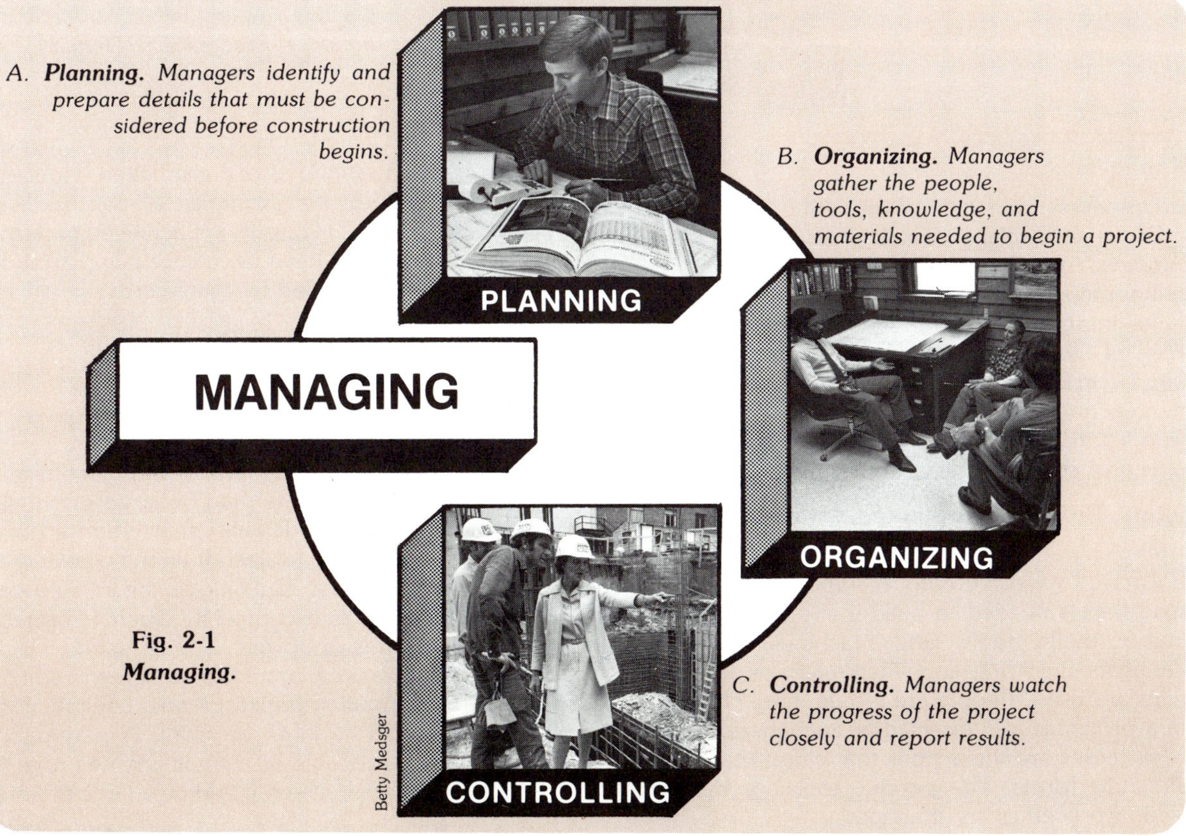

Fig. 2-1
Managing.

A. *Materials, workers, and equipment are some of the **inputs** needed to build a structure.*

C. *The **outputs** are the result of effective use of all inputs and processes.*

B. *Workers use their skills with tools to perform special **processes.***

Fig. 2-2
Inputs + Processes = Outputs

Managers

Top-level managers hire other management employees to plan, organize, and control work. Top-level managers may be company owners. They make major decisions. Middle-level managers also make important decisions. They often direct the work of "first-line" managers. *Superintendents* or *job supervisors* are first-line managers. They manage the production workers. In construction, *production workers* build the structures. They use tools and equipment to do their jobs.

Inputs, Processes, Outputs

Managers in construction manage inputs and processes to achieve outputs. *Inputs* include everything that is needed to build a structure. Cement, lumber, nails, tools, and machines are inputs. Managers direct how and when materials and equipment are used. People are inputs, too. Their work on a project is also directed by managers.

Processes are the actions that cause materials to take new forms. The people, tools, or machines used in processes are managed.

Outputs are the products of construction. Houses, roads, and bridges are examples of construction outputs. See Fig. 2-2. Outputs are achieved because inputs and processes have been carefully planned, organized, and controlled by managers.

Construction Projects

There are two basic kinds of construction projects. One kind is publicly owned. The other is privately owned.

Public projects are planned to satisfy community needs. These projects may be large or small. Constructing a new sidewalk near your school would be a small public project. It would cost only a few thousand dollars. Building a new highway would be a large public project. Such a project would cost millions of dollars.

Private projects are planned to meet the needs of individuals or small groups. Private construction projects may also be large or small. A new patio beside your house would be a small private project. A skyscraper office building could be a large private project.

Fig. 2-3
Construction initiators imagine a finished project on a particular site.

Public and private projects get started in different ways. However, all projects have an *initiator*. This person sees the need for construction and acts on it. All construction projects are planned to meet human needs. See Fig. 2-3.

Planning Construction

Planning a construction project begins many months or years before building starts. *Planning* involves four major activities.

- Formulating
- Researching
- Designing
- Engineering

See Fig. 2-4.

At the start of a project, many questions must be answered. Should the project be built? Can it be built? Do people need the construction project? Will the project add to the well-being of the community?

*A. **Formulating** is done early in planning. It may be done by a group or by individuals.*

*B. **Researching** may involve testing construction materials.*

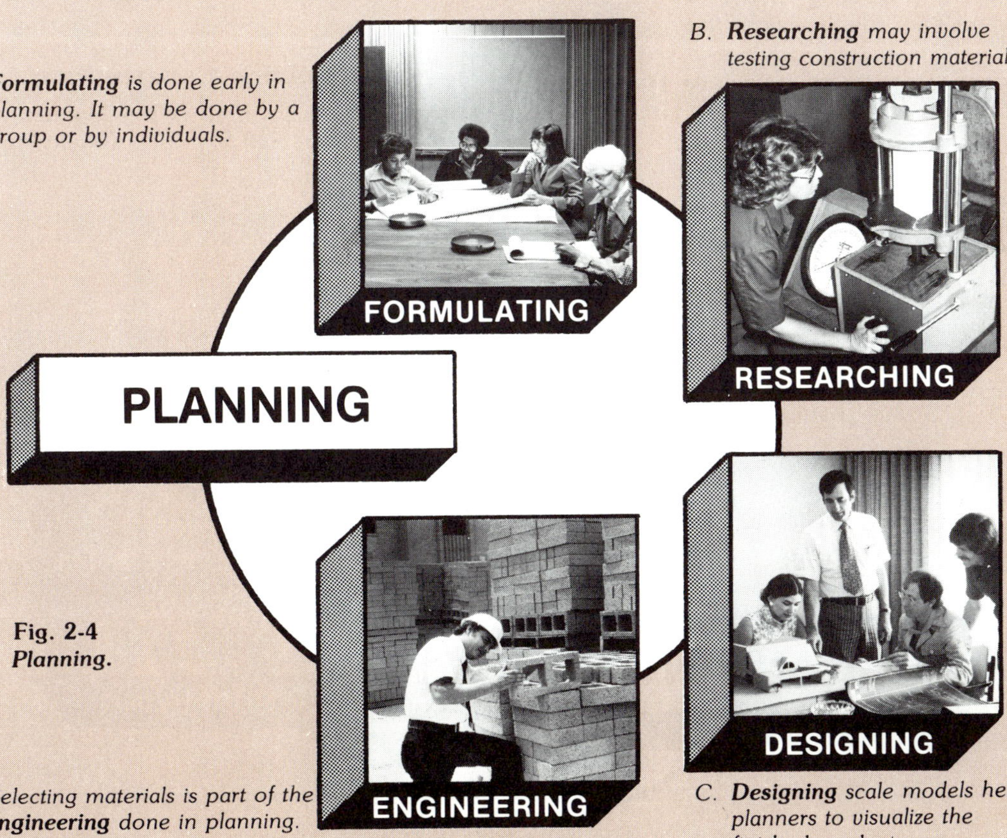

Fig. 2-4
Planning.

*D. Selecting materials is part of the **engineering** done in planning.*

*C. **Designing** scale models helps planners to visualize the finished product.*

Will it make money for the builder? Will it meet the needs of the owner?

Formulating

Answers must be found to such questions. From the answers, the planner can make decisions. Then, from the decisions, the planner sets goals. This beginning process is called *formulating.*

Researching

Project planners must gather facts that will help them make decisions. This information is gathered by *researching.* Some researching is done at the construction site. An example is finding the hardness of soil. This information helps planners determine the kind of foundation needed.

Researching may be done at research centers, universities, or private companies. Often, construction materials are tested. For example, some tests determine the strength of steel or wood beams. Beams are structural members that support much of the weight of a building. Loads are placed on beams to find how much weight they can safely support.

Some researching is done by people who design the project. Finding the location of buried utility lines on a map is an example.

Designing

Designing is creating a plan for construction. It involves finding what purpose a structure is to serve. That is, how will it be used? This is the *design problem. Architects* or *engineers* work to find solutions (answers) to the design problem. Sometimes they work together. Several solutions are created. Sketches (rough drawings) or models may be made. The best design is improved.

Engineering

All the details must be worked out for building the structure. This is the *engineering* part of planning. Engineers, drafters, and estimators work with engineering details. Engineers do most of the planning. For example, they choose the materials that must be used in building the structure. *Drafters* make detailed drawings from sketches and models. Costs are determined ahead of time by *estimators.*

Scheduling is also done. This is determining tasks and the order in which they will be performed. A construction project must be planned from beginning to end.

Organizing Construction

The second major kind of management work is called *organizing.* See Fig. 2-5. Organizing means

A. ***Structuring.*** *Contractors manage by identifying work to be done and the people and materials needed to do it.*

B. ***Supplying.*** *Contractors order supplies, schedule equipment, and hire workers. This is another part of managing.*

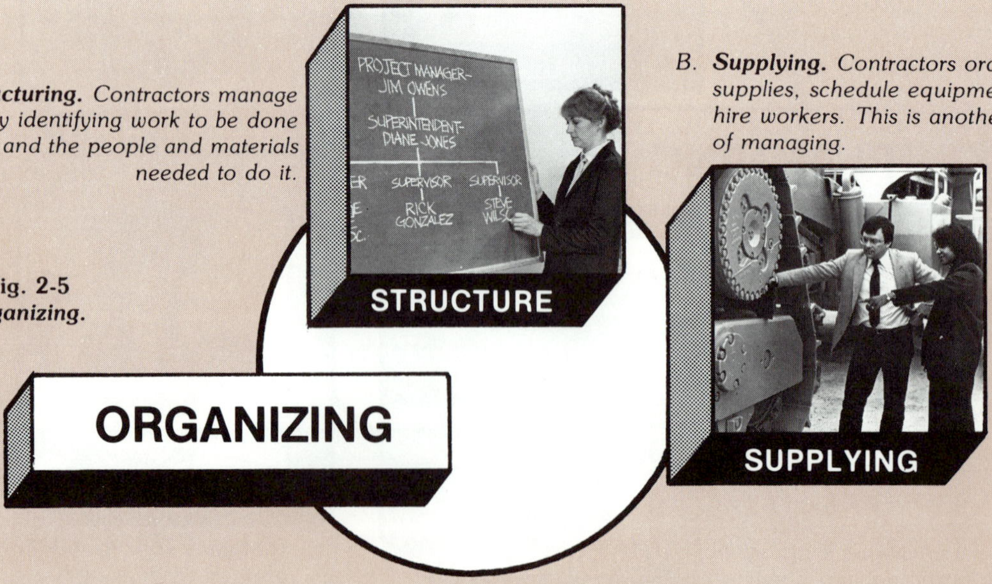

Fig. 2-5
Organizing.

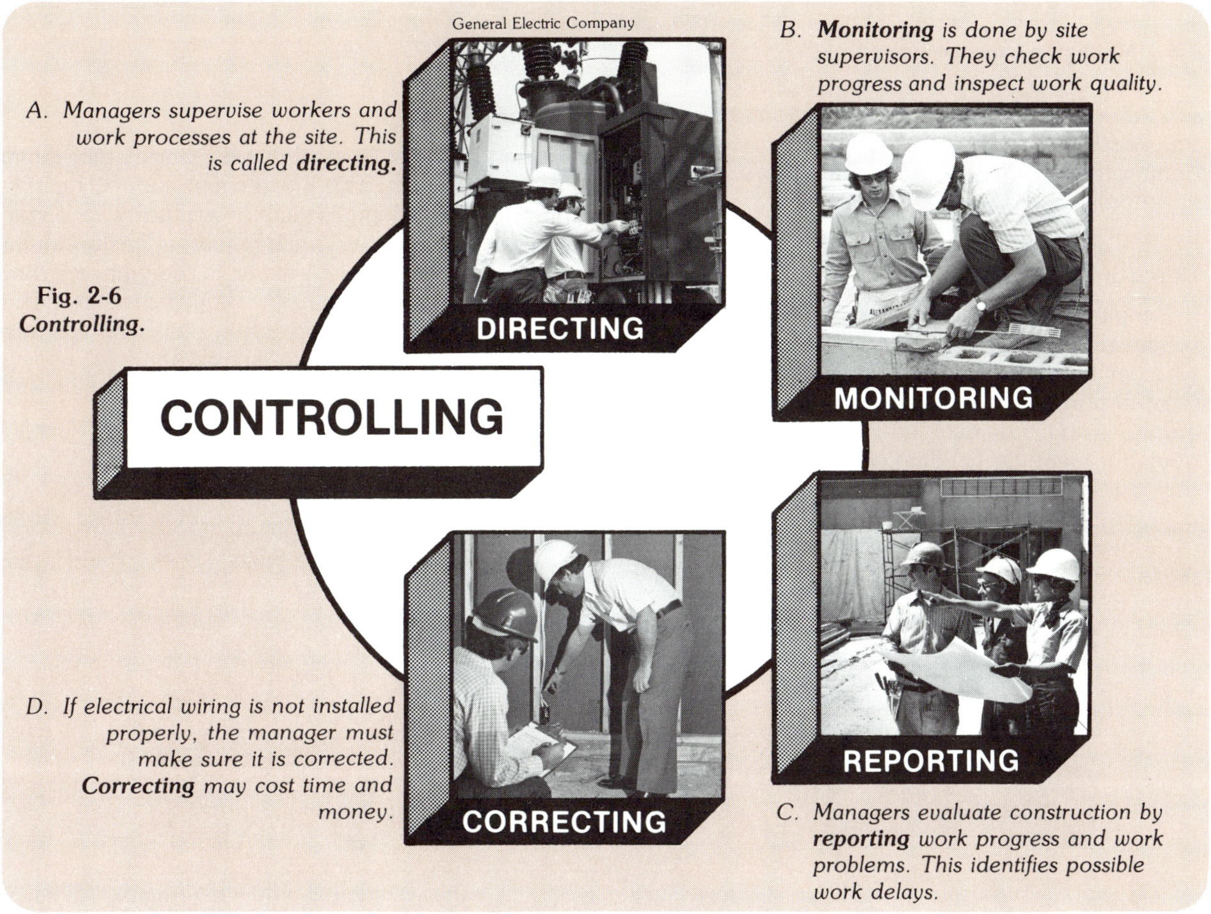

Fig. 2-6 *Controlling.*

A. Managers supervise workers and work processes at the site. This is called ***directing.***

B. ***Monitoring*** *is done by site supervisors. They check work progress and inspect work quality.*

C. Managers evaluate construction by ***reporting*** *work progress and work problems. This identifies possible work delays.*

D. If electrical wiring is not installed properly, the manager must make sure it is corrected. ***Correcting*** *may cost time and money.*

putting in order. It involves *structuring* (arranging the parts of) an organization. Work activities must be determined and planned. Construction *contractors* (builders), for example, must structure or form their own companies. They must decide what the duties will be and what managers and workers will be needed.

Next, contractors obtain all needed equipment and supplies and hire the people. This is called *supplying* the organization. The company may own many tools and machines. It may rent others. Materials are bought from suppliers such as concrete plants and lumberyards.

Some work may be *subcontracted.* This means that a specialized company may do parts of the work. Plumbing and electrical work, for example, are often subcontracted.

Controlling Construction

Controlling involves four major managing activities.

- Directing
- Monitoring
- Reporting
- Correcting

See Fig. 2-6.

Directing

Workers must know what they are to do. Supplies must be ready for use when workers need them. Coordinating people and supplies is called *directing.* Directing helps get work done efficiently. Job supervisors are the key workers. They direct people working with tools, equipment, and materials.

Monitoring

Monitoring is closely watching the work being done. Job supervisors monitor the construction work. Architects and engineers also keep a constant check on the work. *Building inspectors* examine the work to be sure that all *building codes* (rules) are being met. People who take inventory of (count) supplies and materials are monitoring. The timekeeper is a monitor. Monitoring may also be done with machines. A time clock is a machine used for monitoring. Monitors make sure that work is going according to plan.

Reporting

People in charge must know how the work is progressing. *Reporting* lets these people know whether things are going as planned. Reporting is most important when something is wrong. For example, inspectors may find that electrical wiring does not meet the building codes. They will report this to the site superintendent.

Correcting

Changes must be made when work is not going according to plan. This is *correcting*. Correcting may involve minor changes such as replacing damaged lumber. Or, it may involve some costly major changes. For example, architects or engineers may need to redesign some part of the structure. But when work is planned carefully, few changes are necessary. Good planning is important!

Personnel Technology

There are special ways of managing workers in construction. People manage other people to help

Fig. 2-7 ***Personnel technology*** *is managing workers.*

A. *Construction managers are in charge of* ***hiring****. They hire people who will best meet the company's needs.*

B. *The manager arranges on-the-job* ***training*** *for new workers.*

C. *Managers oversee employees as they are* ***working****. They make sure conditions are safe and work is done efficiently.*

D. *Construction managers see that workers who do a good job are moved into better positions, usually with more pay. This is called* ***advancing****.*

E. ***Retiring*** *workers is part of managing. Workers are honored for many years of service with the company.*

them work safely and efficiently. Managers try to make sure workers are treated fairly. They want them to be happy with their jobs. Managers are concerned mainly with five personnel practices.

- Hiring
- Training
- Working
- Advancing
- Retiring

See Fig. 2-7.

The technology that helps us form materials is very exact. But *personnel technology*, the technology of getting workers to cooperate, is not. Sometimes it is hard to find the best ways to encourage people. Still, managers try. Managers want to help workers to do their best.

Summary

A successful building project requires the cooperation of many people. Owners, architects, engineers, supervisors, and workers must all work together. Good management helps to accomplish this.

Construction is much easier when the project is planned carefully. Workers know what to do. They also know how and when to do it. People in construction work to meet human needs.

Terms to Know

architects
building codes
building inspectors
contractors
controlling
correcting
designing
design problem
directing
drafters
engineering
engineers
estimators
formulating
initiator
inputs
job supervisors
managers
managing
monitoring
organizing
outputs
personnel technology
planning
private projects
processes
production workers
public projects
reporting
researching
scheduling
structuring
subcontracted
superintendents
supplying

Think About It!

1. Suppose you are to have a class party at school. What planning, organizing, and controlling tasks would be done by each of the following committees?
   a. Food committee
   b. Decorating committee
   c. Entertainment committee
2. If possible, identify a public construction project and a private construction project presently underway in your community. What are the inputs, processes, and outputs of each project?
3. Managers must solve personnel problems. Some problems may need several solutions. Suggest at least two ways to handle each of the following problems.
   a. A worker who is very friendly wastes time talking.
   b. Several workers do not want to wear their hard hats.

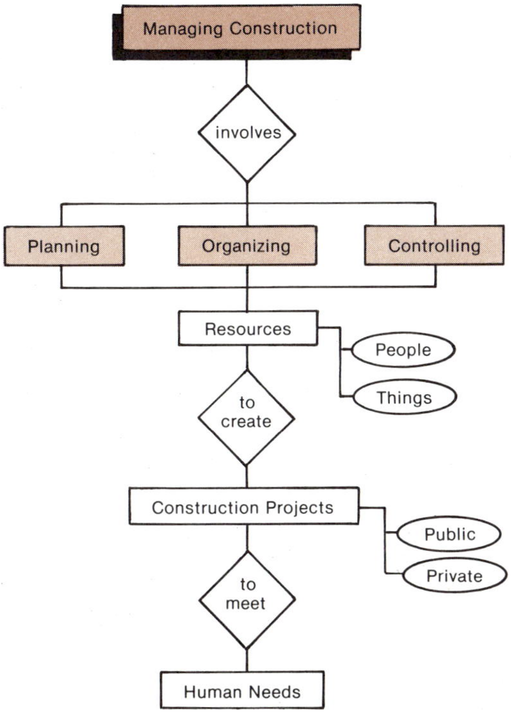

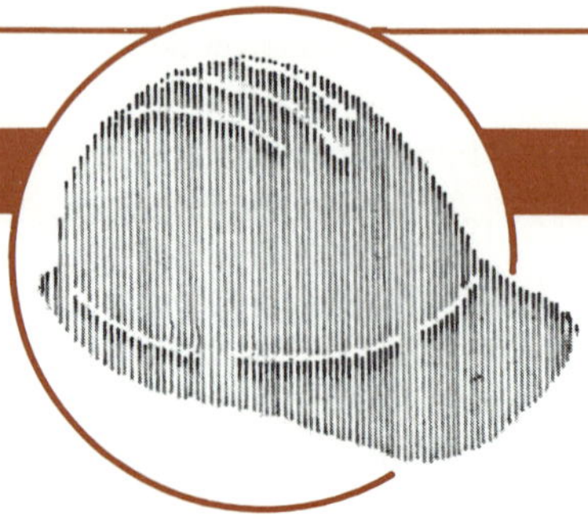

Reading 3

Producing Constructed Works

Producing by Constructing

Construction is an *industry*. It is a managed production system. The system exists to produce and service goods. In the process, people's needs are met. See Fig. 3-1.

Construction planning begins with needs. Project designs are created to satisfy needs. Construction costs are estimated. The work is organized. *Controls* are set up to ensure that project plans are followed. Contracts are made. Only then does actual production of projects begin. There may be a long wait between the time a need is identified and the time it is met.

Fig. 3-1
Construction can make an area more useful and attractive. Here, old structures were torn down and a high-rise apartment house and small park were built.
John D. Cook Associates

Production

Producing goods means changing materials to increase their *value*. The knowledge of how to do this is called industrial production technology. All *raw materials* are taken from nature. Industrial production changes raw materials. It changes them into things that have more value for people.

Manufacturing is done in a plant (factory). Manufactured products are shipped to the locations where they are used. See Fig. 3-2. Constructed

Fig. 3-2
These special heat-resistant bricks are manufactured. They will be shipped to a site, and will become part of a constructed product.
United States Gypsum Company

Tyler Corporation E.I. DuPont DeNemours and Company

Fig. 3-3
Blasting is a construction practice.

products (projects) are built where they are used. Construction involves *assembling* raw materials and manufactured products on a *site*.

Construction production and manufacturing production are similar. Both involve changing materials to increase their value. However, the production practices differ. For example, earthmoving is a production practice of construction. It is not used in manufacturing. See Figs. 3-3 and 3-4. Can you think of other examples?

Construction production practices may be put into three divisions. *Preprocessing* is getting ready to produce. *Processing* is producing. *Postprocessing* is work done after processing.

Preprocessing Materials

Before materials can be processed, they have to be prepared. Preparing materials for processing is called preprocessing.

Preprocessing involves

- receiving,
- handling,
- storing, and
- protecting.

These practices also go on throughout production.

Receiving is done at the site. There may not be room for everything at once. Therefore, materials arrive at the site at different times. They are scheduled to arrive as they are needed.

Fig. 3-4
Pile driving is another construction practice. Piles are driven into the ground to help support a structure.

Fig. 3-5 Wire Reinforcement Institute
Materials may be stored at the construction site. Large pieces of steel mesh are placed in stacks. They can be obtained quickly when needed.

Handling is done as materials are received and counted. They are moved to where they will be used.

Sometimes *storing* is done at this time. Workers may use materials one part at a time. The materials are stored until needed. See Fig. 3-5.

Loss or damage of materials increases project costs. Higher costs usually mean lower profits for

A. **Separating** plywood for subflooring.

B. **Combining** concrete blocks using mortar. This forms a strong wall.

C. **Forming** sheet metal into ducts.

Fig. 3-6 *Materials must be processed in order to make a constructed product.*

the construction company. Therefore, *protecting* is done. Materials must be protected at all times.

Processing Materials

Processing makes materials more valuable. Materials are processed in three basic ways. The ways are

- separating,
- combining, and
- forming.

Separating takes away material. *Combining* adds material. *Forming* changes the shape of material. See Fig. 3-6.

For example, take a piece of paper in your hands. What can you do to change it? You can rip (separate) it into pieces. You can staple two pieces of paper together (combine). Or, you can fold (form) the paper.

Often all three operations are performed on a material. For example, plywood is used in building forms for concrete. The materials are cut to length (separated). Then they are fastened together (combined). They may also be bent into a curved shape (formed).

Postprocessing Materials

A construction project begins its useful life when initial processing ends. Further work may be done after processing. This is called postprocessing. More commonly, it is called *servicing*. Postprocessing or servicing includes

- installing,
- maintaining,
- repairing, and
- altering.

Servicing a home is a good example of postprocessing. See Fig. 3-7. Think about *installing* that can be done in a home. For example, telephones may be installed. And *maintaining* must be done in a home. Windows are washed. Floors are waxed. Walls are repainted. There are many ways of maintaining a home.

Repairing is another common servicing operation. In homes it may include replacing a broken window. Or, it may involve patching a hole in the roof. *Altering* may be done, too. Altering is changing. Suppose a carport is changed into a garage. This is an example of altering.

The skills and knowledge needed for servicing are like those needed for processing. Suppose you put a window in a new home. If later it were to break, you could probably replace it. In the same way, if you wanted to add on a room (alter the house), you would need the same skills needed to build a house.

Millions of people work each year to produce new construction. Millions more work servicing construction projects. Production workers can do servicing work. People who do servicing work can also do production work. The same skills are required.

Fig. 3-7
Installing siding on a home is postprocessing.

The Construction Story

Many kinds of projects are built in the construction industry. Some projects are huge. Dams and manufacturing plants are examples. Some, such as houses or shops, are small by comparison. Some projects cover distances. Highways stretch for miles, and bridges span canyons or rivers.

Structures may be different. But they all have something in common. They are all built by using the same general procedure. The steps in this procedure are performed in order.

The steps briefly described here are used to build a house. But they are much the same as those used to build other structures. Remembering the procedure will help you understand construction production.

1. The site is prepared by clearing away all unwanted objects. The site is surveyed to locate the exact place for the structure. Then earthmoving is done to reach the surface on which the structure will rest.

2. The foundation is set below ground level. It is called the substructure. The substructure will bear the weight of the whole structure.

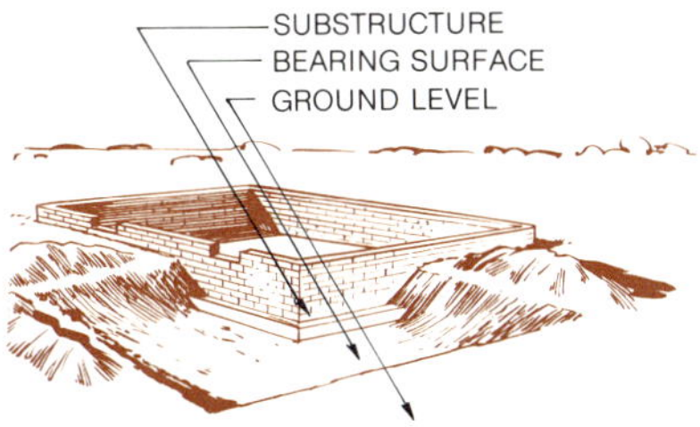

3. The superstructure is built on the substructure. Buildings are enclosed to keep out the weather. Some structures, such as bridges and roads, are not enclosed.

4. Utilities are installed to provide light, water, heat, and cool air. This requires wiring, plumbing, and ductwork.

5. The interior of the structure is enclosed (in a building). Inside surfaces are covered to hide utilities and improve appearance.

6. Surfaces are finished to protect the materials and improve their appearance. This mainly involves painting and decorating.

7. The site is landscaped by smoothing and shaping the land surface. A lawn, trees, and shrubs may be added.

The construction story continues beyond building the structure. Many of the same practices used in building are used again in servicing. If structures are to withstand years of use, they must be kept in good condition.

Summary

There are three stages in the production process. Preprocessing is getting ready to produce projects. Processing is producing projects. And postprocessing is using and servicing projects.

In industry, the form of materials is changed to increase their value. Production requires a material to be changed by separating, combining, or forming.

Construction production is done in a certain order. Understanding the general procedure will help you understand construction production. When you see construction work, you will know what step is being done. This will be true whether the project is a building, highway, or dam.

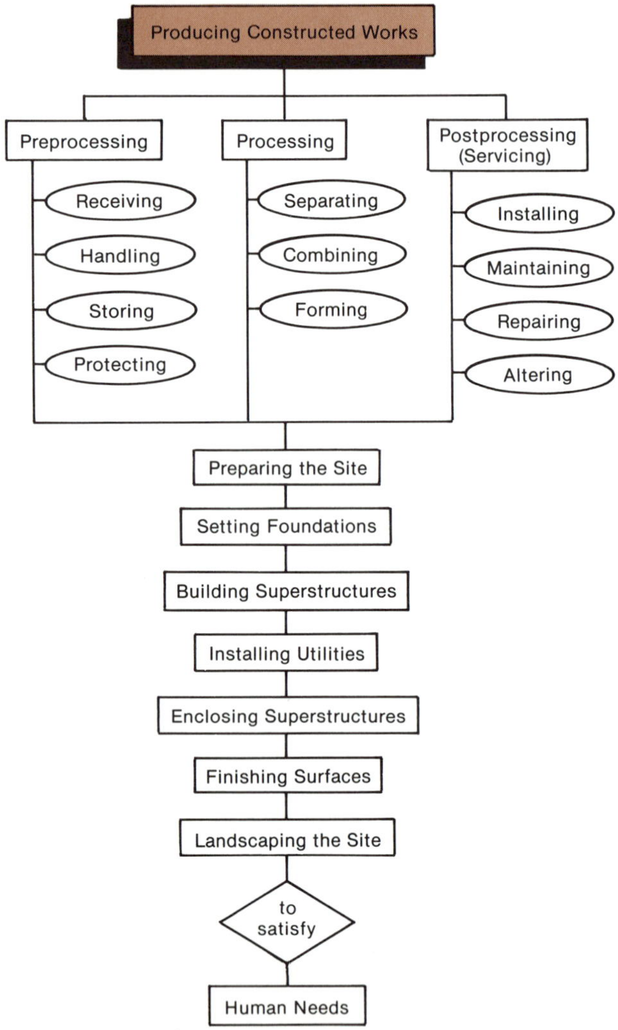

Terms to Know

altering	preprocessing
assembling	processing
combining	protecting
construction	raw materials
controls	receiving
forming	repairing
handling	separating
industry	servicing
installing	site
maintaining	storing
manufacturing	value
postprocessing	

Think About It!

1. Before construction production starts, materials arrive at the site. Other materials arrive later.
   a. What kinds of preprocessing do materials go through?
   b. Why does a contractor not want all the materials delivered to the site at one time?
2. Processing is changing materials in a factory or on a construction site. Name some materials that are processed by
   a. separating. (taking away material or making small pieces from a larger piece)
   b. combining. (putting pieces of material together)
   c. forming. (changing material without adding anything or taking anything away)
3. List six servicing jobs that require the use of construction knowledge and skills. Think about work that has been done to maintain your home.

Reading 4

Selecting and Purchasing a Site

The Importance of a Good Site

Two decisions must be made before construction can begin. One concerns the design. The other is about the site. What will the structure look like? Where will it be located? Structural design is very important. Yet the choice of the site may be more important. A good site is the key to a project's success. The site should be for sale, useful, and low in cost. Getting information about possible sites requires research.

A site is selected with great care. Often it is picked after long, detailed study. The route for a second Panama Canal is an example. An alternate route has been studied for more than fifty years.

Identifying Availability

The site must be *available*. See Fig. 4-1. Often there are many good sites. However, they may be in use by someone else. The owners may not want to sell the land. About one-third of all land in the United States is owned by the government.

Criteria for Selecting a Site

A site is judged by *criteria*. Criteria are standards or tests used as a basis for decision. The exact criteria for choosing a site depend on the type of structure to be built. For example, sites for a canal and a ski resort would be judged by different criteria. Each site must meet different needs.

Hercules Incorporated

Fig. 4-1

A. *Planners of a new factory may consider possible locations all over the world.*

B. *After much research, the desired location is identified. The search for an available site begins.*

Ogden Avondale Shipyard

Fig. 4-2
The ideal site for a ship-building company is near a large body of water.

One criterion for selecting any site is usefulness. The most useful site is the one that best serves the purpose of the project. Sometimes costs are higher for the most useful sites.

Another criterion is a supply of *labor*. Workers must be available to construct the project. They must have the skills needed. They must also have good work records.

If the project is to be a manufacturing plant, it should be located near its *market*. See Fig. 4-2. Suppose a plant will make snowmobiles. The best choice for the site would be in the north, where there is snow. This area is the plant's market. The plant should also be near sources of needed supplies and raw materials. For example, it should be near sources of metal and machine parts.

It is important to select a site that has a good community *environment*. Suppose owners want to build a plant. They want to go where they are welcomed. The area must provide good schools for the employees' children. Good recreational facilities are needed. Sometimes states or communities donate sites to industry. They do this because new jobs will help boost the economy. Long-term, low-interest loans are sometimes offered to attract industry. Often utilities are provided. Some states keep taxes low for companies. Such things help create a good community environment for companies.

There must be good transportation to and from the site. Industries must be able to get their products to market. This requires highways, railways, waterways, and airlines. See Fig. 4-3.

Utilities such as sewage and water systems must be available. Most industry needs large amounts of water for processing, cleaning, drinking, and cooling. Electrical power is of prime importance. Some

Fig. 4-3
Materials must be received and products shipped out. Good transportation facilities are vital to industries.

CBI Industries

Fig. 4-4
Most communities have land restrictions and zoning laws.

plants need large amounts of electricity.They want to locate near sources of low-cost supply.

Climate often determines the location of a site. For example, a ski resort would be located where there is much snow each year. Certain climates are needed for some manufacturing processes.

The physical characteristics of the site are important. Flat, hard sites may be desired. Steep sites may be best for some projects. Drainage must often be considered. Conditions below the ground are considered, too.

Land *restrictions* help determine which site will be chosen. *Zoning* laws tell what kind of structures may be built in an area. Cities are divided into residential, business, and industrial zones. The site must be in the proper zone. See Fig. 4-4.

Selecting a Site

Some sites are easy to select. A site for a home is the least difficult to choose. Cost, physical characteristics, restrictions, and utilities are chief concerns. Buyers choose the site that best meets their *requirements.*

For public projects, elected or appointed officials must make the site decision. Often this is done with the advice of professional *consultants.*

In private industry, top-level managers must always approve the site. *Feasibility studies* are made. These tell whether a choice is practical. They provide data about the best sites. Such studies may be done by the owner. They are often done by the owner's staff or by hired consultants. During such a study, many sites are considered. The sites that are most promising (likely to be chosen) are visited. The choice is narrowed to a few sites. These few are then studied in detail. The owner or the top-level managers make the final choice.

Acquiring the Site

Suppose a site has been selected. Steps must be taken to *acquire* it. Sometimes this is easy. It may take only a few days. But generally it takes much longer. See Fig. 4-5. One way to acquire a site is by negotiation. A second way is by condemnation.

Negotiation

Property for private projects is usually obtained by *negotiation*. Negotiation is a discussion between a buyer and a seller. They are usually assisted by a *realtor*. They try to reach certain *agreements*. They must agree on a price. They must agree on what the seller can remove from the property. Also, they must agree on the date the real estate will change hands. If the buyer and seller agree on these things, they make a contract. See Fig. 4-6.

First the buyer makes a *purchase offer*. The offer states the price the buyer is willing to pay. It sets a

Fig. 4-5
This is a site selected for a dam. It may take months or even years to acquire it.

Fig. 4-7
The federal government used its power of eminent domain to build a new interstate highway across this farm property. The farmer must now cross a busy highway to reach the fields, but a fair price was paid for the land.

Fig. 4-6
People who want to buy a site usually negotiate with the seller before making a contract.

closing date. This date is when the property will change hands. The offer states the amount of *deposit* or *binder money*. This money secures the buyer's right to purchase the property on the agreed terms. These conditions protect the buyer and seller from problems that might arise in the future.

Condemnation

Condemnation is a difficult process. Often a public project needs one or more tracts of land. Some landowners may refuse to sell their land. If it is a government project, the power of *eminent domain* is used. This is the right to obtain real property for public purposes. This can be done legally even if the owner objects. See Fig. 4-7. The owner must be paid a fair price. This is required by the United States Constitution. Of course, condemnation of property is very unpopular. But often it must be done in the *public interest*.

Closing and Recording

The site has been selected. The sales agreement has been made. The buyer must now have the money to make the purchase. Often money is borrowed from a *financial institution*. This may be a bank or a loan company.

A meeting, called a closing, is usually held at the lending institution. At this time the agreement is

Fig. 4-8
At the county office, a record is kept of ownership of all property in the county. Whenever property is sold, the records are changed.

completed. It is made legal. Both the buyer and seller attend this meeting. They both sign the agreement.

Other professional people usually attend this meeting, too. A realtor may be at the meeting. The realtor lists (advertises) and sells the property for the seller. One or more *lawyers* may be present, also. Usually one lawyer represents the buyer. Another lawyer represents the seller. The buyer's lawyer arranges to *record* the transaction. The *deed, abstract,* and *site survey* are taken to the county office or courthouse. These documents identify the property and the legal owners. A *registrar of deeds* sees that the documents are officially recorded or registered. See Fig. 4-8. The buyer — the new owner "of record" — then takes possession of the property.

Summary

The site is a very important part of any construction project. The site must be available. It must be useful. It must also meet the requirements of the project. It must meet many other criteria as well.

Sites are usually obtained by negotiation. The buyer and seller come to an agreement. Sometimes negotiation does not work. Sites for public projects may legally be bought by the government without the owner's consent. The authority to do this is called eminent domain.

Once the sale is agreed upon, it must be made legal. A closing is held, and the site legally changes ownership. Then the land sale is recorded. This process makes the buyer the owner "of record."

Terms to Know

abstract	financial institution
acquire	labor
agreements	lawyers
available	market
binder money	negotiation
closing	public interest
condemnation	purchase offer
consultants	realtor
criteria	record
deed	registrar of deeds
deposit	requirements
eminent domain	restrictions
environment	site survey
feasibility studies	zoning

Think About It!

1. Suppose your community needs a new, larger airport. How could the site be acquired? What criteria need to be applied in selecting the site?
2. Why are realtors, lawyers, and bankers important in a land purchase?

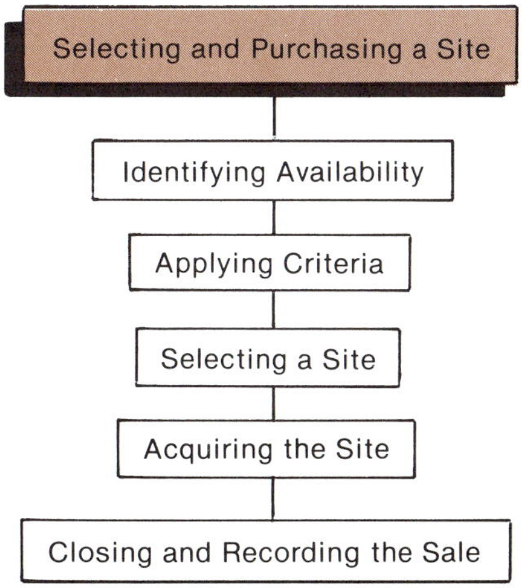

Reading 5

Surveying and Mapping

Why Surveys Are Made

All construction is located on a site. A site has natural and artificial features (characteristics). These features are surveyed or measured. Then they are put on maps. See Fig. 5-1. The maps are used by designers and engineers. The maps are important to people who design construction projects.

Types of Surveys

There are several types of surveys. The name of each describes its function.

Land surveys are made to establish land boundaries. They provide records and maps of the land. They show the proper location and subdivision of lands.

Topographic surveys describe the terrain (surface of the land). These surveys are made to collect data (facts) about natural and artificial features. They describe a site as hilly or flat.

Route surveys are made to determine alignment of routes. Alignment means the line along which a road, for example, will be built. Highways, canals, and pipelines are built along routes. Route surveys are made to collect data about the grades (or changing slope) of the terrain. They help determine the amount of earthwork needed.

Hydrographic surveys describe bodies of water. These surveys collect information about shorelines. They find and record the shape of the land under water.

Fig. 5-1

A. *Before a highway project can begin, natural and constructed features along the site must be accurately located and mapped.*

B. *Cartographers use scribers and other tools to prepare maps. Information for the maps is supplied by surveyors and other researchers.*

American Congress on Surveying and Mapping

Fig. 5-2
Survey information may be obtained by searching records.

Aerial surveys are made from above the surface of the earth. They record the features of large areas of land. Topographic maps may be made from aerial surveys.

Construction surveys describe a construction site. They are made to determine the best location for all the facilities to be built.

Making Surveys

Surveys are made by persons with special knowledge. Construction *consultants* can make surveys. Or, a special surveying company may be hired to make them. Employees of the project owner may do the work. Collecting survey data and making records requires many steps. First, surveyors must do *research*.

Research

Information about a site may be available. Finding this information will save work and money. The facts already known about a site are collected first. Gathering known facts is called research. See Fig. 5-2.

There are many places facts may be found. Most of the country has already been mapped. The United States Geological Survey has done such work. Copies of maps can be requested from their files. Other federal, state, and local agencies have large amounts of surveying data.

Almost all surveys are really *resurveys*. This is a survey of what has been surveyed before. A resurvey is usually done for a new reason. It is done in more detail than earlier surveys. See Fig. 5-3. A resurvey must start from known points on the earth. These points are *location* and *elevation* (height). They are found by research.

Finding Known Points

Horizontal control is locating a known point on the earth's surface. The elevation of known points is the *vertical control*.

Fig. 5-3
Over time, the geography of an area changes. New surveys must be made so that maps will show the most recent site details.

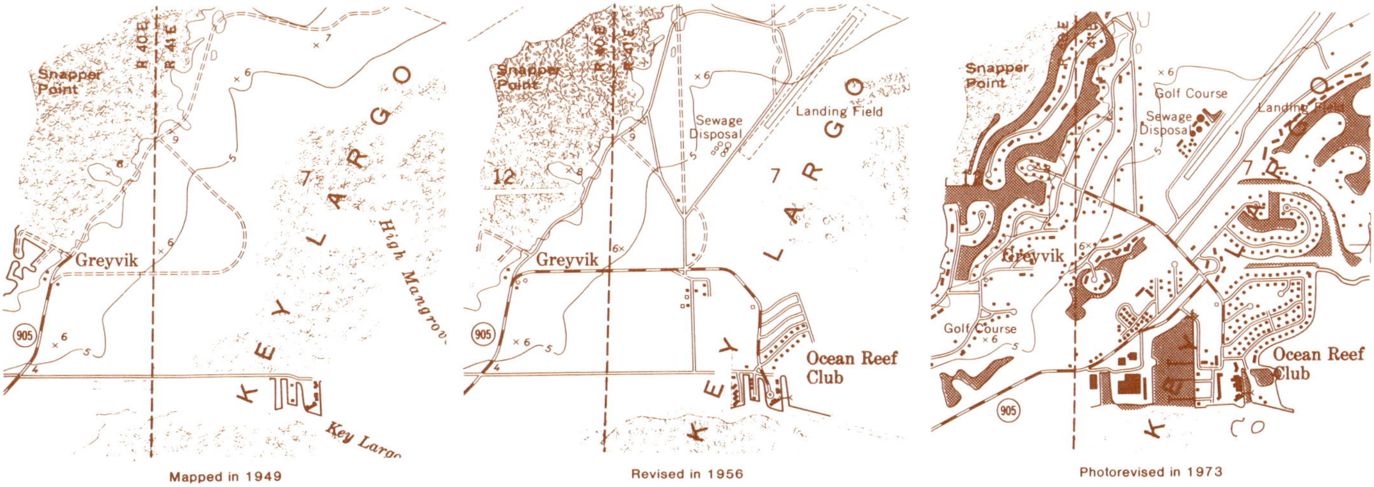

The locations of points (horizontal control) are found by *triangulation*. Accurate surveys have been made of the whole country using large triangles. Each side of the triangle may be several miles long. From the triangle corners, more surveys are made. In this way, millions of other points on the land are located.

Elevations of points are distances above or below certain fixed points. Usually the average sea level is used as a fixed level. For example, the elevation of Lake Superior is about 183.5 meters above sea level.

Systems of Surveying

Two systems of surveying have been used in the United States. The older one is the *metes and bounds* system. The newer system is the *rectangular* system. See Fig. 5-4.

The metes and bounds system was started during Colonial days. It was used to survey the thirteen colonies. Boundary lines were established by natural points such as trees, rivers, and large stones. Sometimes surveyors located and set markers. Today, this system is sometimes used to survey odd-shaped tracts (land sections).

After the Revolutionary War, the rectangular system was set up. It was used in the west, where tracts of land were vast. Land was divided using the meridians and base lines on maps. *Meridians* (longitude lines) run north and south on a map. *Base lines*, or *parallels* (latitude lines), run east and west. Large, unmarked areas were divided into sections one mile square. Sections were further subdivided.

Monumenting

When important surveys find the location or elevation of a point, markers are placed there. Future surveys can start from this point. These permanent markers are called *monuments* or *bench marks*. They are set firmly in place. See Fig. 5-5. Each marker usually has a bronze plate on top. Surveyors' readings are taken from the center of the "X" on the bronze plate.

On a project site, a surveyor works from known points. The points are related to the nearest bench mark. As surveyors locate boundaries, they drive stakes or metal pins into the ground. These stakes or pins locate a specific site or lot.

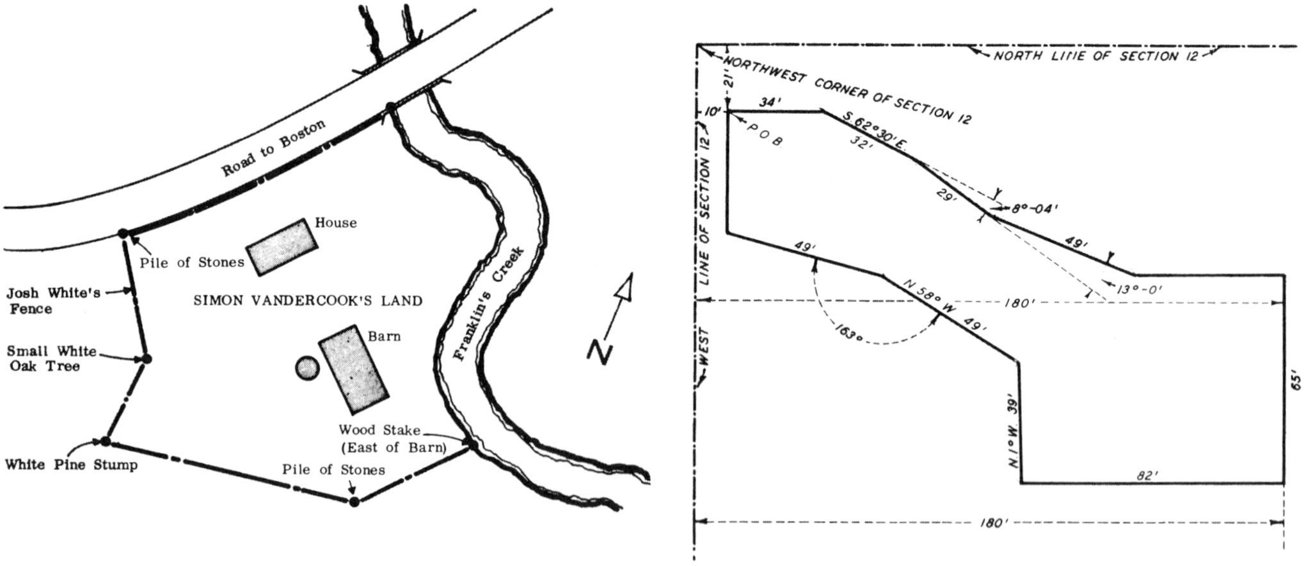

Fig. 5-4

Early metes and bounds surveys were marked by natural and artificial markers, as shown at the left. Now surveys are done more accurately with the rectangular system, right. This one uses the customary system of measurement.

American Congress on Surveying and Mapping

A. *This boundary stone was used 3000 or more years ago to mark a location.*

B. *Today, bench marks such as this are placed in the ground to mark locations and elevations. This marker is located in a parking lot.*

Fig. 5-5

Survey Parties

Most *survey parties* (groups) are made up of two or more persons. See Fig. 5-6. The *chief* plans and supervises the work. The chief, or an assistant called a *recorder*, records findings in a *field notebook*. The instrument assistant sets up and operates the surveying instruments over a known point. A rod used in surveying is held over another point by a *surveyor helper,* the rodperson. Measurements are made using a steel tape called a *chain*. This is done by another surveyor helper, also referred to as the chainperson. Sometimes brush and trees obstruct (block) the view from the survey instrument to the rod. The land is cleared by the axperson, yet another surveyor helper.

Surveying Equipment

The most common surveying instrument is the *transit*. It measures horizontal and vertical angles. See Fig. 5-7. A *theodolite* is another survey instrument. It is a more precise form of a transit. Both are used to locate points.

Fig. 5-6
This survey party is surveying a road. Can you identify the job each person is doing?

Piatt County Highway Department

Fig. 5-7
A transit is used by surveyors to measure horizontal and vertical angles.

Betty Medsger

A *level* is often used to find an unknown elevation from a known one. See Figs. 5-8 and 5-9. A *level rod* is used with a level to measure unknown elevations. Note the level rods in Fig. 5-9.

For many years survey equipment was unchanged. But recently, changes have occurred. A radio-controlled *laser beam* now may be used to find elevations. See Fig. 5-10.

American Electric Power Company

Fig. 5-8
A level is used by surveyors to measure elevations.

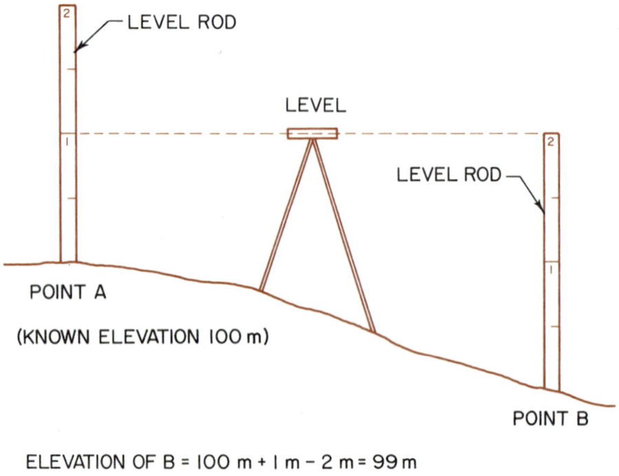

Fig. 5-9
The known elevation here is 100 m (point A). The reading on the level rod is 1 m, and at point B the reading is 2 m. The difference between the two readings is 1 m. Therefore, point B is 1 m lower than point A.

Fig. 5-10
This surveying equipment uses a laser beam to take measurements.

Plotting Data

Data collected during a survey are recorded in a field notebook. The notebook is then taken to an office. There, figuring is done. Data from the survey are used to make a *topographic map*.

Each important point is *plotted*. This locates a point on a map. To show the shape of land, two methods are used. The first is a *grid system*. The second is the use of *contours*.

A simple grid system is often used for small sites. See Fig. 5-11. The site plan is divided into squares to form a grid. Elevations are shown at the corners of squares. The land is flat if adjoining (connecting) squares show the same elevation. If adjoining squares show differing elevations, the site is uneven.

A contour is a line connecting all points of the same elevation. Except in rare cases of overhanging cliffs, contours never cross. Contour lines that are close together mean a steep slope. When contour lines are far apart, the land has a gentle slope or is quite flat. Closed contour lines mean a hill or a low spot of land.

Information from a contour map can be used to make *profiles* of the site. The top part of Fig. 5-12 shows the contours of a small site. The three pro-

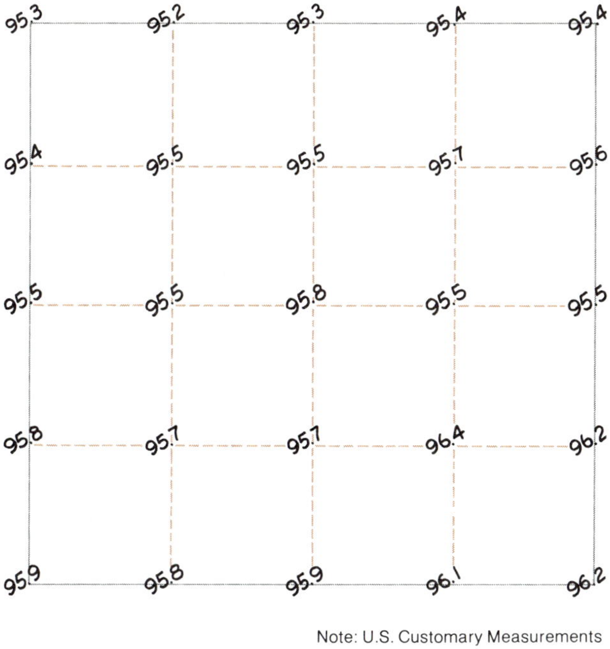

Fig. 5-11
This corner of land is one of the lowest points on the site. Can you find any areas of the site that are flat?

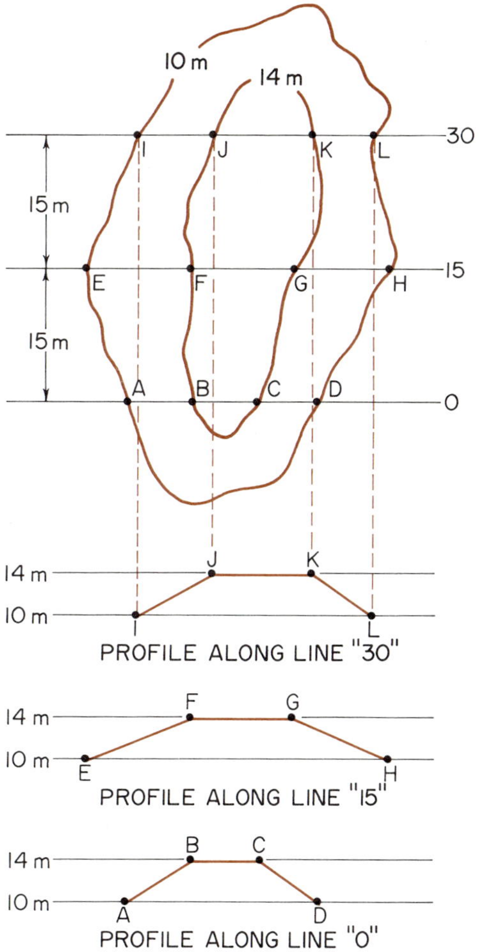

Fig. 5-12
This contour map (top) has been used to make profiles of the site. Note that the lines on the contour map are farther apart near line 15. As you can see on the profile, the land along line 15 has a more gradual slope than other points on the site.

files below the contour map are cross sections of land. Profiles show how the elevation of the site varies.

Surveys of Large Areas

Modern instruments may be used to survey large areas. Aerial photography can be used in such surveys. Photographs taken from the air show features on the ground. Horizontal and vertical controls are needed to plot what the photographs show. Control points are marked before the aerial photographs are made. These methods require expensive, complicated equipment. Two such devices are multiplex projectors and stereocomparographs.

Summary

Surveying is measuring land. Surveying shows the exact natural and artificial features of a site. Surveys are made so maps can be drawn. These maps are used by designers and engineers.

Surveys begin by finding known points. Horizontal locations and elevations are very important. A transit and other instruments help find points. Elevations are found by using a level and rod.

A grid system may be used to describe a small site. Contour lines and cross sections help show the lay of the land.

Terms to Know

aerial surveys
base lines
bench marks
chief
chain
construction surveys
consultants
contours
elevation
field notebook
grid system
horizontal control
hydrographic surveys
instrument assistant
land surveys
laser beam
level
level rod
location
meridians
metes and bounds
monuments
parallels
plotted
profiles
recorder
rectangular
research
resurveys
route surveys
surveyor helper
survey parties
theodolite
topographic map
topographic surveys
transit
triangulation
vertical control

Think About It!

1. Was your community surveyed by metes and bounds? How can you tell?
2. Why does a surveyor begin measuring from a monument or bench mark?
3. Can topographical maps be made without surveying?

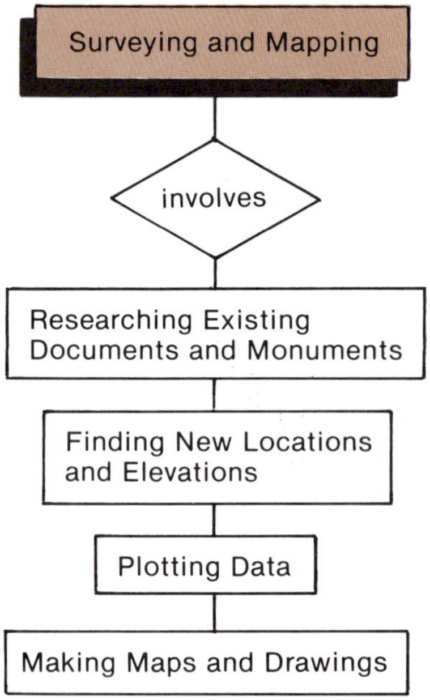

Reading 6

Soil Testing

The Reason for Soil Testing

You have read how land surfaces are measured and described. This reading is about the nature of the land itself.

Rock and soil form the top layer of the earth. Sometimes these are used as building material. They may be used to build structures such as earthen dams. Or, soil may be used to construct playfields.

Generally, the earth provides only *load-bearing surfaces*. This means that a structure's foundation is supported by the earth. For this reason, the earth's surfaces are very important to every construction project.

All structures must have strong foundations. A foundation remains strong and firm only if the earth supports it. Therefore, before designing can begin, it is important to know the nature of the soil. This can be learned by *soil testing*.

Fig. 6-1
If you could cut a cross section of the earth's crust, you would see many layers of rock and soil.

The Earth's Crust

The earth's crust is made up of rock and soil. See Fig. 6-1. Rock is a hard material. Great force is needed to excavate (dig) or remove it.

There are several kinds of rock. *Bedrock* forms a layer of the earth. It may be on or near the surface. Sometimes it is far below it. *Boulders* are large, loose pieces of rock. They are 200 mm or more in diameter. *Cobbles* are small pieces of rock. They are between about 100 mm and 200 mm in diameter.

Soil is made up of loose parts of the earth's surface. It is made of different sized grains. *Gravel* ranges from the size of peas (about 10 mm) to the size of cobbles. Grains of *sand* range from the size of table sugar to the size of gravel. Grains of *silt* vary in size, too. They may be between the size of powdered sugar and the size of sand. Lake and river bottoms are usually silt. Silt is the soil that covers an area after a flood. It actually may be made up of seventy-five percent silt, fifteen percent clay, and ten percent sand. *Clay* has a fine texture, very much like flour. Most soil has a mixture of grain sizes, ranging from fine clay to boulders.

Topsoil is the soil at the surface of the earth. Generally, it is covered with grass, crops, or weeds. It is a mixture of soil and decayed vegetable matter.

Soil Characteristics

Different types of soil have different *characteristics*. Some characteristics depend on the soil particles. Others depend on the action of water on soil particles. These particles are not smooth like marbles. They are varied in size and shape. Sometimes they are rounded. More often they are jagged, with sharp corners. Since the particles are uneven, there are usually spaces between them. These spaces are called *voids*.

Much soil, especially clay, is quite *cohesive*. This means that the grains stick together. In fact, some kinds of soil stick much like cement. These soils are almost as hard to work as rock. Other soils have great *plasticity*. That is, they can be molded very much like modeling clay.

The characteristics of a soil depend a great deal on how much water it contains. At least a small amount of water is in all soils. It is held in the voids between grains. This is called *surface water*. It comes from rains and seepage.

Water also enters soil from pools under the ground. This water fills and flows through the spaces between soil grains. Underground water can be found by digging an open pit. The top of this water is called the *water table*. See Fig. 6-2. The water table is not flat like a surface pool. Its depth varies. It generally follows the surface contour of the land. Like surface water, this water may have a current.

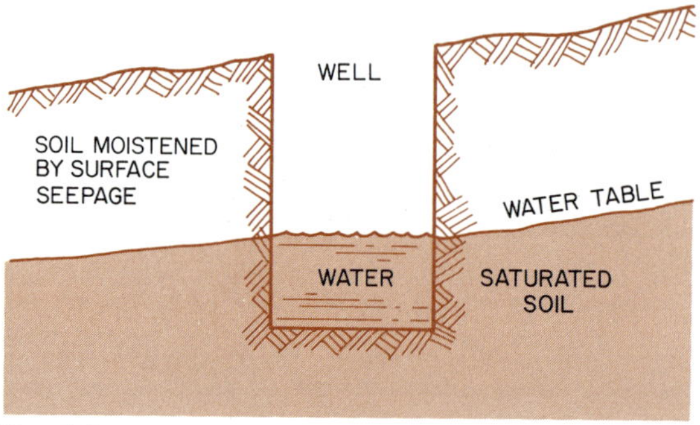

Fig. 6-2
In different parts of the country, the water table is at different depths.

Most soil (especially that which contains much clay) swells as moisture is increased. It shrinks when moisture is taken away. Some clay *slakes* (turns into a soft mass) when it becomes wet. Firm clay may lose its strength when there is water in it.

Climate affects soil. When water freezes, it takes up more space. If the water in soil freezes, the soil also expands. This raises the surface of the ground. When the ice thaws, the soil becomes very wet.

Sometimes there are large voids between grains in soil. Then the soil is called loose. When the voids are small, the soil is called dense. Density can be increased by compressing (putting pressure on) the soil. Most soil is *compressible*. This means that it can be packed down.

Most soil expands or swells when it is dug up. This happens because air gets between the particles. Compressing this soil decreases its volume (size). The soil can even be compressed to less than its natural volume. See Fig. 6-3.

Size of grains, water content, and compressibility affect the soil's strength. They determine whether it will support a structure.

Soil Analysis

Soil specialists may be hired to do a *soil analysis* of the project site. They study the soil. See Fig. 6-4. They make recommendations about how to work with it.

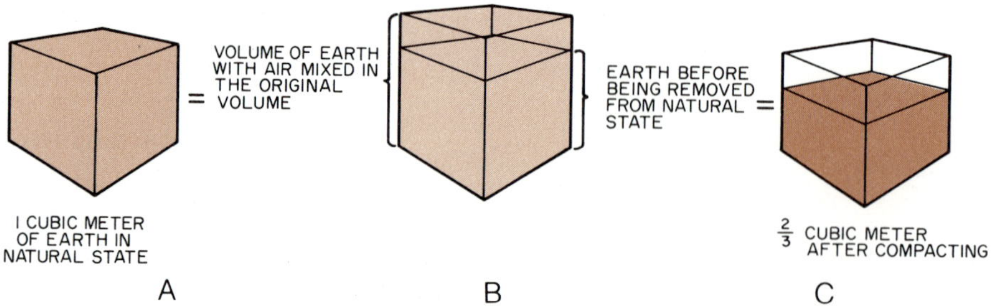

Fig. 6-3
When soil is removed from its original position (A), it becomes loose (B). This is because air enters between soil particles. Its volume also increases. When the soil is packed to support a foundation, its volume becomes (C) less than the original volume.

Betty Medsger

Fig. 6-4
A soil tester can tell much about the soil just by looking at it and touching it.

There are many ways to learn what kind of soil is underground. The cheapest way is to learn what is already known. This is done by checking with government agencies. Nearby property owners and soil laboratories also may have information about the underground soil.

One can learn some things about the soil below the surface by touching and looking at the soil on the surface. This is a low-cost method. But it may not be accurate. Therefore, many tests must be done in soil analysis.

Holes may be bored with drills to see what type of soil comes out. Some drills bring up *samples* that are cylindrical (shaped like round pencils). They are called *cores*. See Fig. 6-5. Cores may be sent to a laboratory for analysis.

Sometimes, large holes are dug in the ground. The soil tester then examines the soil and takes samples at various depths. These holes are called *test pits*.

Field Tests

Two main kinds of soil tests are done in the field. Each measures different factors.

Soiltest, Incorporated

A. The parts are assembled and forced into the earth to gather the sample.

B. The sample is removed for testing.

Soiltest, Incorporated

Fig. 6-5
A split-barrel soil sampler may be used to gather samples for testing.

First, there are *plate-bearing tests*. In these tests, heavy weights are placed on the earth's surface. The tester observes how much the load tilts or leans. This tells the vertical and horizontal stability of the soil. Testers also observe how far the load sinks into the soil. The distance the load sinks is a measure of the *soil stability* and *settlement*.

Second, there are *density tests*. Soil is put into a standard cylinder. Then it is hit a certain number of times with a hammer. This shows how much the soil packs (how compressible it is).

Soiltest, Incorporated

Fig. 6-6
A resistivity meter is used to find out the moisture content of the soil, its density, and its mineral content.

There are also *geophysical tests*. Explosives are sometimes used. Or, the ground's electrical resistance is studied. See Fig. 6-6. Geophysical tests tell how deeply bedrock is buried below the surface.

Laboratory Tests

Soils are also tested in laboratories. See Fig. 6-7. Tests tell how strong the soils are. One type of testing is done to learn how much water is in the soil. See Fig. 6-8. Soil plasticity and soil compressibility are also tested. See Fig. 6-9.

Soil and Construction

The type of soil determines the type of foundation needed. This may affect cost.

The soil under a foundation must be stable enough to hold up the structure. It must not erode or wash away. It also must be kept from *heaving*. Heaving results from freezing or from excess water in the soil.

Soil must not settle too much or unevenly. This would cause the structure to crack. Soil in earthen dams must be placed in such a way that water can drain without eroding the slopes.

During construction, some soils in trenches must be held in place. Supports called *shoring* are used. See Fig. 6-10. Shoring keeps sidewalls from caving in or falling on workers. Firm soil may not need to be shored. Excavating crews must know which kind of soil they are working with. They have to prevent dangerous cave-ins.

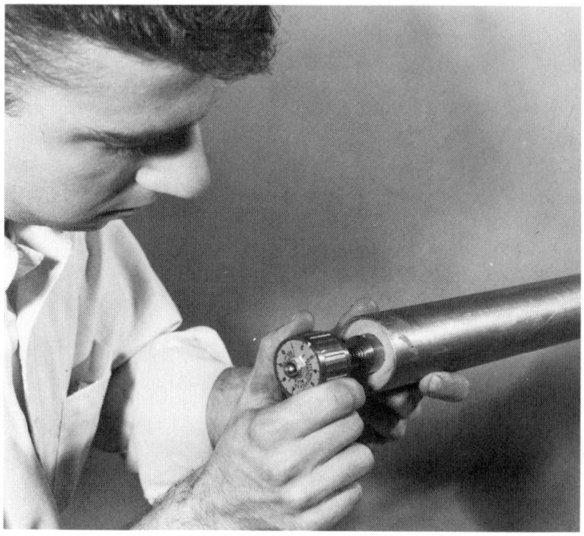

Soiltest, Incorporated

Fig. 6-7
A laboratory technician is testing the strength of a core sample of soil.

Soiltest, Incorporated

Fig. 6-8
A liquid-limit testing device is used to find out how much water is needed to cause the soil sample to become fluid.

Soil Engineering

Construction consultants determine how soil will be affected by the weight of structures. They base their conclusions on soil tests and laboratory reports. Then they decide what must be done to the

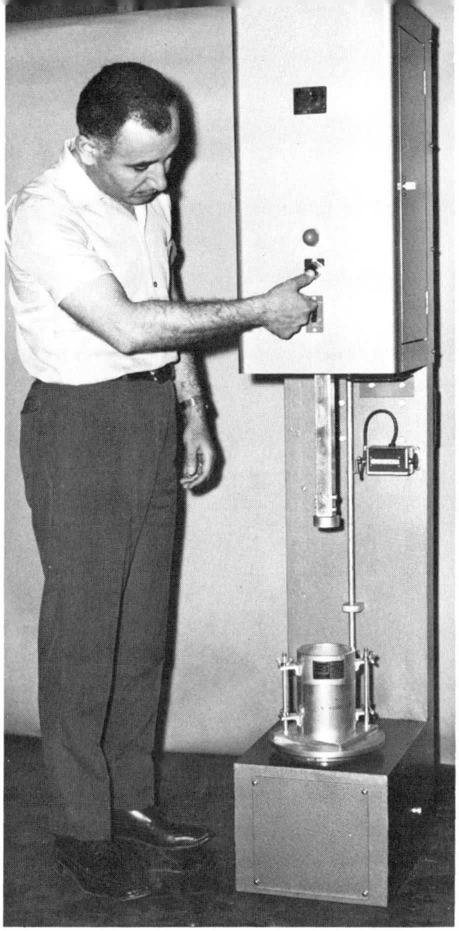

A. *A laboratory soil compactor is being used to measure how much the soil can be compressed.*

Soiltest, Incorporated

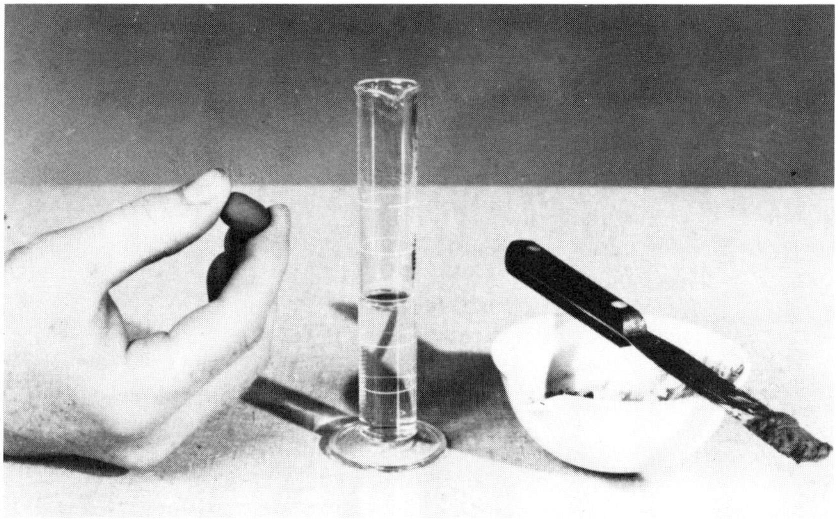

B. *Wet soil is squeezed, then rolled and pulled to test soil plasticity. This is the amount a mass of soil can be molded or pulled without cracking or breaking apart.*

Fig. 6-9

Fig. 6-10

Soil around an excavation is supported to keep it from collapsing.

Schnabel Foundation Company

soil to meet the structure's needs. The need for workers' safety is also considered.

Often, changes are made in the soil. Sometimes soil must be removed. It is replaced with soil that better meets construction needs. Sometimes heavy rollers are used to apply pressure to soil. This reduces the amount of space (the voids) between grains of the soil. Then the soil can hold greater loads. Often the composition of the soil itself is changed by adding different types of soil. Sometimes other substances, such as cement, are added.

Engineers decide how steep the constructed slopes should be. They also decide how to drain off excess water. Engineers must also know how to prevent frost heave.

Engineers determine how much weight the soil can support. They want to keep foundations from collapsing or settling too much. They may even find that *pilings* are needed for support. These are large concrete or steel stakes driven into soft soils. Pilings help support structures. When all soil conditions are known, designing and engineering for the structure begin.

Summary

The earth's crust is made up of rock and soil. Different soils have a variety of characteristics. Soil consultants test the soil. Various tests are used to collect data. These data are studied to learn what needs to be done. Engineers must find ways to provide a good bearing surface. A proper bearing surface is needed for a firm foundation. And a structure cannot work well without a firm foundation.

Terms to Know

bedrock
boulders
characteristics
clay
cobbles
cohesive
compressible
cores
density tests
geophysical tests
gravel
heaving
load-bearing surfaces
pilings
plasticity
plate-bearing tests
samples
sand
settlement
shoring
silt
slakes
soil analysis
soil stability
soil testing
surface water
test pits
topsoil
voids
water table

Think About It!

1. Describe the soil in your community.
2. How do you think the soil in your community affects
   a. water supply?
   b. building foundations?
   c. basements?
   d. roadways?
   e. drainage?
   f. plant growth?

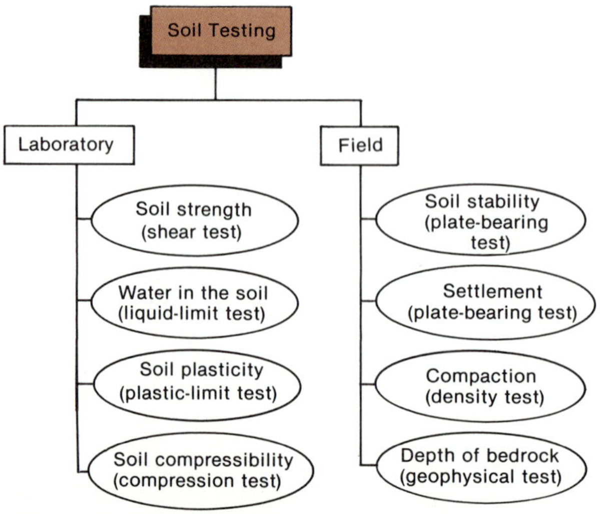

Reading 7

Designing and Engineering Constructed Works

Designing and Engineering

Suppose you wished to build a shopping center. You would need to answer many questions. Where is the right place to build it? How should it look? What materials should be used for its construction? How will the parts be joined together? To help you answer questions such as these, you would employ (hire) an architect or engineer. Finding such answers involves *designing* and *engineering*.

Designing and engineering are important parts of *planning* construction. Planning is part of the managing process. Planning also involves formulating (setting) goals and researching (finding needed information).

Good planning solves construction problems before building begins. Architects and engineers design structures. The designs show the work that must be done. Designing and engineering set a plan of action.

The Design Process

Architects, contractors, skilled workers, and engineers must solve designing and engineering problems. They must approach problems in a reasonable, organized manner. When their work involves design, they generally use the *design process*. This process is a series of steps that designers use to find solutions. You, too, can use the design process. The steps used by most designers are:

1. *Problem Identification*. In this step, a need is described. See Fig. 7-1. Background information must be gathered to write the description. Suppose there is a need for automobile traffic to cross a riverbed. Identification of the problem includes finding many facts. For example, what traffic moves along the river? How many cars will cross the river daily? How deep is the water? How wide is the riverbed? How high might the water rise? This type of information will help describe the details of the problem.

Now suppose there is a need for a home. Again, problem identification will provide details. For example, how many people will live there? What are their interests? Would they use a formal dining room? a game room? What utilities (services such as gas and water) are needed? Gathering these and other related facts will help the designer write a clear statement of the kind of home needed.

Fig. 7-1
Problem Identification. *The amount of heat loss from this building is being tested with a camera that uses special heat-sensitive film. Designers will consider the need for conserving heat when they design new structures.*
Exxon Corporation

Fig. 7-2
Preliminary Ideas. *Building a dam might be one idea for solving a riverbed crossing problem. Many preliminary ideas are sketched quickly in the search for a practical solution.*

Fig. 7-3
Refinement. *Designers develop details for several ideas to find out if the ideas are workable.*

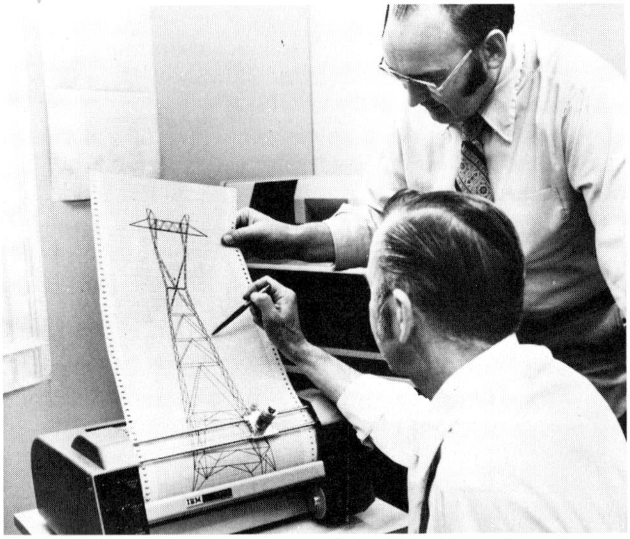

Niagara Mohawk Power Corporation

Fig. 7-4
Analysis. *Designers study the details of the design to make sure no problems exist. Sometimes they use computers to help in their analysis.*

2. *Preliminary Ideas*. After the problem has been identified, the designer thinks of as many solutions as possible. See Fig. 7-2. At this step, it does not matter whether the solutions will work. All ideas are sketched (drawn quickly) or jotted down. This is done as hastily as possible. Even a wild idea may lead to another, more practical idea.

   How might the riverbed crossing problem be solved? The designer might consider building a bridge or tunnel or even rerouting the river. Ideas for solutions have no limits at this step.

   What preliminary (beginning) ideas can be listed for designing a home? The house could have more than one story. Or how about an underground house? Would a house on or in the water be practical? The possibilities go on and on.

3. *Refinement*. Refining means developing and improving details. In this step, several of the best early ideas are refined. Then the designer can judge which design idea is best. See Fig. 7-3. Usually, drawings are made that will specify (tell exactly) dimensions (sizes), shapes, and other design details. For example, several bridge ideas may be drawn in detail. Engineers examine the drawings to see which designs are best. Then the cost of materials is figured for each bridge design chosen.

   When designing a house, the designer finds the cost of utilities for each of several types of homes. The designer also describes many details for each type of home. For example, he or she may select styles of doors and windows. Owners can choose the house designs they like best.

4. *Analysis*. It is important to find out the strength or size of materials needed for each part of the design. This is done by analyzing (studying) the

design. See Fig. 7-4. Analyzing is more closely related to engineering than any other design step.

Suppose a bridge is to be built across the river. The proposed structure is analyzed. Engineers decide the sizes needed for beams. They figure the size and shape of the bridge foundation. Mathematical skills are needed for this work.

In designing a house, analysis is important also. Should the roof be held up by beams or trusses? What size should the main water pipe be? What kind of wire is needed? The demand for details seems endless. Yet, at this point, all the details must be included.

5. *Decision*. By now more than one design has been refined and analyzed. It is necessary for the designer to make a decision. Or, design ideas may be presented to an owner for a decision. A detailed picture may be drawn to show the client what the completed structure and site will look like. The owner can get a better idea of what the final structure will be. Models of designs may be helpful, too. It is easier to picture a design from a model than from drawings. See Fig. 7-5. The decision may be made to
   - accept one design.
   - combine the best features of several designs.
   - turn down all the designs and repeat the design process.
6. *Implementation*. At this step, the owner feels that the best solution has been found. The idea is put into a form that can be used. *Plans* are drawn. The designer supervises this work. *Specifications* are written. The designer checks these carefully. All construction details that cannot be shown clearly on the drawings must be described in the specifications. Drawings and specifications are used as guides for construction. See Fig. 7-6. Minor design changes may be made during construction. However, these are usually costly. They are avoided as much as possible.

Fig. 7-5
Decision. *If the project is a new library or other city building, the designs are presented to the city council. Special features will be explained, then a design will be chosen.*

The Designers

Construction projects may be designed by either an architect or engineer. Sometimes the architect and engineer work as a team. Architects develop plans for buildings, shopping centers, and community developments. Engineers design bridges, dams, highways, and utility systems. The architects in charge of a shopping center design will be responsible for the *master plan*. They will depend on engineers to design certain parts of the project. For example, electrical engineers will design the electrical system. Drainage of the site will be designed by civil engineers. Mechanical engineers will design the framework of the building.

Sometimes a project calls for little designing or engineering. For example, a home buyer may choose a design that has been used before. In such cases,

Fig. 7-6
Implementation. *Plans are drawn before construction begins. They show every part and special detail of the structure.*

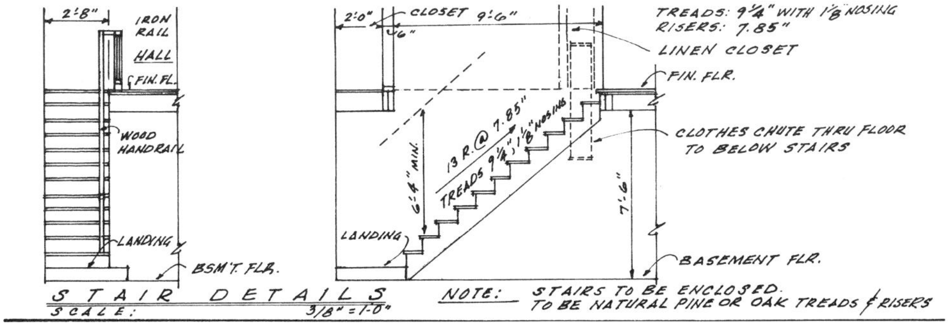

Sears, Roebuck and Company

A skyscraper

Martin Marietta Corporation

A highway

J.F. Santa — United States Department of the Interior

A dam

Lone Star Industries, Incorporated

A bridge

A sports and entertainment center

Control Data Corporation

An earth-sheltered office complex

A home

Fig. 7-7
For all kinds of structures, architects and engineers must consider function, appearance, cost, materials, and strength.

the contractor may make only slight changes for the owner.

Owners and skilled workers sometimes do their own designing and engineering. However, they should do only the work they are sure about. Costly repairs and serious safety hazards (dangers) result from poor designing and engineering.

Some design problems require new solutions. Examples are a solar-powered electrical generating station, a space farm, and a neighborhood transit system. Only trained designers and engineers should try to solve these design problems. Even they may not have had experiences that can help them. Perhaps no one has. It may be difficult or impossible to find background information that will help solve the problems. Designers and engineers must depend on their training and good judgment.

Design Considerations

Whatever design is being developed, some basic *design factors* must be considered.

- Function
- Appearance
- Cost
- Construction materials
- Strength

These factors are common to all designs. They are closely related and overlap. Sometimes they must be considered together. See Fig. 7-7.

The most important part of a design is its *function*. This is the ability to meet stated needs. For example, a foundation must support a building. That is its function. If it will not support the building, the design is poor.

Appearance is important. Structures should be attractive. Sometimes, however, people think this factor is more important than other factors. When this happens, the effectiveness of one or more other factors may be lost. An ideal project has a good appearance. But it must also work well. And its cost must be reasonable.

Almost all design projects begin with a *budget*. The budget is an outline of how project money is to be spent.

The use of materials and the strength of a design are important to any project. For example, which material is better for a door hinge — leather or steel? Steel is better. Why? A steel hinge gives more strength. It lasts longer. It also holds the door in position.

Summary

Designing is part of planning. In general, designing is finding new solutions to problems. Architects and engineers solve design problems. As they design, they consider many design factors. The final design is usually presented in the form of drawings and specifications. These are used by builders to complete the project. The final product of a designer is a completed project that satisfies a given need. This project may be a building, a bridge, or a community.

Terms to Know

analysis	implementation
budget	master plan
decision	planning
design factors	plans
designing	preliminary ideas
design process	problem identification
engineering	refinement
function	specifications

Think About It!

1. You want to design your "dream home" (an ideal home from your point of view). Explain what you will do at each step of the design process.
2. How do architects and engineers work together?

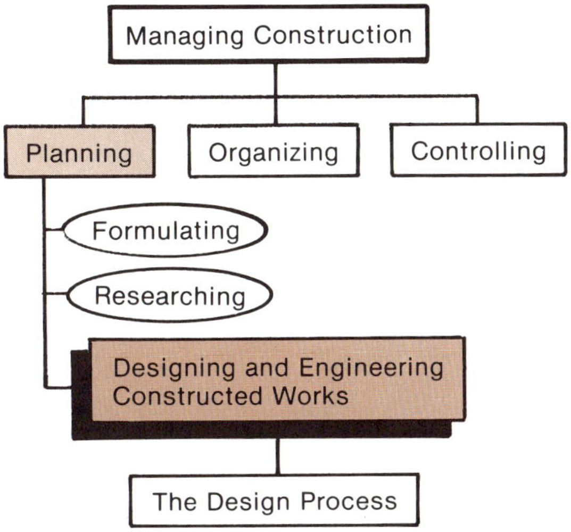

Reading 8

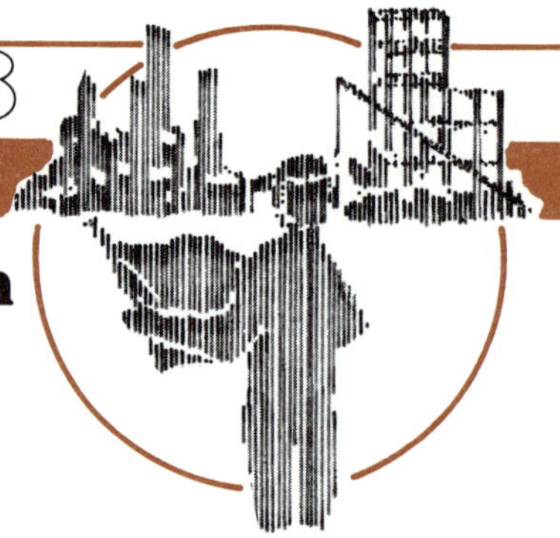

Identifying the Design Problem

Problem Identification

The first step in the design process is to define the problem clearly. What kind of structure is to be designed? Why is the structure necessary? What are its *needs* and *requirements?* Knowing the answers to questions such as these helps define the design problem. Designers call this step of the process *problem identification.*

Identifying a design problem includes three basic steps.

Fig. 8-1
A way was needed for many vehicles to cross a river quickly. A large bridge was built to satisfy that need.

National Semiconductor Corporation

- Identifying needs
- Gathering information
- Judging information

Identifying Needs

Designs are based on needs. See Fig. 8-1. A bridge is designed to span (extend across) a canyon. People need it for more direct travel. A dam is designed to control the flow of water. People need the power created at the dam. A camp is designed to serve the needs of campers.

Suppose a scout troop needs a camp. The need might be stated as follows:

> A camp is needed on Lake Indian to provide a recreation area for Scout Troop 403. It should have eating, sleeping, and activity areas. It should be reasonable in cost.

By writing a *design statement,* the designer clearly defines the design's purpose. Design statements should be written for all proposed structures. A design statement for a house might be written as follows:

> A home is needed on the eastern edge of Rocky Creek. It should cost no more than $60,000. It should house four people.

Gathering Information

Design statements do not give many details. More information is needed before designing can begin. For example, the designer of the scout camp must know many things. How many people will use the camp on a typical outing? In what activities will the scouts be involved? How many parking spaces will be needed? How many people of each sex will use the camp? How many boats are to be housed? Is a permanent lodge required? Answers to questions of this type must come from the people responsible for the project. They must decide what features are needed.

Suppose a site is donated to the scout troop. The problem then involves a study of the site. See Fig. 8-2. How can the designer take advantage of the best features of this site? Perhaps a parking lot and lodge are to be constructed. Where should they belocated? A bridge is needed across the creek. Where should it be built?

Information must be collected about the site. What is its size and shape? Where are trees located? Is the land flat or hilly? What are the normal high and low levels of the lake? What other facts might a designer need to know?

Much information is also needed to design a home. Has the family chosen a style for the house? What activities do the family members enjoy? Do they want a basement? Can a basement be dug on the chosen site? How many cars does the family own? Should the garage be attached to the house? To plan well, the designer needs to know a great deal about the family. See Fig. 8-3.

Fig. 8-3
The designer meets with clients to gather information about their needs and wants.

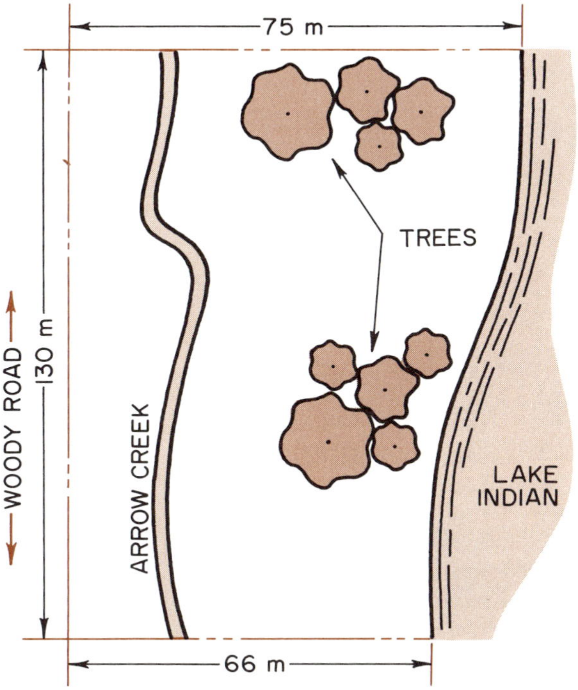

Fig. 8-2
The proposed site for the scout camp.

Before the structure can be designed, information must be gathered about the site. The appearance of the site may influence the structure's design. Will the topography (lay of the land) be suitable? Do zoning laws allow the construction of the desired type of structure? Are utilities and services available? Is the site for sale? What is the cost? See Fig. 8-4.

Suppose you are looking for a home. When you find the right place, then what? You buy it. First, you make an *offer*. You tell what you are willing to pay. If the owner wants more money, you *negotiate* (discuss). If you reach an agreement, you must *close* the deal (make it legally binding). Finally, the sale must be recorded. This is done at a county office (a courthouse, for example). There, a Registrar of Deeds records (registers) the legal documents (papers) that confirm the sale. This ensures that a copy of the records is always available.

Early in the design process, the owner and designer must consider costs. The planned structure should not cost more than the buyer can afford. Buyers often borrow money. The amount paid for borrowing money is called *interest*. Sometimes interest is the buyer's greatest cost.

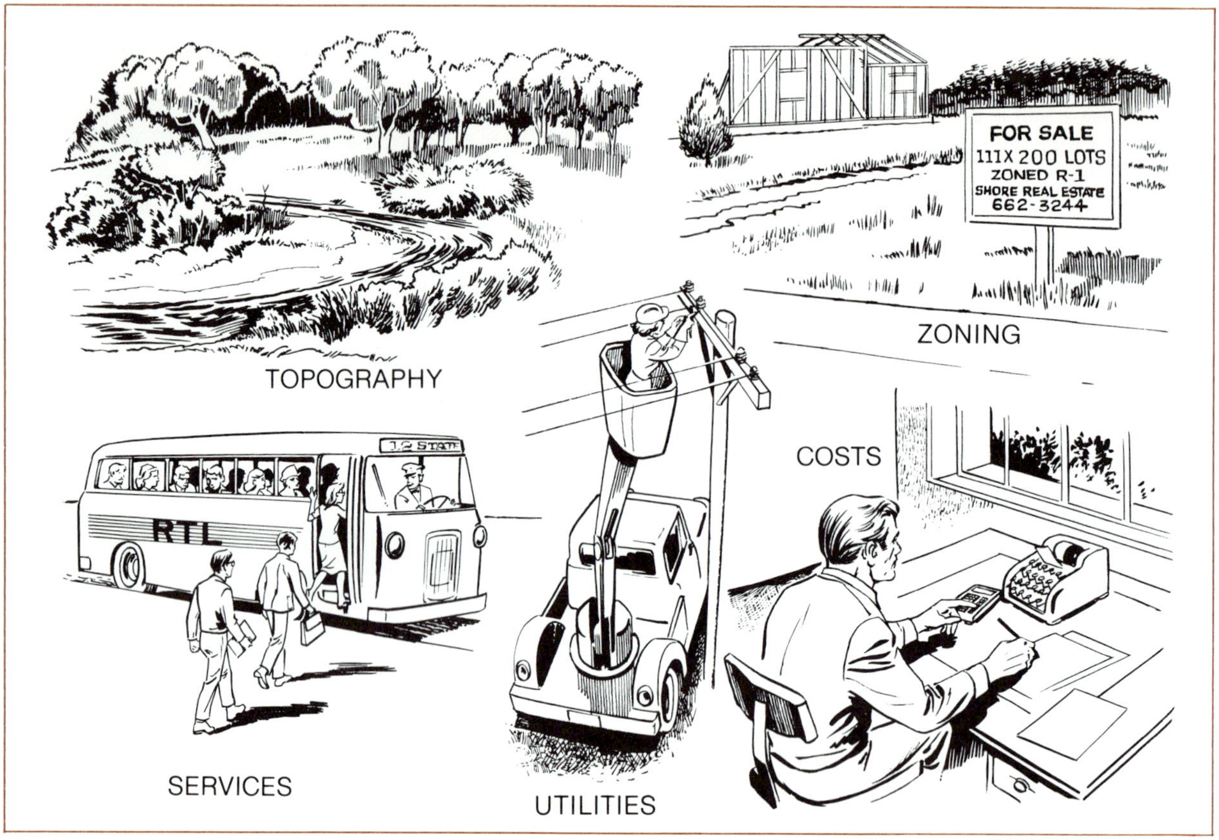

Fig. 8-4
Many factors must be considered in selecting a homesite and designing a home.

Judging Information

The information gathered about a design problem must be *evaluated* (judged). Evaluating information gives the designer a better understanding of the problem. Graphs may be drawn to help judge information. They show past records and what may be expected in the future. For example, look at the graph in Fig. 8-5. The number of scouts in Troop 403 has increased each year. The designer must plan room for continued increase. Graphs showing ages of scouts may also help the designer understand the problem. Why might a rainfall record be helpful in designing a camp?

Information about a home design must also be judged. A home designed to include all the family's interests may cost too much. The family must decide which interests are most important. For example, would the family prefer a swimming pool or a garden? Judgments such as these must be made carefully.

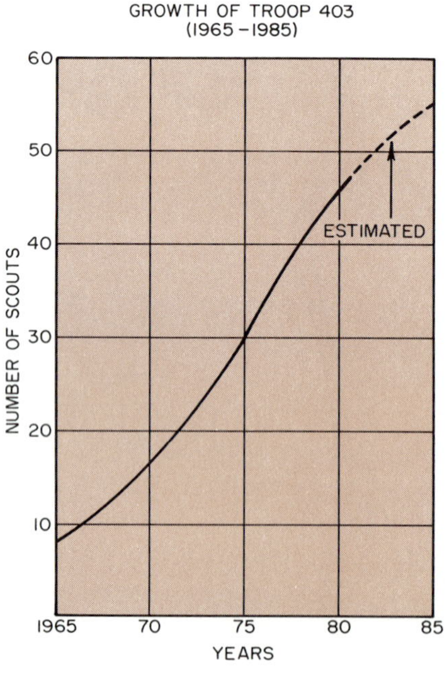

Fig. 8-5
The number of scouts in Troop 403 has increased every year since 1965. It will probably continue to increase.

Summary

The three basic steps of problem identification are identifying needs, gathering information, and judging information. These steps give designers information they need to solve the design problem.

Terms to Know

close
design statement
evaluated
interest
needs
negotiate
offer
problem identification
requirements

Think About It!

1. Sometimes a designer must ask clients personal questions. (How much money can you spend for this project? Are you planning a large or small family?) Why must a designer evaluate answers to such questions?
2. If a design is based upon faulty information, that may happen after the project is completed? (Suppose information gathered about family size was incorrect or was poorly judged.)

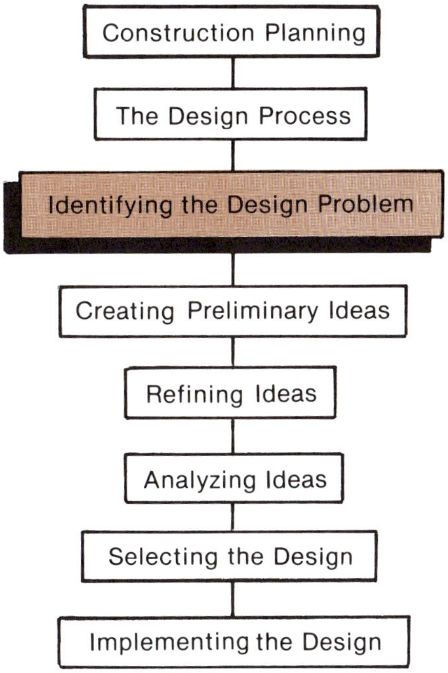

Reading 9

Creating Preliminary Ideas

Preliminary Ideas

How can the design problem be solved? Finding the solution begins with exploring different ideas. See Fig. 9-1. This is the second step in the design process. Considering several beginning or *preliminary ideas* is likely to result in a better final design. Thinking about many ideas opens up several paths to follow. All ideas considered must be ways to satisfy the stated need.

Preliminary Ideas and Creativity

Preliminary ideas are those the designer thinks of as he or she first studies the problem. The designer lists or sketches many design ideas. Ideas are not detailed. The main concern is finding solutions. The designer may seek ideas that are better than anyone has thought of before.

The designer is free to consider many ideas at this point. This is the most creative step of the design process. One new design idea often leads to others. Perhaps a new idea will improve older methods of construction. Or, it may be a new way of solving a design problem.

An engineer may be asked to solve the problem of supplying water for a city. The engineer may list ideas people have already thought of, such as a well, dam, or pipeline. Or, the engineer may list unusual ideas. For example, one idea might be to design a new roof for every home in the area. The new roofs would catch water. The water could then be stored and used in the homes. This new idea might work. All ideas are good at this time. It does not matter if they seem impossible at first.

To think of new ideas, designers must be creative. Most people can be creative. They must give themselves a chance. To be creative, it is necessary to keep an open mind. For example, if you say you cannot create new designs, you probably will not. If you try, you may be surprised. Designers, too, must keep an open mind to be creative.

Fig. 9-1
The problem of designing multiple-family housing may have several different solutions.

Recording Preliminary Ideas

First ideas should be listed or drawn as they come to mind. Otherwise they may be forgotten. Drawings can be made quickly without special tools. Such drawings are *freehand sketches*. Some drawings may be made to explain ideas or show parts of ideas. These may not look like the items they represent. These drawings are *diagrams*. Both freehand sketches and diagrams help to detail ideas. See Fig. 9-2. *Notes* (bits of information) can be added to the drawings.

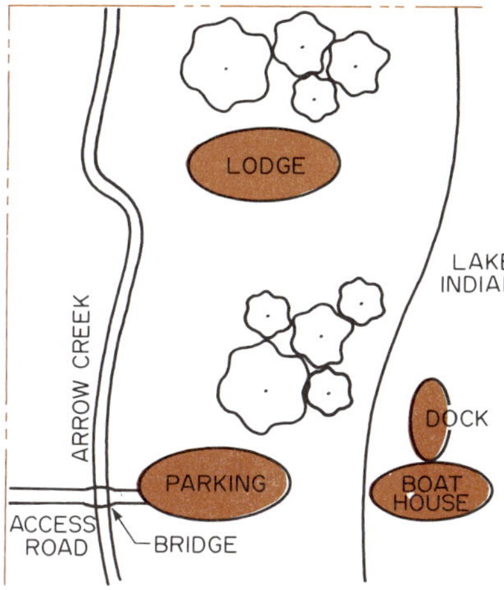

Fig. 9-2
Preliminary idea #1 for the site plan of the scout camp.

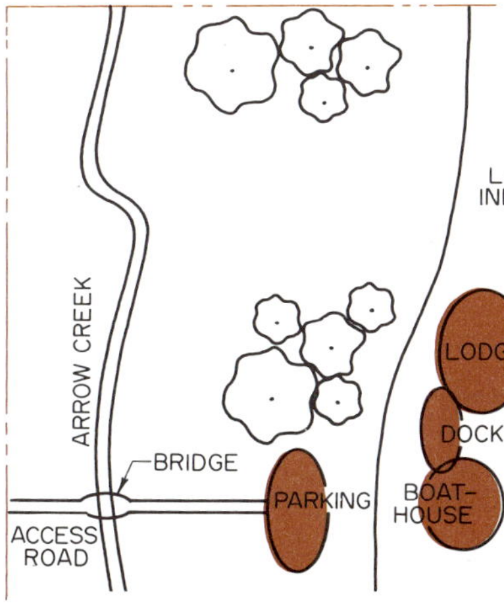

Fig. 9-3
Preliminary idea #2 for the site plan of the scout camp.

Making drawings and notes helps the designer find other ideas. Ideas are recorded as quickly as they come. This frees the designer's mind to think of other ideas.

The designer keeps all the work sheets (drawings and notes). They are useful records of design ideas. The designer may refer to them at any time during the design process.

In Reading 8, the need for a scout camp was identified. Beginning ideas for the camp design involve a *general layout* of the various parts of the camp. Such features as the entry road, parking lot, lodge, and water facilities are considered. See again Fig. 9-2. General areas may be sketched on diagrams. The designer can see how areas relate to each other. A different plan is shown in Fig. 9-3. According to this plan, the lodge would be built over the water. Sketching *general relationships* helps the designer see ways to combine parts of the design.

Each part of the design plan may have several solutions. For example, the designer may consider a variety of layouts for the lodge. One layout design is shown in Fig. 9-4. It shows the general relationships of the major activities that will take place in the lodge. Several such diagrams are made. Each shows a different design. Then the designs are compared. The best features of each design will provide a basis for drawing floor plans. See Fig. 9-5.

Fig. 9-4
A preliminary idea for the lodge, in diagram form.

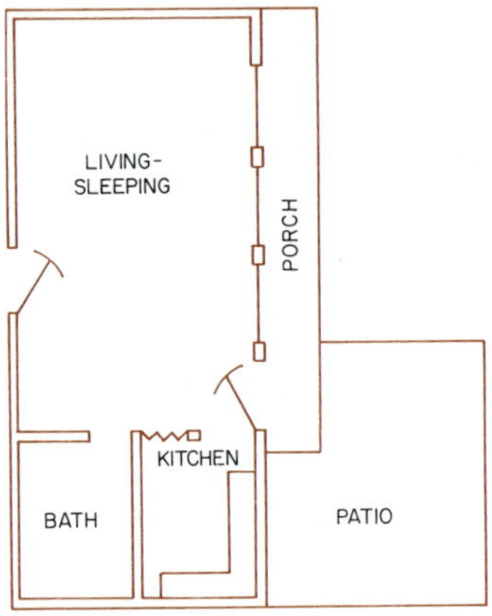

Fig. 9-5
A preliminary floor plan of the lodge.

This procedure can be followed for developing beginning ideas for all designs. You can use it, too. It can help you locate your home on a lot. It can help you plan your dream home.

Planning Living Spaces

Many activities will take place in a home. They will determine the design of the home. When planning the living spaces inside the home, the designer must

- identify family activities.
- determine area relationships.
- plan traffic flow.
- plan rooms.

Identifying Activities

Some activities take place in all homes. These are *basic*. Eating, sleeping, relaxing, and bathing are basic activities. Other activities depend on personal likes. These are *nonbasic*. Hobbies are nonbasic activities. Many nonbasic activities require a special room, such as a den or a workshop. A home is planned to meet a family's needs for basic and nonbasic activities.

Determining Area Relationships

Some activities are quiet. Others are noisy. When a home is planned, areas for noisy activities should be separated from areas for quiet activities. See Fig. 9-6. Sleeping and studying are quiet activities. Areas for these activities should be placed close together. Noisier areas are those in which people work, eat, and play. They, too, should be grouped together.

Sketching these general areas can be helpful in determining area relationships. Such sketches are called *space relationship diagrams*.

Planning the Flow of Traffic

People must be able to move easily from one area of the home to another. This movement is known as *traffic flow*. See Fig. 9-7. Which areas will people walk through most often? Perhaps halls are needed in these areas. The area used for traffic flow generally takes up about ten percent of the total living space.

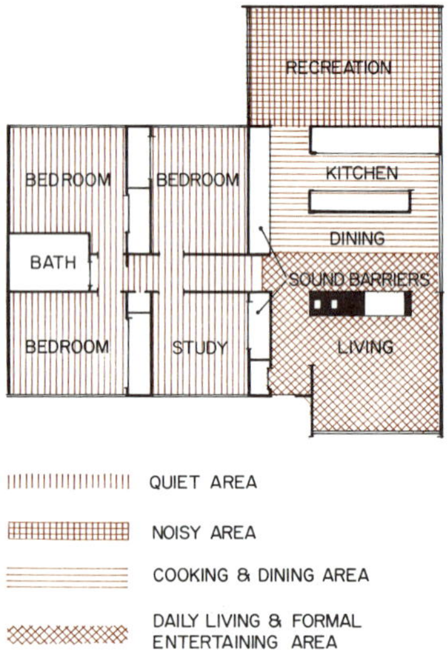

Fig. 9-6
This design is planned well. Different types of activities are in different areas.

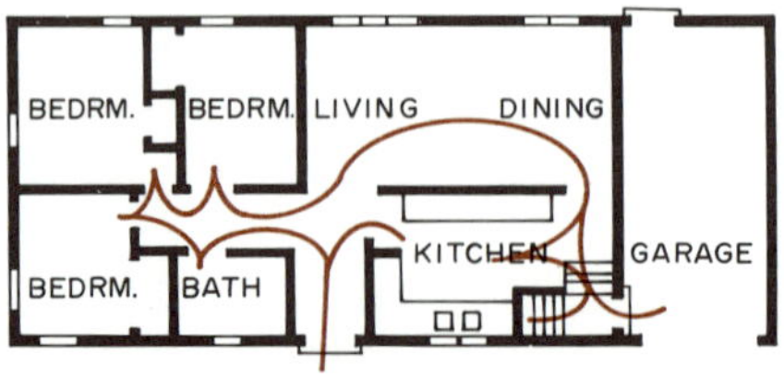

Fig. 9-7
A well-planned design allows for traffic flow. People can move from room to room without unnecessary steps.

Fig. 9-8
This room was planned for active use. Durable floor covering, overhead lighting, acoustical ceiling tile, and space for the game table help meet the room's purpose.

Planning Rooms

When rooms are planned, many things must be considered. Most important, perhaps, is the way in which the rooms will be used. The purpose of each room will largely determine the size of the room. It will also influence the planning of doors, windows, lighting, and other features. See Fig. 9-8.

Size is important. Rooms must be large enough for the planned activities. Rooms are most often planned to be slightly longer than they are wide. The width should be about two-thirds the length. For example, a room that is 6000 millimeters long should be about 4000 millimeters wide.

Another factor that influences room planning is furniture. If furniture is not considered during the planning stage, the new owners may be disappointed when they move into the home. Space for the piano or that special easy chair may not be available.

Summary

The most creative step in the design process is developing preliminary ideas. These ideas can be unusual. There should be little concern about whether a preliminary idea will work. Each new idea may suggest others.

Preliminary ideas should be sketched or diagrammed on work sheets. General notes can be added to outline the ideas. It is extra work to develop several choices. But it is worth it.

Designers carefully examine the space relationship diagrams. In the next design step, the best preliminary ideas will be refined into detailed drawings.

Terms to Know

- basic
- diagrams
- freehand sketches
- preliminary ideas
- general layout
- general relationships
- nonbasic
- notes
- space relationship diagrams
- traffic flow

Think About It!

1. Preliminary means "before the main business." Why should designers sketch and write down preliminary ideas before working out the exact details?
2. Can you create other preliminary ideas for the scout camp? Try diagramming one.

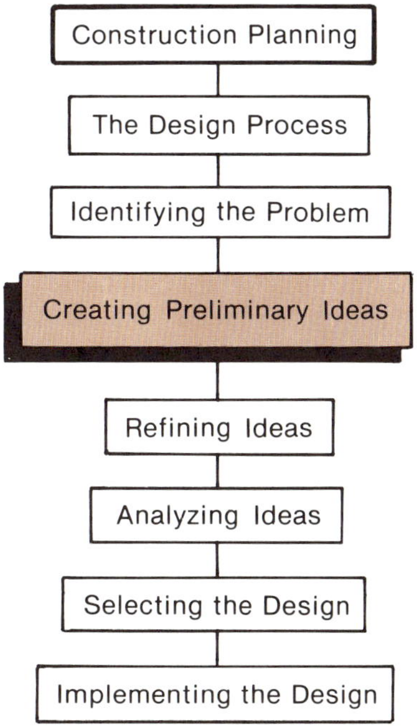

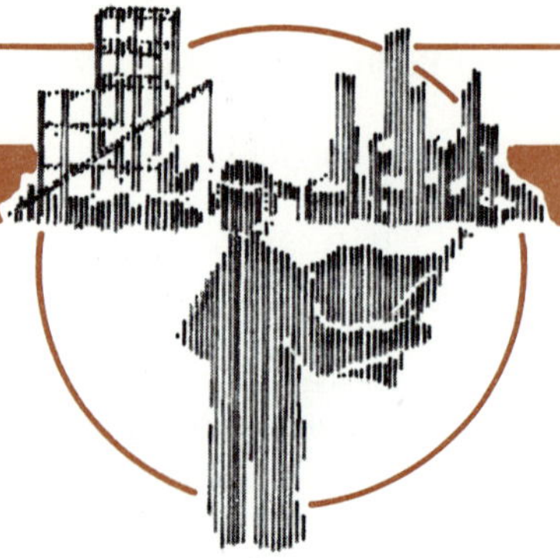

Reading 10

Refining Ideas

Refining Preliminary Ideas

Many preliminary ideas for solving a design problem may seem workable. The designer selects several of the ideas and improves them. This step is called *refining ideas.* It is the third step in the design process. Refining makes ideas more complete.

Drawing to Scale

When making freehand sketches of beginning ideas, the designer pays little attention to shapes and sizes. For example, a parking lot may be sketched as a general area on the site plan. Because the actual sizes are not shown, the sketch may give wrong ideas. The parking lot, as sketched, may be too small. Or there may be too much parking space allowed.

Beginning ideas need to be refined. They should be drawn so that sizes can be compared. This is done by *drawing to scale.*

In scale drawings, small measures represent larger measures. Areas or parts are shown in the correct size relationship to each other. When a scale drawing is made, the designer may find that parts of the design idea will not fit together. The idea may be rejected (turned down). Sometimes ways may be found to improve the idea. For example, the designer may see that the parking lot, as planned, will take up space that would be good for recreation. If the parking lot can be made smaller, the design may still work. If not, the designer will reject the design.

Selecting Ideas

Previous steps of the design process involved much creative expression. It did not matter if the ideas would not work. Refining requires making judgments about the ideas. It involves making decisions about which ideas will work best. Designers should have reasons for the decisions they make.

Designs are selected carefully. Usually, designers refine several of their first ideas. Then they compare them. Sometimes designers combine the best features of each idea. If a solution is chosen too quickly, some good ideas may be lost.

As they work, designers should refer to their first ideas, notes, and sketches. They should also keep referring to their original definition of the problem. In refining, designers try to find the best *workable solutions* to design problems.

Finding the Best Solution

At any point during the design process, designers may find that a design will not work. If this happens, they turn to other beginning ideas. They repeat this procedure until they find the best solution.

Suppose an engineer selects the most likely location for a bridge. The site is studied in detail. It is found that developing this site will be very costly. The engineer must consider other sites. There may be a site that serves the purpose at a lower cost. Suppose, however, that the cost of building a highway to the new site will be higher than the cost of the bridge. If so, perhaps the extra money needed to develop the first bridge site would be well-spent. Limits on cost or time may cause an architect or engineer to reject preliminary ideas.

The same general procedure is followed in designing a home. A variety of ideas must be refined. Many may be rejected.

Rejection is not failure. It is a way of finding the best solution. The best ideas from the rejected solutions may be worked into the new design. Studying many ideas helps the designer make the best choice.

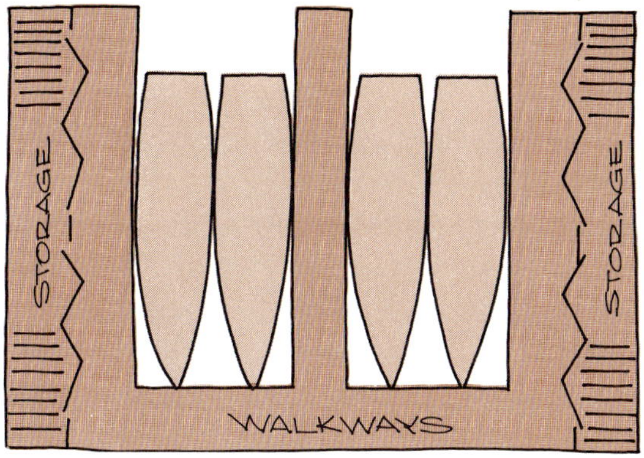

Fig. 10-1
A preliminary sketch of a boathouse floor plan.

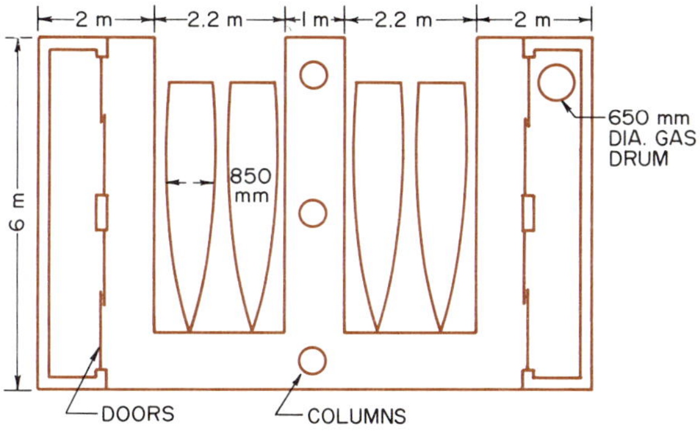

Fig. 10-2
The boathouse floor plan, drawn to scale.

Determining Sizes and Shapes

Refining involves finding the sizes and shapes of areas and structures. For example, how can the boathouse design for the scout camp be refined? When the problem was identified, it was decided that four boats would be kept in the boathouse. One preliminary idea is shown in Fig. 10-1. This sketch gives a rough idea of the general layout. It shows how the boathouse might be designed. This is not a scale drawing.

A scale drawing of the boathouse floor plan is shown in Fig. 10-2. This drawing shows the actual measurements of the boats and the items to be stored. The width of walkways is also given. All these *dimensions* are used to determine the dimensions of the boathouse.

Just as boat sizes affect the planning of a boathouse, furniture sizes affect the planning of a home. Suppose a king-size bed must fit in a bedroom. The bedroom must be more than large enough to contain the bed. If other furniture is to be added, the room must be even larger. As with the boathouse,

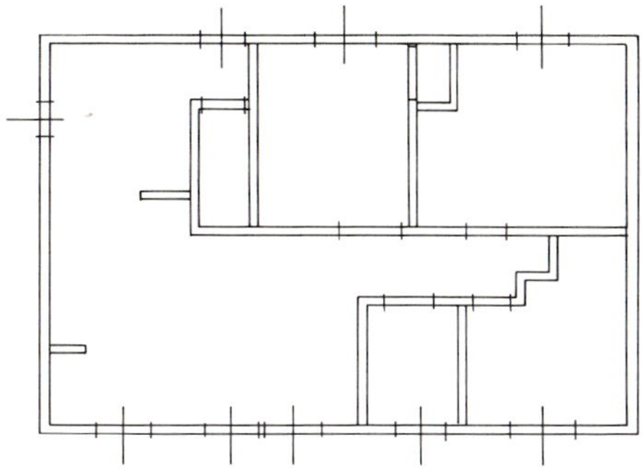

A. *The basic plan for a home.*

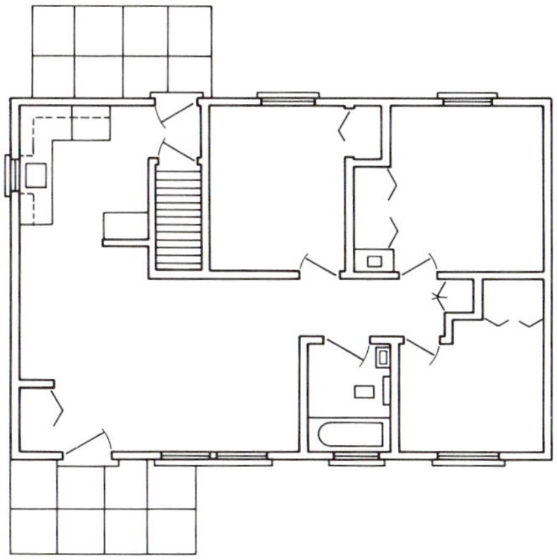

B. *Details for doors, windows, stairs, cupboards, and bathroom are added.*

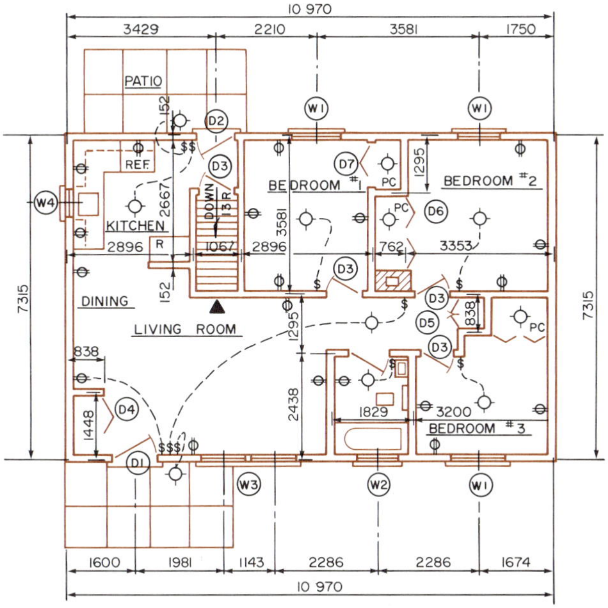

C. *Dimensions and electrical symbols complete the plan.*

Fig. 10-3

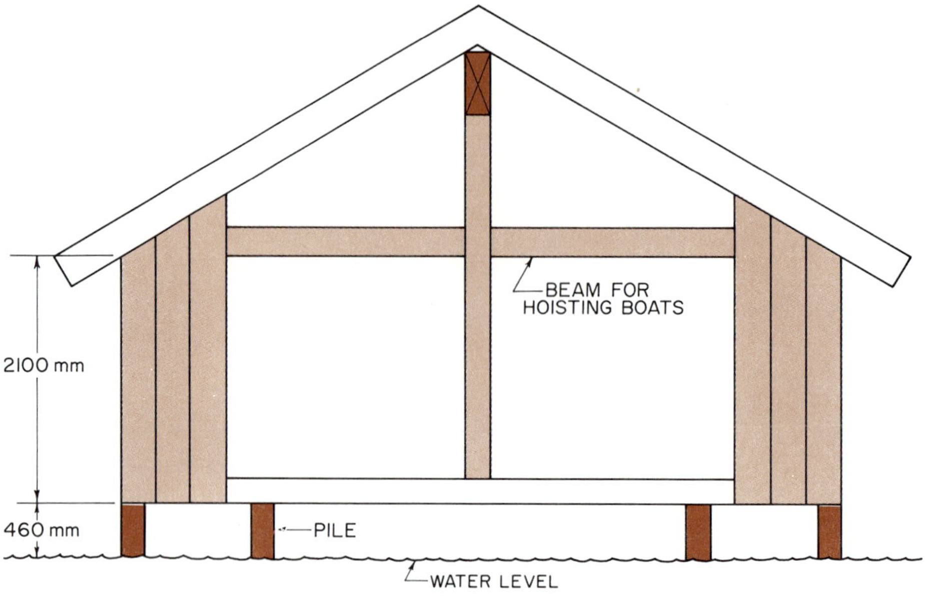

Fig. 10-4
A refined drawing of the exterior of the boathouse.

many factors must be considered in making the floor plan of a house. Refining may change the floor plan. As further details are known, rooms may be made larger or smaller. Other refinements are required. See Fig. 10-3. Dimensions — such as roof height, floor area, storage space, and parts of the structure — may need adjustment. This can be done as the design is improved.

The floor plan is basic to all plans. Even so, it is just the beginning point for a workable design. Thought must also be given to the desired type of roof, outside walls, and general exterior. An exterior refinement of the boathouse is shown in Fig. 10-4.

As the design is improved, details of construction become important. For example, how will the boats be hoisted (lifted) out of the water? What material should be used to build the boathouse? How much will construction cost? As refining continues, answers must be suggested for each of these questions.

Summary

Refining involves judging beginning ideas. These ideas are improved, drawn to scale, and studied. The designer tries to get a good idea of the size, shape, and details of a design. Refinement drawings are not finished plans from which a project can be constructed. They are general drawings. They show important details that should be refined further.

While refining ideas, designers may see problems that cannot be solved easily. Then they select other ideas to improve. Designers should also try to work the best features of the rejected ideas into the new designs. Rejection is a natural step in finding the best solution to a problem.

Designers should not put all their efforts into a single idea. They should refine several workable ideas. They will then be in a better position to select the best design. Before making a final decision, designers must study each design. They will study its function (use), strength, and cost. This is called analyzing ideas. It is the next step of the design process.

Terms to Know

- dimensions
- drawing to scale
- refining ideas
- workable solutions

Think About It!

1. When a design idea is refined, exact sizes become important. What dimensions will you need to study if you are designing a family room?
2. Suggest some things that could go wrong if a designer does not carefully work out the family room sizes and dimensions.

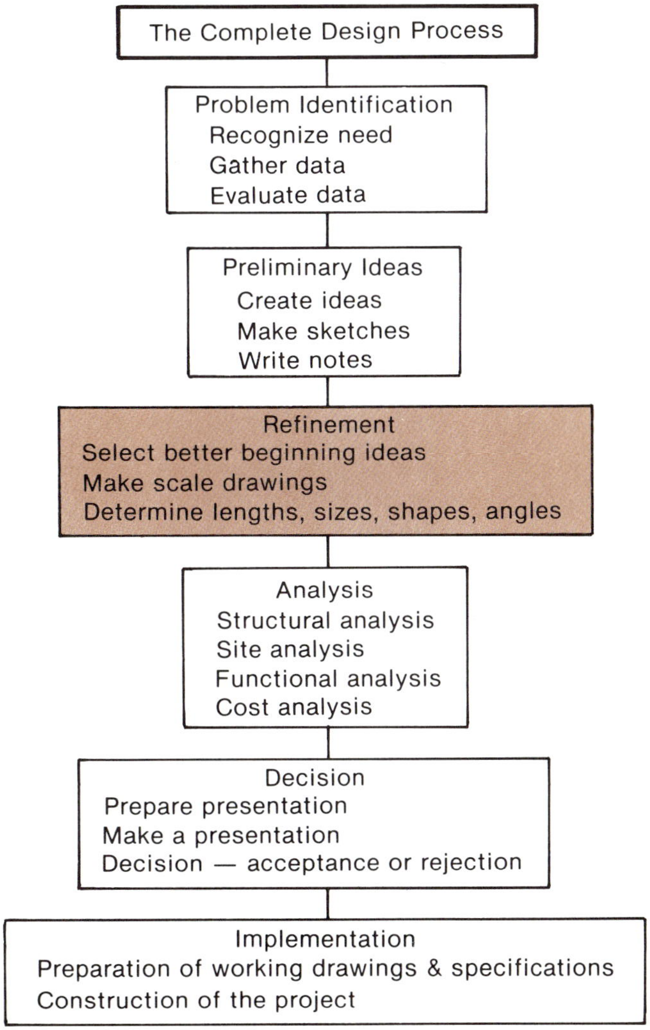

Reading 11

Analyzing the Architectural Design

Analyzing Preliminary Designs

After preliminary designs are refined, they are *analyzed*. Studying designs in detail helps architects and engineers find problems. See Fig. 11-1. Does the design meet the purpose of the project? A *functional analysis* answers this question. Does the design take advantage of features of the site? Studying the site as it relates to the design is *site analysis*. Are all parts of the structure strong enough? A *structural analysis* uncovers weak points in the structure's design.

Functional Analysis

Functional analysis shows whether the design will work. The designer can see if the project will do what it is supposed to do.

A scout camp design must provide all the necessary structures and features. Will the lodge be large enough for the troop? Will one diving board be enough? Is the lodge far enough from (or close enough to) the lake? Perhaps the fishing dock is too small for several people to fish at one time. Can boats be lifted out of the lake easily? See Fig. 11-2. All these questions must be answered.

An architect must answer the same kinds of questions when planning a house. Is the front entrance large enough? Are there enough bedrooms? Is the stairway to the basement wide enough? Is the garage large enough to store tools and garden equipment?

Questions of function must be asked for every work of construction. For a small structure, only a

Fig. 11-1

Architects and engineers study the refined drawings in detail.

Fig. 11-2
Boats must be stored dry, in a suspended position. Designers analyze the method used to lift boats from the water.

few questions may need to be answered. But analyzing the function for a very large construction project can be complex. Many questions must be answered. Designers must be sure that what they design will do the job.

Site Analysis

The site chosen for a structure must be studied carefully. Now is the time to identify problems that can ruin an otherwise good project. Analyzing the site can help improve the project. It will ensure better use of the site.

Some areas of a site are better suited for certain activities than others. Think about a parking lot for the scout camp. It should be located on level ground, near the main road. This will make parking easy. The access road (from main road to lot) will not be too long. Also, little *earthmoving* will be needed to prepare the site.

It is important to make a site analysis when choosing a site for a road. A roadway must not be too steep or have big dips. The roadway site must be prepared. Suppose you wish to back a boat trailer to the lakeshore. This will be difficult without site and roadway preparation. Study Fig. 11-3. The

Fig. 11-3
Designers must analyze the shape of the land to be used for boat launching. A steep incline can cause damage to vehicles.

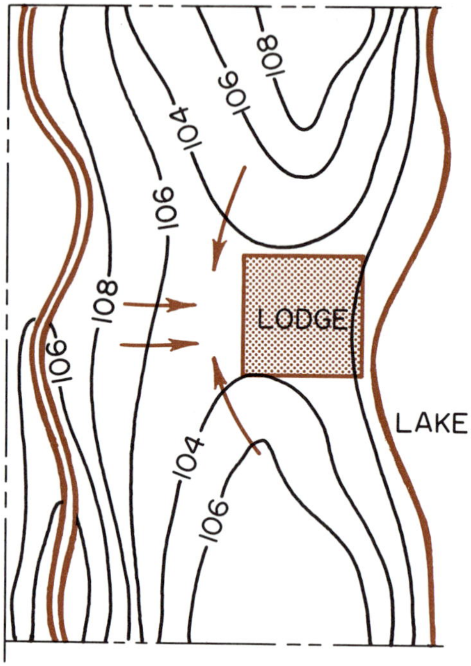

Fig. 11-4
Designers analyze the site by studying contour maps. If the lodge were located in this position, it could be damaged by draining water.

hill may be too steep. Some earth must be moved to make a gradual incline (slope).

It is very important to do site analysis before building a structure. People planning a house may want it to have a basement. But what if there is solid rock just under the surface of the site? It will cost too much to make the *excavation* (hole). Or, what if the site has a drainage problem? For example, refer to Fig. 11-4. The scout lodge should be on higher land. If the lodge is placed on low land, there could be trouble in the future. Water collects in low areas. It may damage the building and the things inside.

Every type of construction project is placed on a site. Each site must be analyzed. A good analysis helps designers find the best possible solution to the design problem.

Structural Analysis

The design of the structure is analyzed. Each part must be the right size. Each must be strong enough to serve its purpose. Structural analysis can be done by an architect or engineer.

Making a structural analysis requires special knowledge. Suppose a diving board is to be built for the scout camp. The engineer must be sure it can withstand hard use. See Fig. 11-5. The engineer figures the maximum load (weight or force) that can

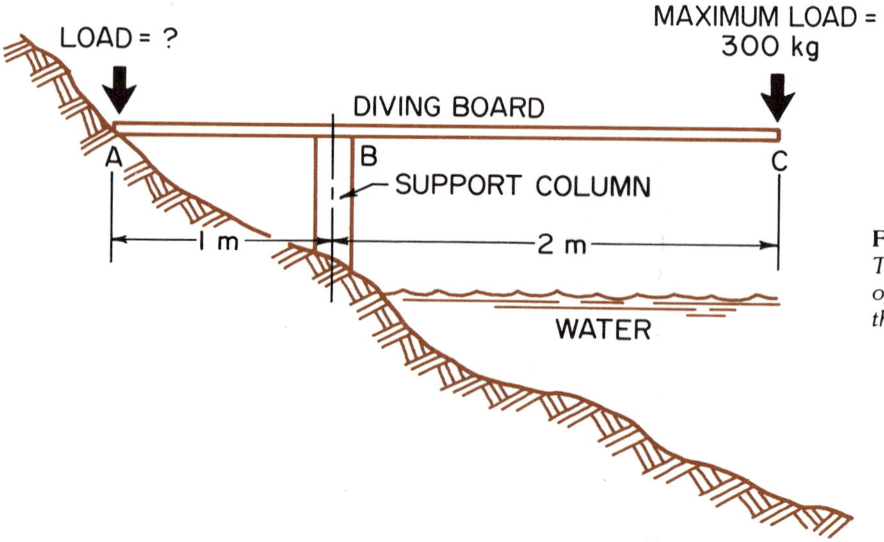

Fig. 11-5
The engineer analyzes the design of the diving board to make sure the board can be used safely.

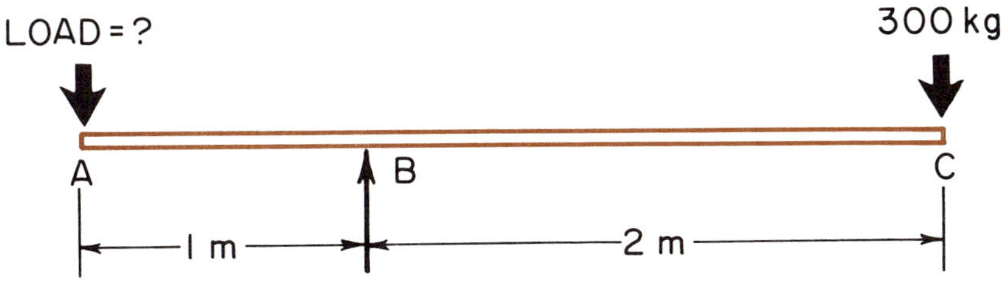

Fig. 11-6
Engineers use mathematics and physics as their tools to analyze designs such as that of this diving board.

be applied to the end of the board. Suppose a swimmer jumps at point C. The load at point A must be enough to make up for the load at point C. The load at point A holds the board in place when the swimmer jumps. The board is supported at point B. This makes it work like a lever. Engineers use knowledge of mathematics and physics to analyze the design. See Fig. 11-6. Note the safety factor. Engineers plan structures to support more weight than is expected. In this case, the safety factor is two. This means that the board can support twice the expected maximum weight. It ensures that the anchor on the shore end will hold. Rock or concrete could be used to hold the board in place. A dead weight (mass) of 1200 kilograms (kg) would hold the board.

Structural analysis is important in building construction. All materials used to build a house must be the correct size. Each part must be strong enough to carry its load. For example, floor joists (beams) must support the weight of the floor and everything on it. Study the two sizes of floor joists shown in Fig. 11-7. When weight is applied, the lighter (narrower) joist bends more than the heavier (wider) joist. This bending is called *deflection*. The lighter joist under a finished floor will not support people and furniture. A floor joist should not deflect more than one unit for every 360 units of length (1 to 360). This means the part should not bend more than about 14 millimeters in a 5000-millimeter span.

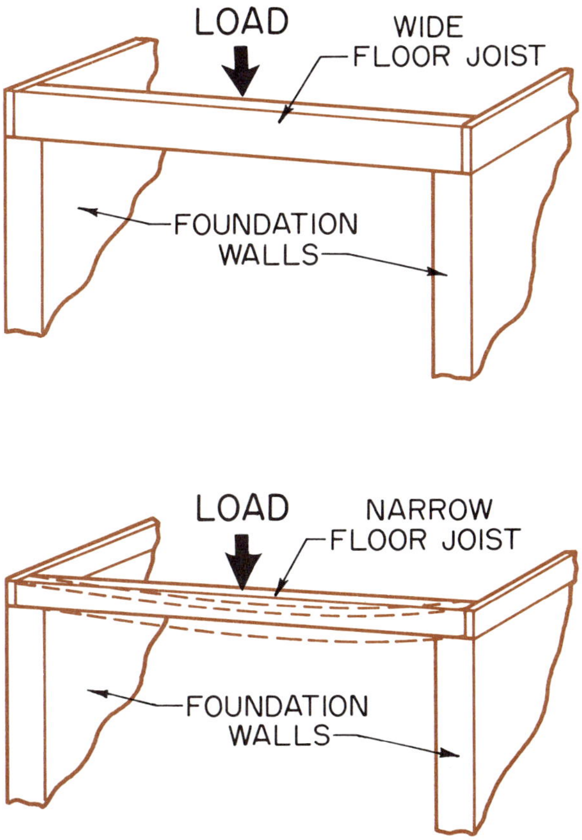

Fig. 11-7
A floor joist must be strong enough to carry the load.

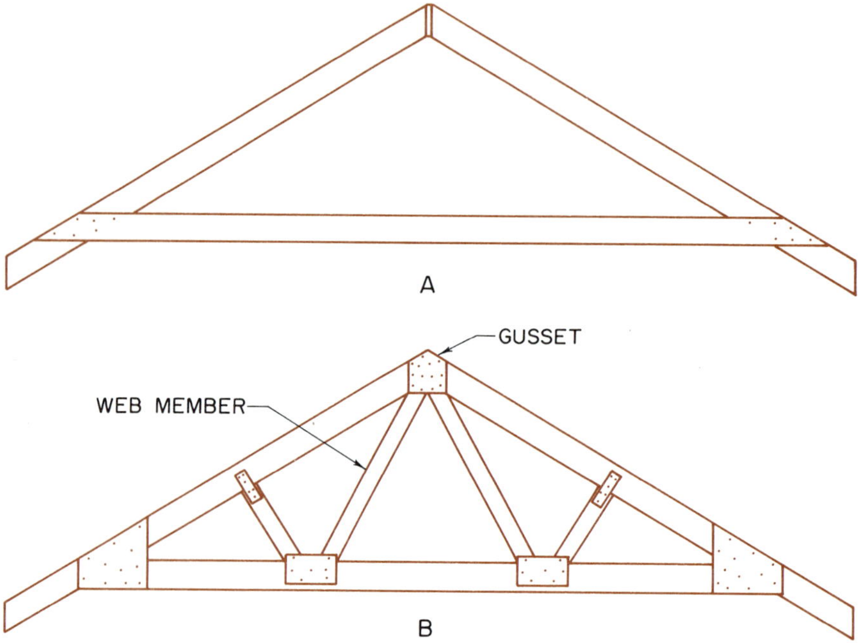

Fig. 11-8
Conventional roof framing (A) is not as strong as a roof truss (B).

Roof members of houses need to be strong. Not long ago, the cross section of a roof looked like the one shown in Fig. 11-8A. Today, engineers design *roof trusses*. See Fig. 11-8B. Note the web members and gussets. These features help the roof support more weight. They also help keep the ceiling from sagging.

Engineers must know the maximum load that each part will be required to support. Then they can determine the size of parts. This is structural analysis. See Fig. 11-9.

Tektronix, Incorporated

Fig. 11-9
Engineers often use computers when making a structural analysis.

Summary

Analyzing is studying in detail. Architects and engineers analyze designs. They must determine whether designs will work under certain conditions. The function of the project is analyzed. The design must suit the purpose. The site, too, is studied carefully. The design must use the features of the site in the best possible ways. Every part of a structure is carefully designed. Each must be strong enough to carry its load.

Sometimes problems are found. A refined preliminary design may not work. If it does not, the designer may have to return to other design ideas. A workable solution is found by analyzing the architectural design.

Terms to Know

analyzed
deflection
earthmoving
excavation
functional analysis
roof trusses
site analysis
structural analysis

Think About It!

Suppose that you have been asked to buy and install recreation equipment for a neighborhood playground.

1. What function problems must you think about?
2. What site problems do you need to examine?
3. What special structural problems might require study?

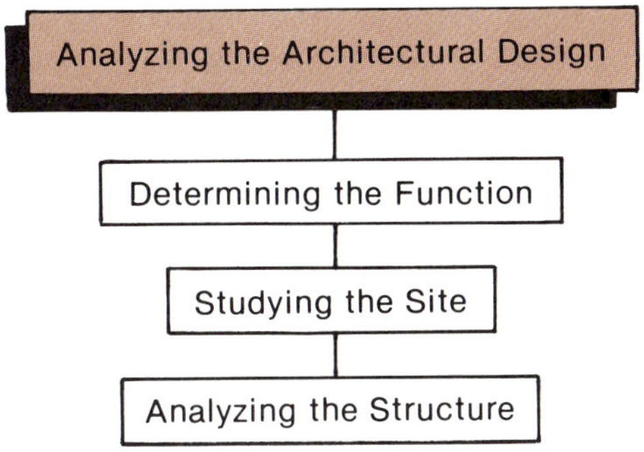

Reading 12

Analyzing the Engineered Design

Assuring Project Success

Project success must be assured before construction can begin. Engineers must be satisfied with the design. It must serve the purpose of the project in the best possible way. To make sure it will, engineers *analyze* the design. They study each of its parts to find what will work best. It is hard to analyze a whole bridge. But engineers do this. They study it part by part. A bridge is only as strong as its parts.

Some engineers are involved in the entire design process, including the analyzing step. Others work mostly in analyzing.

Designs are analyzed in several ways. A *functional analysis* is done. This helps find whether a project will actually work well. A *site analysis* is also done. This tells whether the site is suitable for the structure. *Structural analysis* is done to find whether the structure is strong and safe. *Cost analysis* tells whether the project can be paid for.

Fig. 12-1
Imagine the work required to build the Glen Canyon Project! Each part had to be analyzed and planned carefully. Engineers had to make sure the design would work.

F.S. Finch — U.S. Department of the Interior

Functional Analysis

Finding out whether a project design will work is a complex task. It often takes a long time.

Let's look at how a huge project in Arizona, the Glen Canyon Project, was analyzed. The general plan was studied. This was done to determine whether the project could be built successfully.

The project was proposed years before it was built. It included a dam, a bridge, and a power plant. See Fig. 12-1. The usual design steps were followed. The problem was described in detail. A number of possible design solutions were created. The most promising of these were refined. Then a more careful analysis began.

The proposed Glen Canyon bridge was to be the world's highest single-span steel arch. Could it stand all the expected wind and other conditions? Could the structure be built at such height?

The dam was to rise 216 meters. It would require over four million cubic meters (m^3) of concrete. Could so much concrete be placed in the canyon? Could a dam that high hold the water behind it?

The generating plant was expected to generate enough energy for one and a half million people. Could this much energy be distributed without too much loss of electricity? Could turbines withstand the pressure of a 200 meter waterfall?

Answers could only come after much detailed study. Who could answer such questions? Who could determine whether the design would work? Engineers were to provide the answers.

As they studied the project, engineers used much data. For example, they searched weather records. What were the highest winds ever recorded in the area? What was the average rainfall? the least rainfall? the greatest rainfall? The drainage area had to be outlined.

New information was needed, too. No one knew how much traffic would use the bridge. No roads had been built in the area yet. The project would cause a city to develop. This city's size and needs had to be estimated.

No such bridge or dam had ever been built. Therefore, some new production technology was needed. Could available materials be used? Could the water be held back to allow construction? See Fig. 12-2. Would bypass tunnels be needed? Would a temporary dam overflow before the new dam was ready?

A functional analysis has to look even beyond construction. For example, how fast would the lake silt up? That is, would the waters flowing into the lake behind the dam carry much soil? If so, as the water slowed, silt would drop. The lake would become a large mud puddle. The dam would not serve its function well.

Site Analysis

Many details about the site are analyzed. The bearing surface for the dam is bedrock. It is carefully studied for hidden cracks that will allow leakage. During construction, excess water must be run around the dam site. Tunnels will be *bored* to carry this

Fig. 12-2

Two temporary dams (cofferdams) were built to keep water away from the construction site.

A.E. Turner — U.S. Department of the Interior

A.E. Turner — U.S. Department of the Interior

Fig. 12-3
Here, a tunnel has been built to allow water to flow around the construction site. On another site, a ditch might be dug to route water away.

water away. See Fig. 12-3. An analysis determines where and how this will be done. The site is also analyzed to find the best place to *stockpile* materials. See Fig. 12-4.

A great many other site details must be studied further. Again, engineers know the technology needed to provide answers. They can tell whether water currents will *erode* the rock. They can find how much foundation area is needed. And they can tell how the site should be blasted. That is, they can tell how explosives should be used to loosen rock.

Everything that is built rests on a foundation. Engineers who design foundations must be schooled in *soil mechanics*. They must also learn *hydraulics*.

Fig. 12-4
Deciding where to locate roads and stockpile materials requires careful site analysis.

A.E. Turner — U.S. Department of the Interior

Kerr-McGee Corporation

Fig. 12-5
Engineers keep a close check on construction progress. Here, engineers are surveying a tunnel to be sure it is being built as planned.

This is the study of how flowing water can be used to produce energy and how the flow affects land. This knowledge helps engineers analyze sites. Surveying and soil testing are parts of analyzing. Engineers must do much analysis to make sure sites will be used well. See Fig. 12-5.

Structural Analysis

Civil engineers are the *skeleton* analysts of industry. They must make sure that no structural part fails. See Fig. 12-6. They work with electrical and

Fig. 12-6
Even the tiniest structural details must be analyzed and planned carefully.

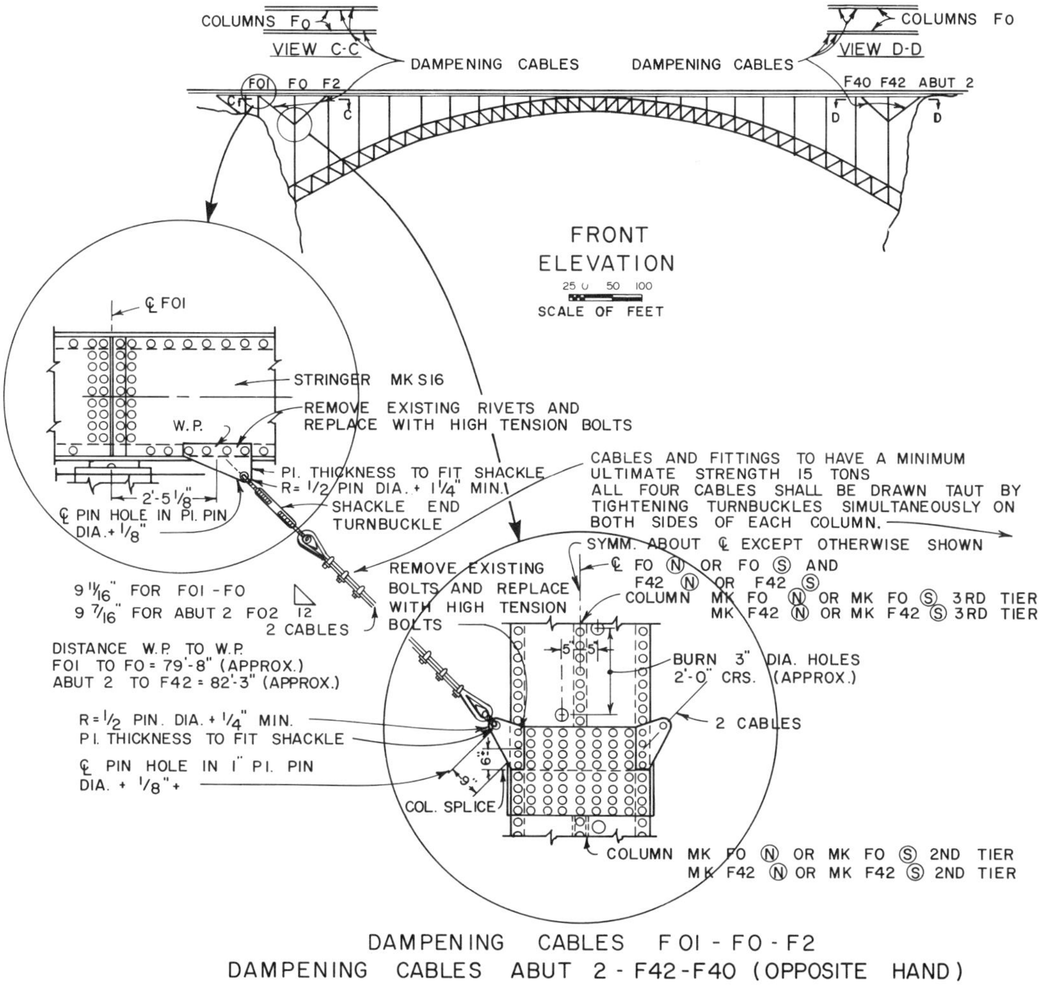

mechanical engineers. Together, the engineers control the quality of structures. They also assure that *systems* in the structures work well.

Structural analysts must know the nature of construction materials. They also must know how these materials react under various conditions. They must know, for example, how materials bend or drain. With such knowledge, engineers make certain structures will be safe and efficient.

Cost Analysis

Engineers are concerned about project costs from start to finish. Detailed costs are figured when the design is analyzed. One major cost is for the site. Other costs are for materials and labor.

Engineers may seek the help of such persons as bankers, accountants, purchasing agents, and lawyers. Every possible cost must be known. And each must be the best use of funds.

Summary

Engineers analyze designs to see if they will work. They analyze each design to make sure the function, site, and structure will be adequate. They also study each design to know what it will cost.

Engineers use very specialized construction technology to analyze designs. They must know construction materials, processes, and costs. With their special knowledge, they help make designers' ideas into completed projects.

Terms to Know

analyze	site analysis
bored	skeleton
cost analysis	soil mechanics
erode	stockpile
functional analysis	structural analysis
hydraulics	systems

Think About It!

Suppose you want to build a heavy construction project such as a bridge, dam, or highway. Select just one project and answer these questions.

1. What problems of function need to be considered?
2. What site problems will you need to study?
3. What special structural problems will you need to study?
4. What cost factors must be considered?

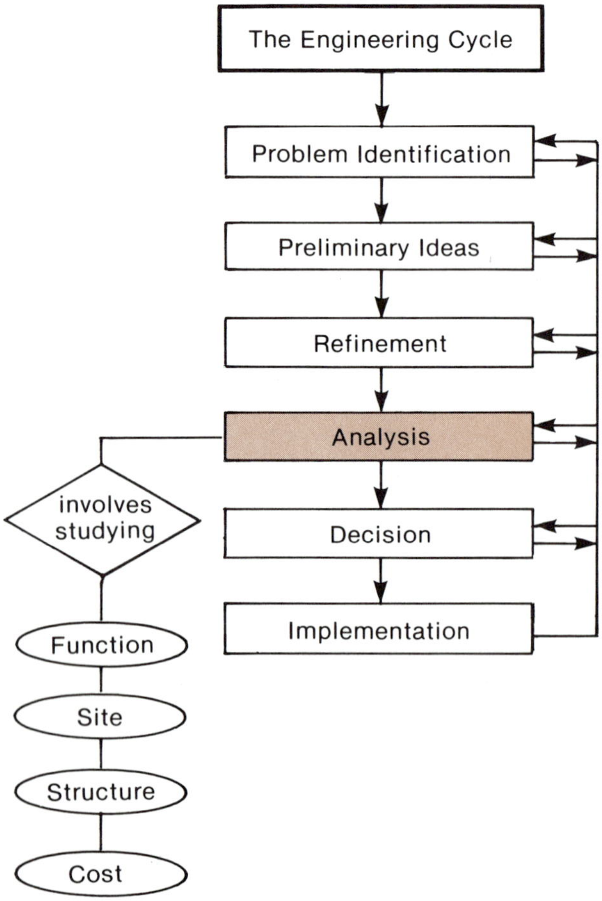

Reading 13

Selecting the Architectural Design

Making the Decision

Which of the refined designs is best? The structure must meet certain needs. Which design does this best? To find the answers to the questions, each design solution is analyzed. Each detail is studied. Poor solutions are discarded. Remaining solutions may all be good ones. Now a decision must be made. This is the process of *selecting*. It is the *decision* step of the design process.

Selection must be made carefully. If designs are not analyzed well, the best one may be rejected. Or, a poor design solution may be selected. This reading is about the final selection process. It is also about the *contracting* process.

People Who Make the Decision

There are many types of construction projects. Therefore, different people may be involved in choosing a design. Sometimes the designer (architect or engineer) selects the final plan. In many cases, a *client*, usually the owner, makes or helps make the design decision. Sometimes a project will serve citizens in a community. See Fig. 13-1. Then city council members make the final selection. Often they must approve the designer's choice.

The designer of an outdoor patio probably makes the design decision. The designer recommends one design to the client as the best solution. For a more complex project, such as a house, the decision may be made by the architect. Or, the architect may present several designs to the client. The client then selects a design. For a public housing project, the designer creates several designs. One or two designs are presented to a city council. The city council considers costs and many other factors. Then it reaches a decision.

Decision by the Designer

The designer may select the design. An architect, for example, must be very familiar with the design problem. To find a solution, the architect prepares several sketches and scale drawings. Pictorial views and even scale models are made. See Fig. 13-2. All these things help the architect compare the best designs. Next, the advantages and disadvantages of each design are listed. This list guides the architect in arriving at the final decision.

Fig. 13-1
Making a design decision is not easy. After the presentation, city council members may ask the designer many questions.

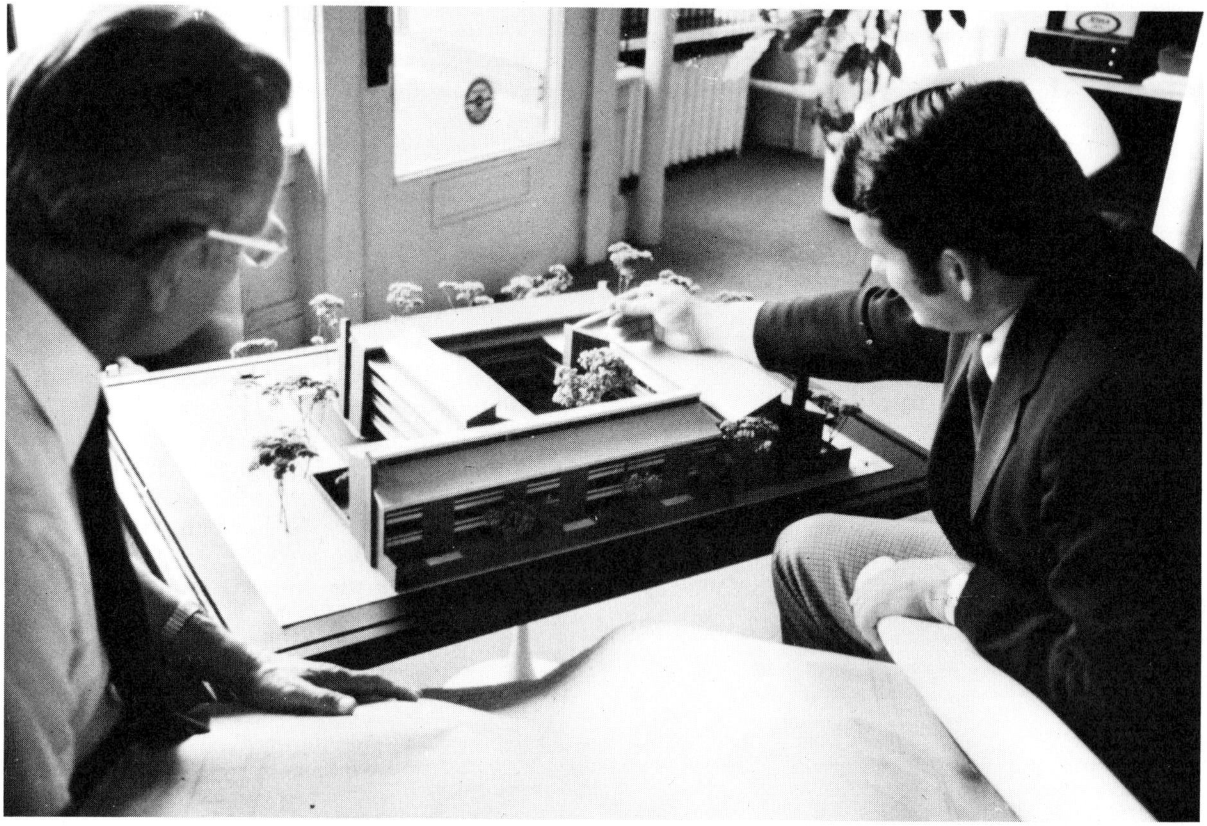

Fig. 13-2
To solve a design problem, the architect may prepare scale drawings or models.

Sketches, drawings, and models are made. These help the designer choose the best design solution. There is no *presentation* to other people. The decision is made by the designer alone.

Decision by Others

The design may be picked by one person or by a group. The one person (client) is usually the owner. The group may be a city council or a company's board of directors. Or, it may be a similar deciding committee from a larger group. In any case, the problem must be approached in much the same way as just described. But the designer must make a presentation of the ideas. He or she must provide the client or group with all the information about the design. This information gives a sound basis for making the decision.

The designer must explain the design ideas step by step. One design may be better than another. However, it may cost more to build. The design explanation may convince the client or group to increase the budget. The more expensive solution may be the best one in the long run.

A deciding committee is often made up of people who do not understand the technical details of a design. They come from different backgrounds. Therefore, all details of the design problem must be presented clearly. The designer must prepare materials that help identify all points to be considered.

Presentation of the Design

A presentation may be given to one individual. It may be given to a small group. For very large projects, 100 or more persons may attend the design presentation.

Successful presentations often depend on the use of *visual aids*. See Fig. 13-3. The designer may use models, flip charts, pictures, and slides. Diagrams or graphs may also help explain the design.

A set of working drawings may be presented. If the project is a simple one, detailed drawings are often presented. The drawings are usually supported with sketches, diagrams, or models.

Some projects are complex. For example, consider an airport addition. Detailed drawings for such a project would take a long time to prepare. This would be expensive. For this reason, complex projects are presented only in general form. Detailed drawings are not made until after the decision is made to build.

Architects may construct a scale model of an entire city development. Most people can understand a model more easily than a complex set of drawings. Sometimes models cannot be used. For example, if the group is quite large, a model may be difficult to see. Then pictorial drawings can be used. These show the appearance of a finished project. At times, photographic slides are used to describe a design solution. Cost of construction, tables of figures, and other information can be shown easily on a graph. Such visual aids save decision-makers many hours of study. These aids make important information easy to understand. This helps the client or group make a wise decision. The designer must keep the decision-makers in mind when developing and presenting the aids.

General Dynamics Corporation

Fig. 13-3
Visual aids are used to help people understand complex information.

Plans for Contracting

After making the design decision, the client or group must arrange for the work to be done. This step is called *implementation*. For a small project, like a home, all planning has already been done. A complete set of drawings has been prepared. Now, the specifications are prepared. For a large project, such as a subway or skyscraper, detailed drawings are made after the design decision. Other engineering activity also begins. All drawings and specifications must be made ready for the contracting process. This all takes money.

Sometimes the client or group has money on hand. More often, money is borrowed from a bank or other *lending agency*. See Fig. 13-4. The client or group makes special arrangements to obtain the money from the lending agency. Then a contract is *negotiated*.

If the project is a home, the owner may make a contract with the architect. The architect can supervise (oversee) the entire project. Often the owner makes a contract directly with a *general contractor*. An owner may act as the contractor by arranging for subcontractors. Some owners even build their own homes. But most owners have work done through a contractor.

Summary

The design process involves a decision step. A design must be accepted, changed, or rejected. Some designers pick a design for their clients. Most designers must develop and present several good designs. Then other persons make the final selection.

Loan Application

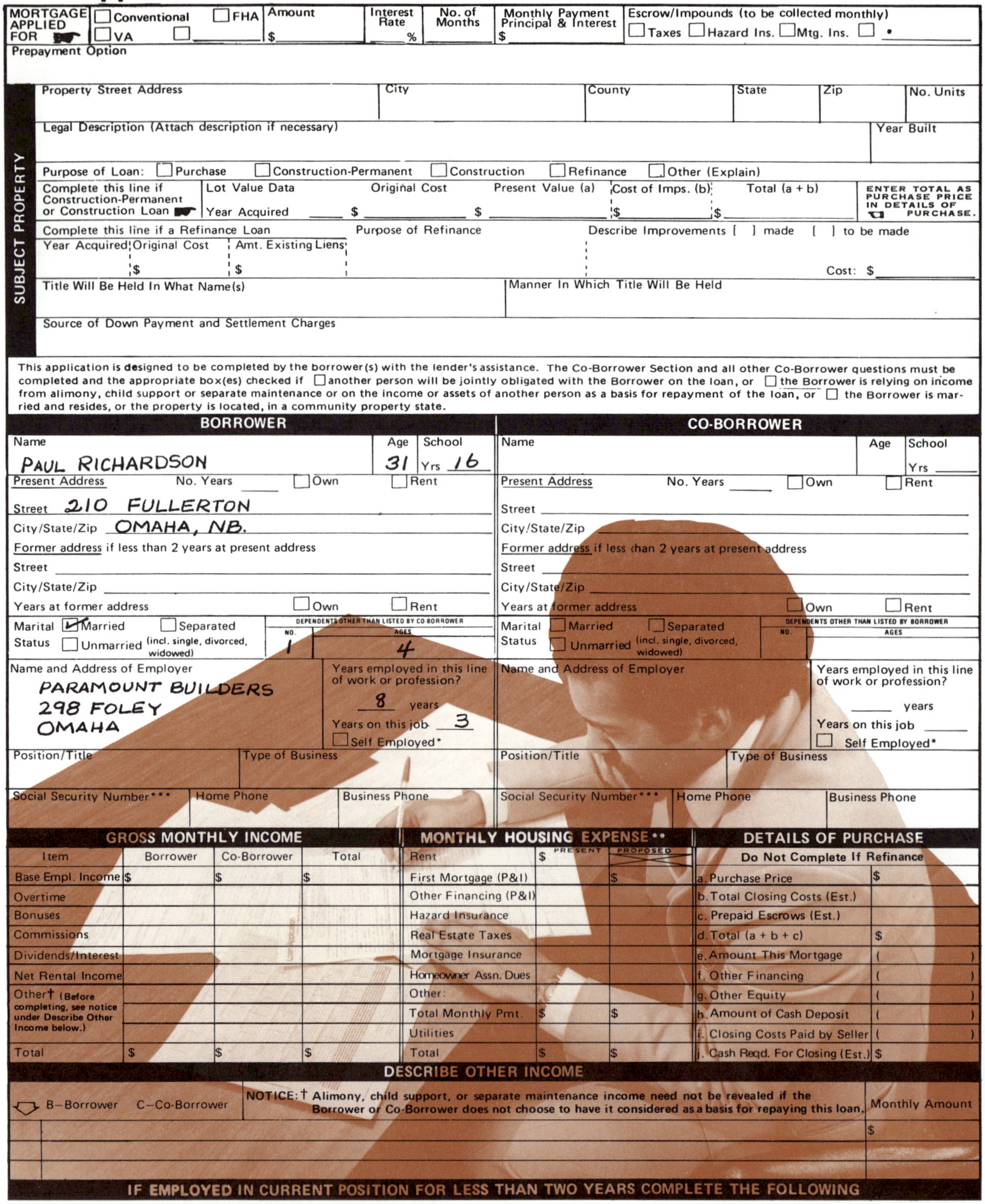

MORTGAGE APPLIED FOR ☐ Conventional ☐ FHA ☐ VA ☐ ______ | Amount $ | Interest Rate % | No. of Months | Monthly Payment Principal & Interest $ | Escrow/Impounds (to be collected monthly) ☐ Taxes ☐ Hazard Ins. ☐ Mtg. Ins. ☐ ______

Prepayment Option

SUBJECT PROPERTY

Property Street Address | City | County | State | Zip | No. Units

Legal Description (Attach description if necessary) | Year Built

Purpose of Loan: ☐ Purchase ☐ Construction-Permanent ☐ Construction ☐ Refinance ☐ Other (Explain)

Complete this line if Construction-Permanent or Construction Loan | Lot Value Data | Year Acquired | Original Cost $ | Present Value (a) $ | Cost of Imps. (b) $ | Total (a + b) $ | ENTER TOTAL AS PURCHASE PRICE IN DETAILS OF PURCHASE.

Complete this line if a Refinance Loan | Purpose of Refinance | Describe Improvements [ ] made [ ] to be made

Year Acquired | Original Cost $ | Amt. Existing Liens $ | Cost: $

Title Will Be Held In What Name(s) | Manner In Which Title Will Be Held

Source of Down Payment and Settlement Charges

This application is designed to be completed by the borrower(s) with the lender's assistance. The Co-Borrower Section and all other Co-Borrower questions must be completed and the appropriate box(es) checked if ☐ another person will be jointly obligated with the Borrower on the loan, or ☐ the Borrower is relying on income from alimony, child support or separate maintenance or on the income or assets of another person as a basis for repayment of the loan, or ☐ the Borrower is married and resides, or the property is located, in a community property state.

BORROWER

Name: PAUL RICHARDSON | Age: 31 | School Yrs 16

Present Address | No. Years ______ ☐ Own ☐ Rent

Street: 210 FULLERTON

City/State/Zip: OMAHA, NB.

Former address if less than 2 years at present address

Street

City/State/Zip

Years at former address ☐ Own ☐ Rent

Marital Status ☑ Married ☐ Separated ☐ Unmarried (incl. single, divorced, widowed)

DEPENDENTS OTHER THAN LISTED BY CO-BORROWER — NO.: 1 — AGES: 4

Name and Address of Employer: PARAMOUNT BUILDERS, 298 FOLEY, OMAHA

Years employed in this line of work or profession? 8 years

Years on this job: 3

☐ Self Employed*

Position/Title | Type of Business

Social Security Number*** | Home Phone | Business Phone

CO-BORROWER

Name | Age | School Yrs ______

Present Address | No. Years ______ ☐ Own ☐ Rent

Street

City/State/Zip

Former address if less than 2 years at present address

Street

City/State/Zip

Years at former address ☐ Own ☐ Rent

Marital Status ☐ Married ☐ Separated ☐ Unmarried (incl. single, divorced, widowed)

DEPENDENTS OTHER THAN LISTED BY BORROWER — NO. — AGES

Name and Address of Employer

Years employed in this line of work or profession? ______ years

Years on this job ______

☐ Self Employed*

Position/Title | Type of Business

Social Security Number*** | Home Phone | Business Phone

GROSS MONTHLY INCOME

Item	Borrower	Co-Borrower	Total
Base Empl. Income	$	$	$
Overtime			
Bonuses			
Commissions			
Dividends/Interest			
Net Rental Income			
Other† (Before completing, see notice under Describe Other Income below.)			
Total	$	$	$

MONTHLY HOUSING EXPENSE**

	PRESENT	PROPOSED
Rent	$	
First Mortgage (P&I)		$
Other Financing (P&I)		
Hazard Insurance		
Real Estate Taxes		
Mortgage Insurance		
Homeowner Assn. Dues		
Other:		
Total Monthly Pmt.	$	$
Utilities		
Total	$	$

DETAILS OF PURCHASE

Do Not Complete If Refinance

a. Purchase Price	$
b. Total Closing Costs (Est.)	
c. Prepaid Escrows (Est.)	
d. Total (a + b + c)	$
e. Amount This Mortgage	( )
f. Other Financing	( )
g. Other Equity	( )
h. Amount of Cash Deposit	( )
i. Closing Costs Paid by Seller	( )
j. Cash Reqd. For Closing (Est.)	$

DESCRIBE OTHER INCOME

B—Borrower C—Co-Borrower

NOTICE: † Alimony, child support, or separate maintenance income need not be revealed if the Borrower or Co-Borrower does not choose to have it considered as a basis for repaying this loan.

Monthly Amount $

IF EMPLOYED IN CURRENT POSITION FOR LESS THAN TWO YEARS COMPLETE THE FOLLOWING

Fig. 13-4

In order to borrow money, it is necessary to fill out a loan application.

Design ideas must be presented very clearly. A good design may be rejected if it is presented poorly. The designer must be able to present the solution in understandable form. Visual aids are very useful in this process.

After the design is selected, it must be implemented. For a residence, usually the owner arranges to get the work done. A contract is negotiated and signed. This completes the design cycle.

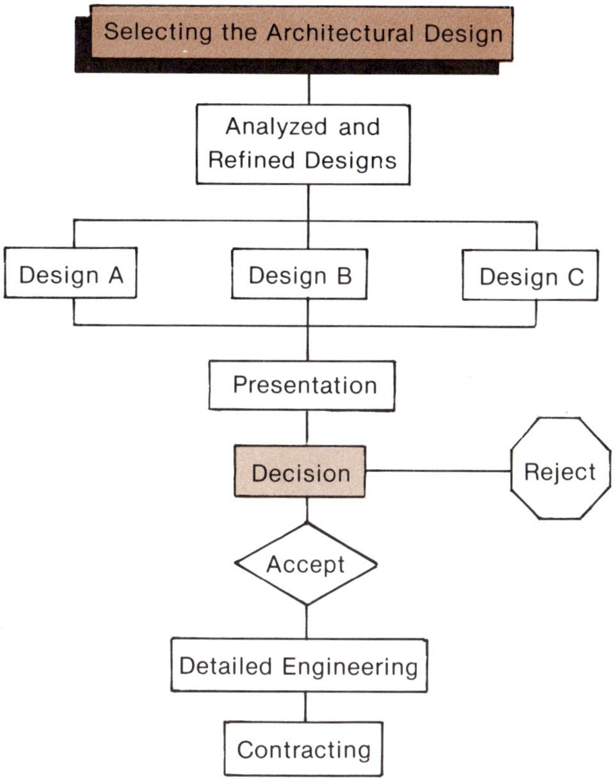

Terms to Know

client	lending agency
contracting	negotiated
decision	presentation
general contractor	selecting
implementation	visual aids

Think About It!

1. Who would make the decision about a design for a new school building in your community?
2. What experience and training would you need to build your own home without a contract? Do you think you would ever try this?

Reading 14

Selecting the Engineered Design

Conoco Incorporated

Fig. 14-1
A petro-chemical plant must be designed carefully. People within the plant must approve the design.

The Decision Step

Selecting the best design for a complex construction project carries great responsibility. A good choice leads to satisfaction with the project. The project will serve its purpose well for years to come.

Engineers refine several designs. Each design has been analyzed in detail. Each seems to solve the design problem. Which one will be chosen? One design may have more advantages than the others. Selecting the engineered design is called the *decision step* of the design process.

People Who Make the Decision

Large, complex projects are usually built for private or public groups. A number of people help select the best engineered design.

In private projects, selection may be made by people in a company. See Fig. 14-1. Sometimes, a company employs its own engineers. As soon as the managers approve the project, engineers begin their work.

In public projects, engineered designs may have to be approved by a department of the government. For example, the highway department may approve a bridge design. Other public projects may only be of great local interest. See Fig. 14-2. The choices may be presented at public hearings. Later, public officials or voters may make a final choice.

People Who Help Make the Decision

Sometimes, decision-makers employ *technical consultants*. These consultants may be experts in certain fields of engineering. Their opinions on structure, site, water supply, or other matters can help the group make a choice. Lawyers and bankers may also be consulted. They give facts about legal and financial concerns.

Fig. 14-2

Some public projects affect many people. If so, the process for selecting the design may take years.

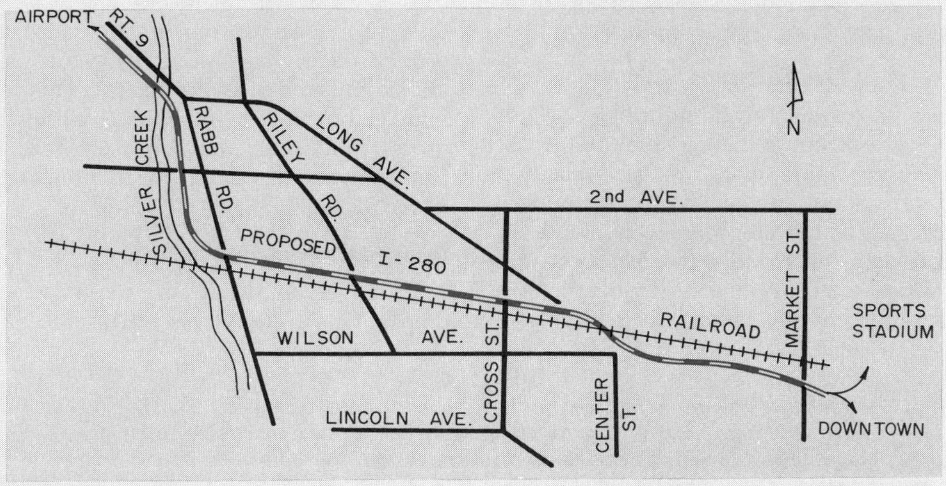

The dotted line shows the planned route for the I-280 freeway, which will connect downtown Smithville to northwestern Benjamin County.

Officials recommend route for controversial I-280 plan

Last Tuesday, officials of the state and city transportation departments recommended a route for the controversial Interstate 280.

The route is known as "Alternative XY," and includes connector streets to the airport and to the sports stadium.

If Alternative XY is accepted, 500 people will have to move from their homes, and 25 businesses will have to relocate. Some other alternatives would affect more residences, and some would affect fewer. Fewer businesses would be disturbed by XY than by any other alternative.

Transportation planners selected XY for several reasons. It is designed for smooth traffic flow onto freeway ramps. Natural surroundings would be least affected by XY than any other alternative. Also, XY makes better use of existing rail yards and is more easily worked into future airport plans.

The opinions of the experts may agree with the designer's judgment. If so, the decision-makers can feel more secure about a design they favor. When experts' opinions vary, the decision-makers must be more cautious in making a choice.

Various governmental agencies may have to approve a decision. See Fig. 14-3. These approvals are required to protect the public interest.

Deciding

Deciding on an engineered design often is a difficult task. One problem is the complex nature of engineered designs. Members of the selection group may not have engineering backgrounds. Design details may be hard for them to understand.

Another problem has to do with cost. Even the cheapest proposal may cost millions of dollars. Paying the interest on money loaned for the project may be expensive. Sometimes the best design costs the most to construct. But the structure may cost less to operate or to maintain. It may last longer. It may be more efficient in other ways. In the long run, the most expensive project may be the cheapest. All these considerations make it difficult to reach a design decision.

A good design presentation helps an approval group reach a decision. Therefore, each design choice must be presented fairly. The designer must explain the choices simply and clearly. The approval group must reach an unbiased (fair) decision. A wise decision is based on an honest presentation of facts.

Presentation of the Design

Visual aids may help explain a design's *technical details*. One common aid is a *model*. See Fig. 14-4.

Fig. 14-3
A number of governmental agencies may have to approve a project. Building a golf course may alter natural water drainage channels. Streets may have to be closed and zoning changed. These kinds of changes must be approved by the proper agencies.

Wheelabrator-Frye, Incorporated

Fig. 14-4
Scale models help people see the project details. This helps them select a design.

Designers must know how to make models and use them for presentations. Using the model, they can point out and explain the project's design features to the approval group.

Flip charts, diagrams, or similar aids also help engineers present their designs. People generally understand these aids better than complex drawings. Visual aids can save a decision-maker hours of study.

Some working drawings may be presented. They help explain the project. These drawings do not contain all the details for construction. Most often they are used along with visual aids.

The Decision

The design process leads to a decision. Either one design is finally accepted, or all are rejected. Usually, a design is selected at this time. Other times, the project is set aside. The complete process may have to start all over again.

After a choice has been made, further details of the design are drawn. A contractor is found. Then, a contract is negotiated and signed. Contracting will complete the design cycle. During construction, the engineer will check the work being done. Construction must be completed as designed.

Summary

Selecting the best design is very important. A wrong choice can lead to years of dissatisfaction.

Design ideas must be presented very clearly. A good design leads to a satisfactory project. But even a good design may not be accepted if it is not understood. Visual aids such as models can help people understand designs. A good presentation can help them make wise design decisions.

Terms to Know

decision step
diagrams
flip charts
model
technical consultants
technical details
visual aids

Think About It!

1. Name several local projects that were probably designed by engineers. For each, identify which person or group probably selected the design.
2. Select one of the projects you identified in question 1. Then list some facts that you, as a decision-maker, would have needed to know to select the best design.

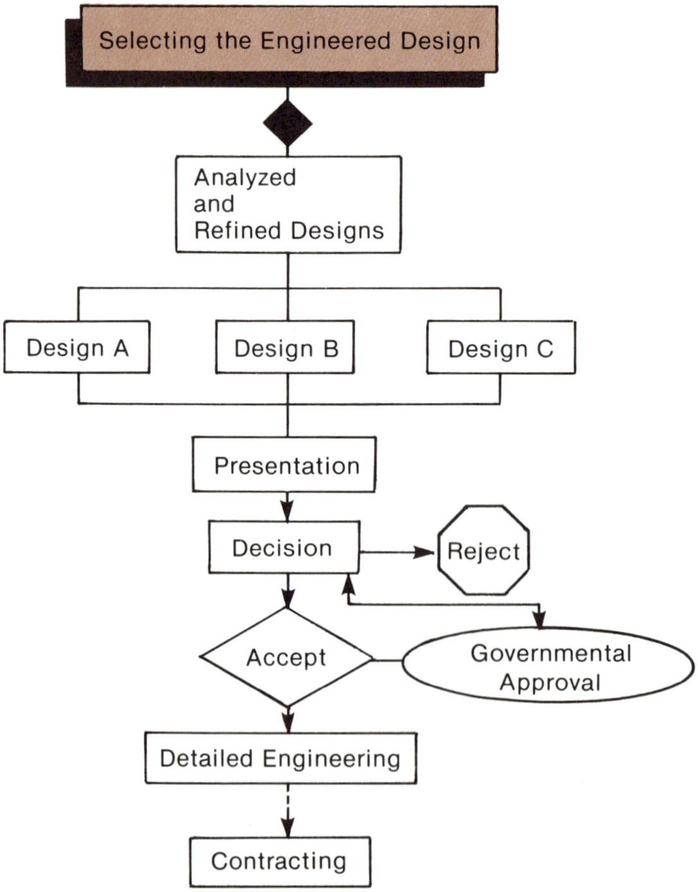

Reading 15

Metricating Construction

Conversion to Metric Measure

The United States is the last major country of the world to *convert* to metric measurement. Over ninety percent of the world's population uses metric units. Within a few years, the metric system will be the system of measurement in the United States. Measurement will be based on the *International System of Units*. (This is abbreviated SI). Using metrics will have a great effect on the construction industry. This reading is about SI Metric units and how using them will change construction.

The Customary System

Our present system of measurement developed over thousands of years. We call our system the English or *customary system*. The reason for this name is that much of the system came to us from Great Britain. We use it all the time.

Most people in the United States are more familiar with customary units than with metric units. See Fig. 15-1. Because of this, customary units may seem easier. Actually, they are very complex. There are many different terms to memorize. Difficult fractions must be used. Working with customary units is not easy.

The World Economy

The United States is a leader in the world *economy*. American materials, parts, and products are sold all over the world. Other countries produce

Fig. 15-1
Common units of measure.

Unit of Measure	Common Metric Units and Their Symbols		Common Customary Units	
Length or Distance	millimeter (mm) centimeter (cm)*	meter (m) kilometer (km)	inch foot	yard mile
Mass or Weight	milligram (mg) gram (g)	kilogram (kg) metric ton (t)	ounce pound ton	
Volume	milliliter (mL) liter (L) cubic centimeter (cm³) cubic meter (m³)		fluid ounce cup pint cubic inch cubic foot	quart gallon cubic yard
Area	square millimeter (mm²) square meter (m²) hectare (ha) square kilometer (km²)		square inch square foot square yard	acre square mile
Temperature	degrees Celsius (°C)		degrees Fahrenheit (°F)	

*Centimeters are not ordinarily used in construction.

goods, too. In world trade, there is much competition. Countries try to produce goods that other countries will want to buy. The United States buys many goods from other nations.

Most countries use the metric system. Their products are sold in metric units. But buying and selling is complicated if countries use different systems of measure. Measures must be changed from one system to another. Trade is easier when countries standardize (use the same system of) measurement. And when trade is improved, it helps the world economy.

0.001	m =	millimeter
0.01	m =	centimeter
0.1	m =	decimeter
1.0	m =	meter
10.0	m =	dekameter
100.0	m =	hectometer
1000.0	m =	kilometer

Fig. 15-2
Metric measures of length.

International System of Units (SI)

SI Metric units will provide a worldwide, unified system of measurement. Nations will agree on all of the *standards* of the system. They are working toward this goal now.

Refer again to Fig. 15-1. *Meter, gram,* and *liter* are basic metric units of measure. These terms may be used with *prefixes* to show units smaller or larger than themselves. For example, the prefix "milli" means one thousandth $\left(\frac{1}{1000}\right)$. The word "**milli**meter" means one thousandth of a meter. The prefix "kilo" means one thousand (1000). A "**kilo**gram" is one thousand grams.

Metrics is a decimal system. Units are related by *powers* of ten. You can change a unit of measure just by moving the decimal point. See Fig. 15-2.

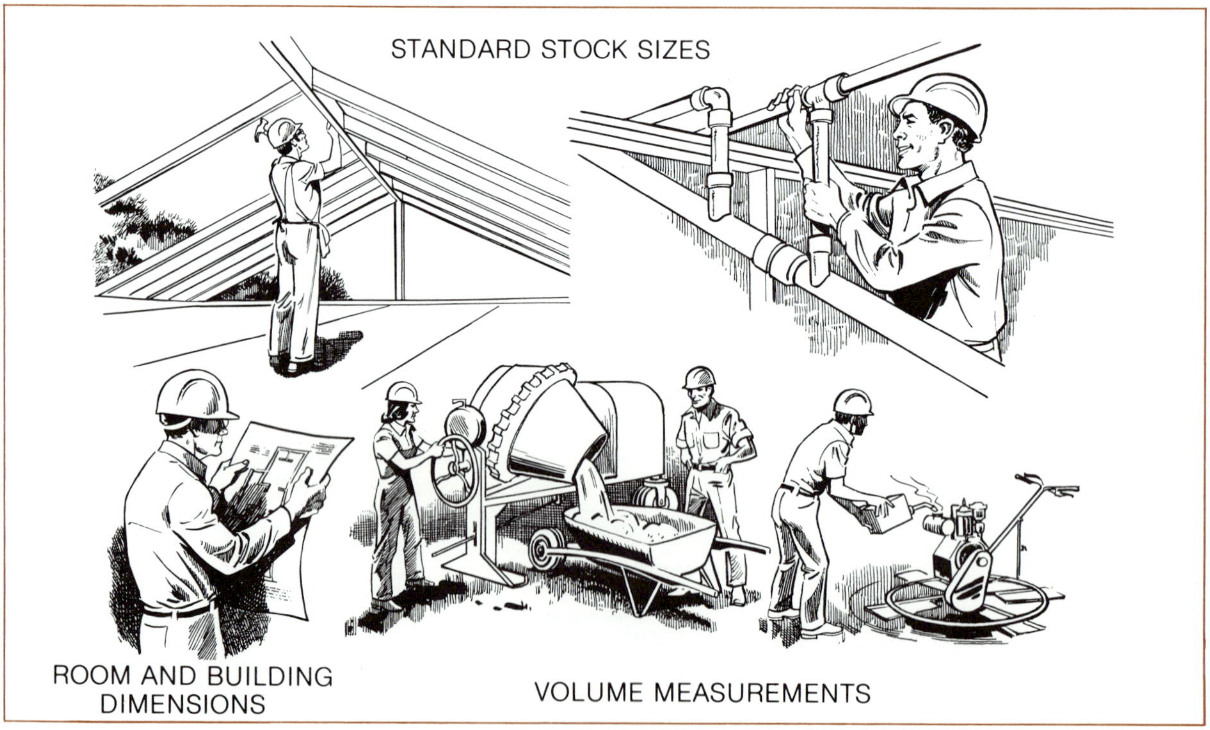

Fig. 15-3
Changing to the metric system will affect the construction industry in many ways.

just by moving the decimal point. See Fig. 15-2.

There are many other SI units. Few persons will ever use all of them. The metric measures most people will need are simple to learn.

Effects on Construction

Life in the United States will change as conversion to metrics takes place. And construction will change. See Fig. 15-3. A few of the changes that will be made in construction technology are listed here.

1. The unit for measuring lengths will be the *millimeter* (mm).
2. In architectural planning, the standard *module* (unit) for planning will be 100 mm. This module may be referred to as M.
3. Products such as lumber, bricks, doors, and windows will be manufactured to new dimensions. The dimensions will be based upon the 100 mm module.
4. Spacing of studs and other building members will be based upon a *planning grid* of 300 mm (300 - 600 - 1200 mm spacing).
5. Panel sizes for plywood, gypsum board, and laminates will be 1200 mm × 2400 mm. This is very close to the present size, 4′ × 8′.
6. Area measure will be in *square meters* (m^2).
7. Land area will be in *hectares* and distance in *kilometers*.
8. Volume and capacity measure for earth (removal or fill), concrete, and space will be calculated using the *cubic meter* (m^3).
9. Climate conditioning and plumbing specialists will calculate rates of flow in *liters per second* (L/s), heat flow in *watts* (W), and pressure in *pascals* (Pa).
10. Many kinds of tools will be changed or converted to metric. Drafters will need new drawing scales. Skilled workers will need new folding rules and tapes.

In 1978, the construction industry in Canada began changing to metrics. This has been done without many problems. In the long run, changing to metrics will make construction work easier to do.

Summary

The United States is changing its system of measurement. Soon all the measurements you use will be metric. Converting to metrics will cause changes in the construction industry. But the metric system is easy to learn and use. All workers in construction will soon begin to "think metric" and "do metric."

Terms to Know

convert	meter
cubic meter	millimeter
customary system	module
economy	pascals
gram	planning grid
hectares	powers
International System of Units (SI Metrics)	prefixes
kilometers	square meters
liter	standards
liters per second	watts

Think About It!

1. Why is the United States changing to metrics?
2. Suggest ways that using the metric system will affect construction (other than those listed in this reading).

Reading 16

Writing Specifications

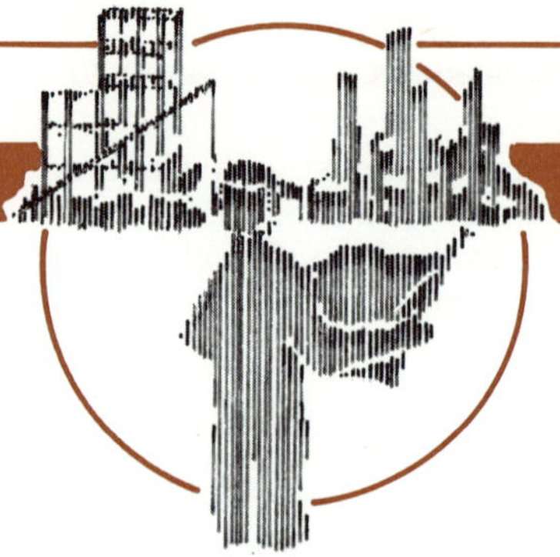

Construction Specifications

Before construction can begin, the materials and work needed for the project must be specified. To specify means to limit. It means to be precise. For example, "a specific day" means just one day. In the same way, *construction specifications* limit work and material choices for a project.

Specifications are written instructions. They give details that are easier to describe with words than with drawings. Together with drawings, specifications tell the contractor what materials to use. They also tell how the job is to be done.

Construction specifications are important. If drawings and specifications disagree, the builder (contractor) follows the specifications.

Fig. 16-1

Specifications are prepared by people who know about construction materials, methods, costs, and regulations.

Computer Consoles, Incorporated

Specifications – Content and Uses

Specifications describe the work to be done. See Fig. 16-1. They give the limitations set by the owner. They specify deadlines, procedures, and work standards. The required fixtures, materials, and equipment are specified. Specifications also give instructions about all parts of the job. For example, they may tell how plumbing and wiring are to be done.

Specifications include the general conditions and special conditions of the project. *General conditions* tell which tasks are most important. They tell the way the work is to be done. See Fig. 16-2. *Special conditions* explain how difficult tasks are to be done. They include items not covered in the general conditions. Such items as protecting existing trees and cleaning the site and structure may be listed in the special conditions. The contractor's guarantees for the work may also be listed.

Several factors influence the way specifications are written. Specifications must go along with standards set by the construction industry and by code-approval authorities. The specifications must meet the architect's and engineer's design requirements. They must also meet the owner's budget. Specifications must be in line with the work rules, methods, and labor rates. The contractor, lending agencies, lawyers, and suppliers also play a part in specifications.

Contractors refer to specifications to figure out how much it will cost them to do a job. Then they make a *bid*. A bid is an offer to do work at a certain price. After a bid has been accepted, the specifications become part of the *contract*. The contract is a signed agreement between owner and contractor.

Sometimes *contract documents* are combined into a book. This book is also referred to as "the specifications." It is divided into four sets of documents.

SPECIFICATIONS

General Conditions

1. The building permit and all other permits and inspections required shall be obtained and paid for by the contractor.
2. A portion of the site will be allotted the contractor for the prosecution of the work.
3. All water and electricity used on the work will be furnished by the owner.
4. All surplus materials to be removed shall become the property of the contractor and shall be removed from the premises.
5. The contractor shall pay for all materials, labor, tools, and transportation necessary for the completion of the work.
6. The contractor shall maintain such insurance as will provide protection from claims under Workmen's Compensation Acts.
7. The owner will be responsible for maintaining fire insurance and liability insurance.
8. The contractor must present a claim for extras before performing any extra work.
9. The owner has the right to require the contractor to furnish bonds covering the faithful performance of the contract.
10. The owner may withhold final payment and any retained percentages until entirely satisfied that the contractor has paid all bills.
11. The contractor must submit a list of subcontractors for the architect's approval.
12. The contractor shall at all times keep the premises free from accumulation of waste materials or rubbish caused by the contractor's employees or work.
13. Specific payments will be made to the contractor after completion of the floor, foundation, and roof.

Fig. 16-2
The general conditions that relate to a project, such as those shown here, are included in the specifications.

- Bidding requirements
- Contract forms
- General conditions
- Project specifications

You will learn about bidding and contracting in other readings.

Kinds of Specifications

There are four major kinds of specifications.

- Outline specifications
- Standards specifications
- Guide specifications
- Project specifications

Outline specifications are made at the beginning of a project, when preliminary drawings are being made. These specifications give only the main facts about the materials and how they are to be used.

Standards specifications are made up by companies. In catalogs, companies tell about the use of their own materials. If one company's product is named in a standards specification, the words "or equal" may follow the name. See items C and D in Fig. 16-3. This means that another company's product may be used if it is of the same kind and quality. But this can be done only with written permission from the architect or owner. The American Society for Testing

SPECIFICATIONS

Lathing and Plastering

Materials

A. Metal Lath: 3.4 lb. copper-bearing steel, 3/8″ diamond mesh, galvanized, or coated with rust-inhibiting paint after fabrication.
B. Tie Wire: No. 18 W & M gauge, weighing not less than 475 lb. per 1000 ft.
C. Expanded Corner Beads No. 1 "Milcox," 26 gauge, galvanized (or equal).
D. Casing Beads or Plaster Stops: "Milcox" No. 66, 24 gauge, galvanized (or equal).
E. Portland Cement: ASTM C 150, Type I.

Fig. 16-3
Typical materials specifications for one feature of the work to be done.

and Materials has standards specifications for many materials. In addition, *Sweet's Catalog* gives specifications for many kinds of work and materials.

Guide specifications provide a model for writing project needs. They are forms with blanks. The blanks can be filled in to show what is needed for a job. They are a guide the writer uses to organize specifications.

Project specifications are the finished specifications for a particular construction project. They are usually made up of outline specifications, standards specifications, and guide specifications.

When Specifications Are Made

Usually, specifications are written after most decisions about materials and equipment have been made. Because of time limits and deadlines, specifications are usually begun soon after the first drawings are started.

Specifications are written first for parts of the project that will probably not change. For example, the first specifications may be for general conditions of the project, earthwork, and concrete work. As drawings are made for the rest of the project, more specifications are written. *Specifications writers* work with the drafters. That way, specifications can be completed soon after the final drawings.

Fig. 16-4
The specifications writer must keep up with new materials and their uses.

Specifications Writers

In a small office, the person who makes the working drawings may also write the specifications. Larger firms have persons whose main job is to write specifications. They also keep a *specifications file* up-to-date.

Specifications writers must have special knowledge and experience. They must understand construction practices and materials. They must know the newest materials and ways of using them. See Fig. 16-4. These writers need to know where to get information they do not have. They must be able to find state, county, village, and city *building codes* or *ordinances*. The specifications must agree with these regulations. Writers often seek information from persons who are trained in construction and in writing standard specifications. With experience, specifications writers can draw up specifications quickly and accurately.

How Specifications Are Made

There are at least four ways to prepare specifications. One way is to work from a set of old specifications. This can be done only if the old set was for a job much like the new one. The old set is rewritten to suit the needs of the new project.

A second way is to gather several sets of specifications. The writer uses the "cut and clip" procedure. The sections needed for the new job are cut from the specifications. They are clipped to a page. Additions are inserted where needed. The writer does not have to prepare information that has already been written.

The "paragraph procedure" is another way of writing specifications. The writer selects cards from a specifications file. Each card has a single statement paragraph on it. The statement may tell about work standards, materials, or methods. The writer gets all the file cards needed to make the specifications. Then, the writer puts the cards in order and writes the specifications from the cards.

Another method is to work from guide specifications. These were described earlier in this reading. They are forms with blanks for the writer to fill in and use as a guide.

Bidding Requirements

- Invitation to Bid
- Instruction to Bidders
- Form of Bid Proposal

Contract Forms

General Conditions

- Supplementary General Conditions
- Special Conditions

Specifications

Division 1 — General Requirements

- 0170 Alternates

Division 2 — Site Work

- 0211 Demolition
- 0221 Site Grading
- 0222 Excavation and Backfilling
- 0274 Irrigation System
- 0281 Soil Preparation
- 0282 Lawns
- 0284 Trees and Shrubs

Division 3 — Concrete

- 0310 Concrete Formwork
- 0320 Concrete Reinforcement
- 0331 Heavyweight Aggregate Concrete
- 0341 Precast Concrete Panels

Division 4 — Masonry

- 0410 Mortar
- 0422 Concrete Unit Masonry

Division 5 — Metals

- 0551 Metal Stairs
- 0570 Ornamental Metals

Division 6 — Carpentry

- 0621 Wood Trim
- 0622 Millwork

Division 7 — Moisture Protection

- 0711 Membrane Waterproofing
- 0751 Built-Up Bituminous Roofing
- 0762 Metal Roof Flashing and Trim
- 0790 Caulking and Sealants

Division 8 — Doors, Windows, and Glass

- 0812 Aluminum Doors and Frames
- 0820 Wood Doors
- 0839 Revolving Doors
- 0852 Aluminum Windows
- 0860 Wood Windows
- 0870 Finish Hardware
- 0885 Glass and Glazing

Division 9 — Finishes

- 0913 Gypsum Lath
- 0915 Plastering Accessories
- 0916 Plaster
- 0926 Gypsum Drywall System
- 0927 Gypsum Drywall Finishing
- 0931 Ceramic Tile
- 0941 Cast-in-Place Terrazzo
- 0957 Wood Parquet Flooring
- 0965 Resilient Flooring
- 0990 Painting
- 0995 Wall Covering

Division 10 — Specialties

- 1013 Chutes

Division 11 — Equipment

- 1132 Audio-Visual Aids
- 1197 Stage Equipment

Division 12 — Furnishings

- 1240 Carpets and Mats
- 1250 Drapery and Curtains

Division 13 — Special Construction

- 1385 Swimming Pool

Division 14 — Conveying Systems

- 1420 Elevators

Division 15 — Mechanical

- 1510 Basic Materials and Methods
- 1521 Water Supply Piping
- 1522 Domestic Hot Water System
- 1526 Soil and Waste Piping
- 1529 Roof Drainage System
- 1530 Plumbing Fixtures and Trim
- 1580 Air-Tempering System
- 1590 Refrigeration
- 1595 HVC Controls and Instruments

Division 16 — Electrical

- 1610 Basic Materials and Methods
- 1630 Electrical Distribution System
- 1640 Lighting Fixtures
- 1651 Telephone Equipment
- 1671 Motors and Motor Controls
- 1695 Lightning Protection System

Fig. 16-5
The Construction Specifications Institute's outline for specifications.

Specifications Format

Specifications are set up in the order in which the job is to be done. They are made up of many sections. Each section is kept in order. Everything about one part of the job is kept together. For example, everything about plumbing is in one section. Electrical work is in another. Organizing specifications in this way makes it easier for a contractor to *estimate* (figure the cost of) a job. It also helps determine who is going to do each part of a job. A sample set of specifications sections and divisions is shown in Fig. 16-5. Sections and divisions such as these are often used in the construction industry.

Summary

Construction specifications are written instructions to a builder. They give specific information about the materials and work needed for a project. Specifications and drawings are used by contractors to estimate and bid for a job. They become part of a contract after a bid has been accepted.

There are four major kinds of specifications. These are outline, standards, guide, and project specifications.

Specifications are started when the first drawings are begun. By working with the drafters, specifications writers can complete the job soon after the final drawings are made.

There are four methods specifications writers use to write specifications. They may rework an old set of specifications. Or, they may cut and clip several sets of specifications to get one new set. Another method is writing from cards in a specifications file. The fourth method is using guide specifications.

Specifications are put in sections. The sections are arranged in the order in which the job is to be done. This makes it easier for a contractor to estimate a job. It also helps determine who is going to do each part of the job.

Terms to Know

bid
building codes
construction specifications
contract
contract documents
estimate
general conditions
guide specifications
ordinances
outline specifications
project specifications
special conditions
specifications file
specifications writer
standards specifications

Think About It!

1. Why is it necessary to have written specifications in addition to project drawings?
2. What are some things you would need to know in order to write construction specifications?

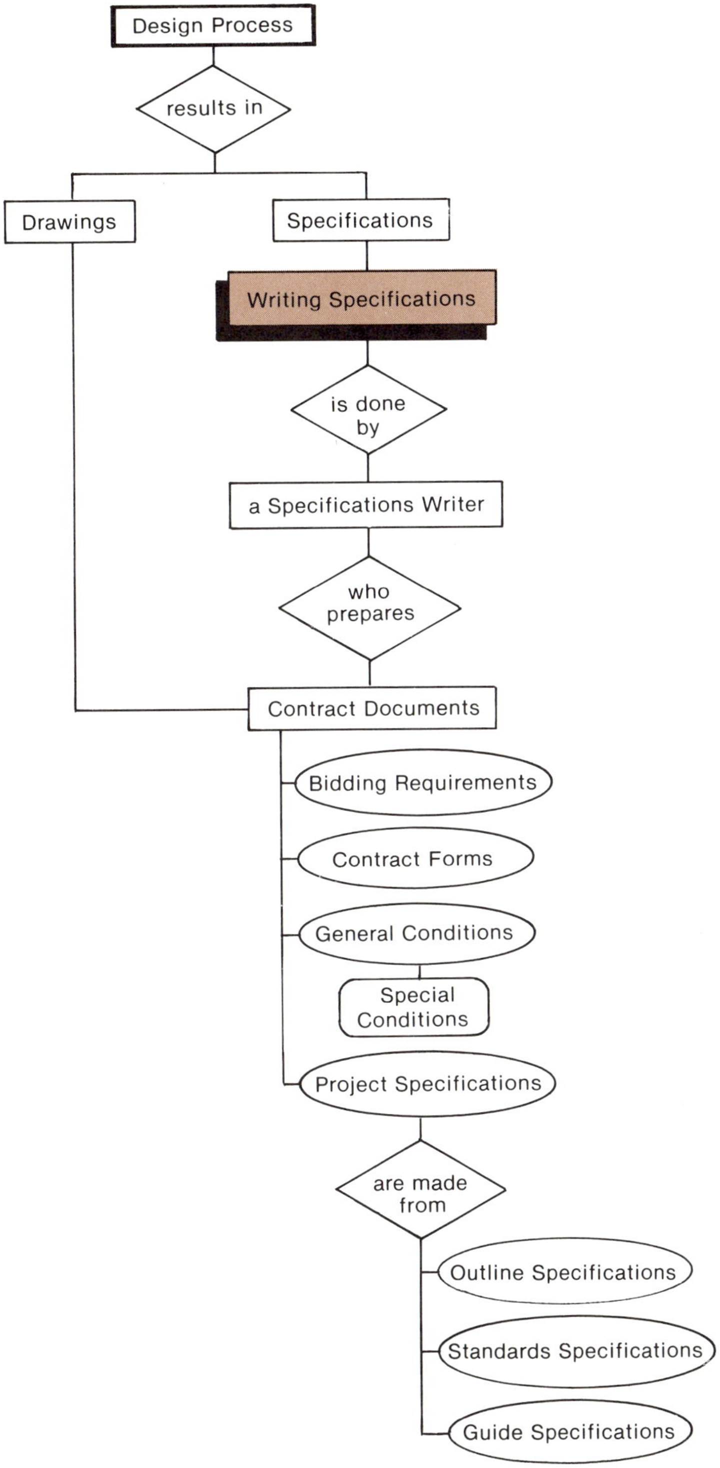
Design Process
results in
Drawings
Specifications
Writing Specifications
is done by
a Specifications Writer
who prepares
Contract Documents
Bidding Requirements
Contract Forms
General Conditions
Special Conditions
Project Specifications
are made from
Outline Specifications
Standards Specifications
Guide Specifications

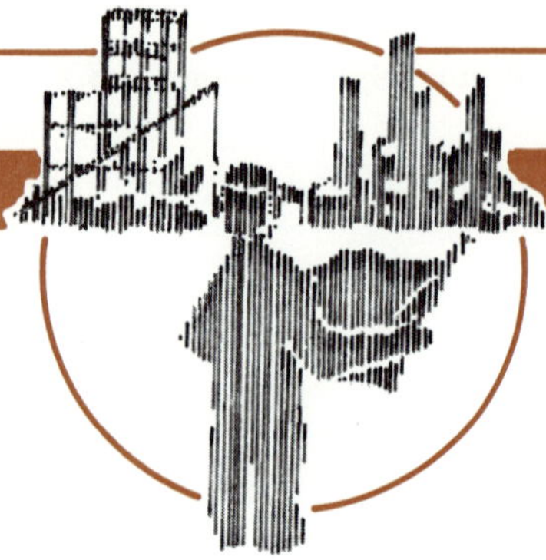

Reading 17

Preparing Architectural Drawings

Working Drawings

After a refined, analyzed design is approved, final drawings are made for construction. The architect may make these drawings. Or, the architect may have the final drawings made by a drafter. Final drawings show exactly how a structure is to be built. They show where it is to be placed on a site. Plans drawn for construction are called *working drawings*. They are fully detailed and dimensioned.

A house is not generally a large construction project. But it requires working drawings. A client (buyer) orders a design and approves the refined drawings. Then working drawings are made for approval. Sometimes a client chooses a design for which working drawings have already been made.

Larger construction projects are very complex. A large apartment building or a skyscraper would need many working drawings. Such drawings are not made until after a decision to build. Design plans show only the general appearance of large projects. Plans are made of the site layout. Floor plans are prepared. Enough details are provided to give the design a finished look. After the proposed design is approved, detailed working drawings and specifications are prepared.

Builders need information to construct a house. Working drawings, together with specifications, provide this information. Sample working drawings are given on pages 88-92. Refer to these drawings as you read.

The Plot Plan

A *plot plan* is a drawing of a lot as seen from above the ground. See Fig. 17-1, page 88. The plot plan shows where a house is to be located on the lot. Often other information is given. *Building lines* are shown. These indicate the distance from the house to the front and sides of the lot. The plan may also show walks, driveways, trees, and other features.

The plot plan should give the elevation (height) of the finished first floor level. This is the distance above the *datum level*. The datum level is the basic level used for finding other heights. Often it is part of the *building code*. The building code contains the rules for building in an area. It sets the building standards.

Floor Plans

The most important working drawing is the *floor plan*. All other drawings depend on it. The floor plan is a *horizontal section* (cut) through the building. That is, it appears that the building has been cut and the top part removed. The "cut" is below the tops of the windows and doors. See Fig. 17-2. A simple floor plan shows the locations of walls, partitions (inside walls), doors, and windows. Most floor plans show more detail. Dimensions are usually given.

Today, most dimensions are given in U.S. Customary measures. Dimensions are stated in feet and inches. For example, a measurement of three feet, four inches is shown as 3′-4″. But soon the construction industry in the United States will use the metric system. In this system, all floor plan dimensions are given in millimeters. The same distance (3′-4″) would be given as 1016 mm. Which system is used in Fig. 17-2?

Floor plans for a large building are often on several sheets. Information about electrical and plumbing systems is generally placed on separate sheets. *Climate conditioning* systems (heating and cooling) are also presented separately. For a small structure, all this information may be placed on one floor plan.

Exterior Elevations

An *exterior elevation* shows the finished appearance of one side of the outside of the building. See Fig. 17-3. Elevations are made of the front and rear of the structure. Left and right sides may also be shown as elevations.

Each elevation is based upon the floor plan drawing. That is, all dimensions used for elevations match those on the floor plan. All surface materials are shown on the elevation. Enough exterior elevations are drawn for the builder to be able to build the house.

Structural Details

Structural sections are drawings of views that best show important construction features. These are usually *vertical sections* of the building. They show the structure from the foundation to the roof. See Fig. 17-4. Sections are often drawn to a larger scale than other working drawings. They show how parts fit together.

Finish Details

Because working drawings are drawn to scale, they are small. Many details cannot be shown. Some need more complete drawings. These are drawn separately. They are called finish *detail drawings*. There are two kinds of detail drawings.

- Standard
- Custom

Both are important. The architect determines which parts of a structure should be detailed.

Standard details are usually repeated many times in a building. A typical standard detail might show a window of standard type and size. Designers and contractors become very familiar with such details.

Custom details (one-of-a-kind) show the solutions to special problems. For example, a certain house may need special kitchen cabinets. A custom detail drawing will be made. A specially designed staircase would also require a custom detail drawing. Items that most often need detailing are stairways, fireplaces, cabinets, and other custom features. Custom details may also be made of electrical or mechanical features.

Foundation Plans

A foundation supports the weight of a structure. The foundation may or may not include a basement. A *basement plan* shows the basement wall and *footing* layout. The footing is the bottommost part of the structure. It is also the widest part. When no basement is planned, the drawing is called a *foundation plan*. It shows the footings and the foundation walls. The foundation and basement are sometimes shown on one plan. See Fig. 17-5.

Many types of foundations and floor systems are in use today. Two types are used most often in house construction. One is *wood floor construction*. This type carries the construction load above grade (ground level). It is often used in houses with basements. The other type is concrete floor or *slab construction*. Houses with no basements often have slab floors. The footings carry the wall load. The soil under the slab carries the ground floor load.

The floor plan can be used as a guide in drawing the foundation plan. A foundation and basement plan shows support columns, room partitions (if any), footings, and other details. Look again at Fig. 17-5.

Joist Layout and Roof Plan

Joist details are often drawn for wood floor construction in housing. These details include placement, size, and spacing of joists.

The roof plan shows the type of roof and its slope. A typical roof section is given on the plan. It shows the placement of principal roof members (rafters or trusses).

Electrical Plan

An *electrical drawing* is a plan view (floor plan) of the building. It is usually made on a simple tracing of the floor plan. It shows the electrical service entrance and meter. All electrical outlets and switches are shown on this plan. Special *symbols* represent these items. All features are designed to meet local electrical codes and standards. These codes and standards are set up by the National Fire Protection Association. Electrical plans are often included on the floor plan. Look again at Fig. 17-2.

Climate Conditioning Plan

New housing should be designed to save energy. Drawings are made to show the details of all *heating, cooling, and ventilating* equipment. These details are often shown by special symbols and notes.

Supply and return lines are planned carefully. They should fit in with structural members. Sometimes structural members must be cut to allow for these lines. Then they should be strengthened by special supports.

Plumbing Plan

The *plumbing plan* is a plan view of the building. It is made on a simple tracing of the floor plan. It shows all plumbing lines and fixtures above and below ground. It maps out soil and waste lines. It shows gas lines, waterlines, and special plumbing fixtures. Symbols are used to show locations of sinks and of appliances such as dishwashers.

Summary

Working drawings and specifications are very important in construction. Together, they tell the kinds of materials to be used and how work must be done. They provide all the information needed to build a structure.

There are many types of working drawings. The floor plan is the most important one. Other working drawings are based on the floor plan.

Think About It!

1. Why is the floor plan the most important kind of working drawing?
2. What would happen if a builder tried to build a house without the necessary working drawings?

Terms to Know

basement plan	electrical drawing	horizontal section	structural sections
building code	exterior elevation	joist details	symbols
building lines	floor plan	plot plan	vertical sections
climate conditioning	footing	plumbing plan	wood floor construction
custom details	foundation plan	slab construction	working drawings
datum level	heating, cooling, and ventilating	standard details	
detail drawings			

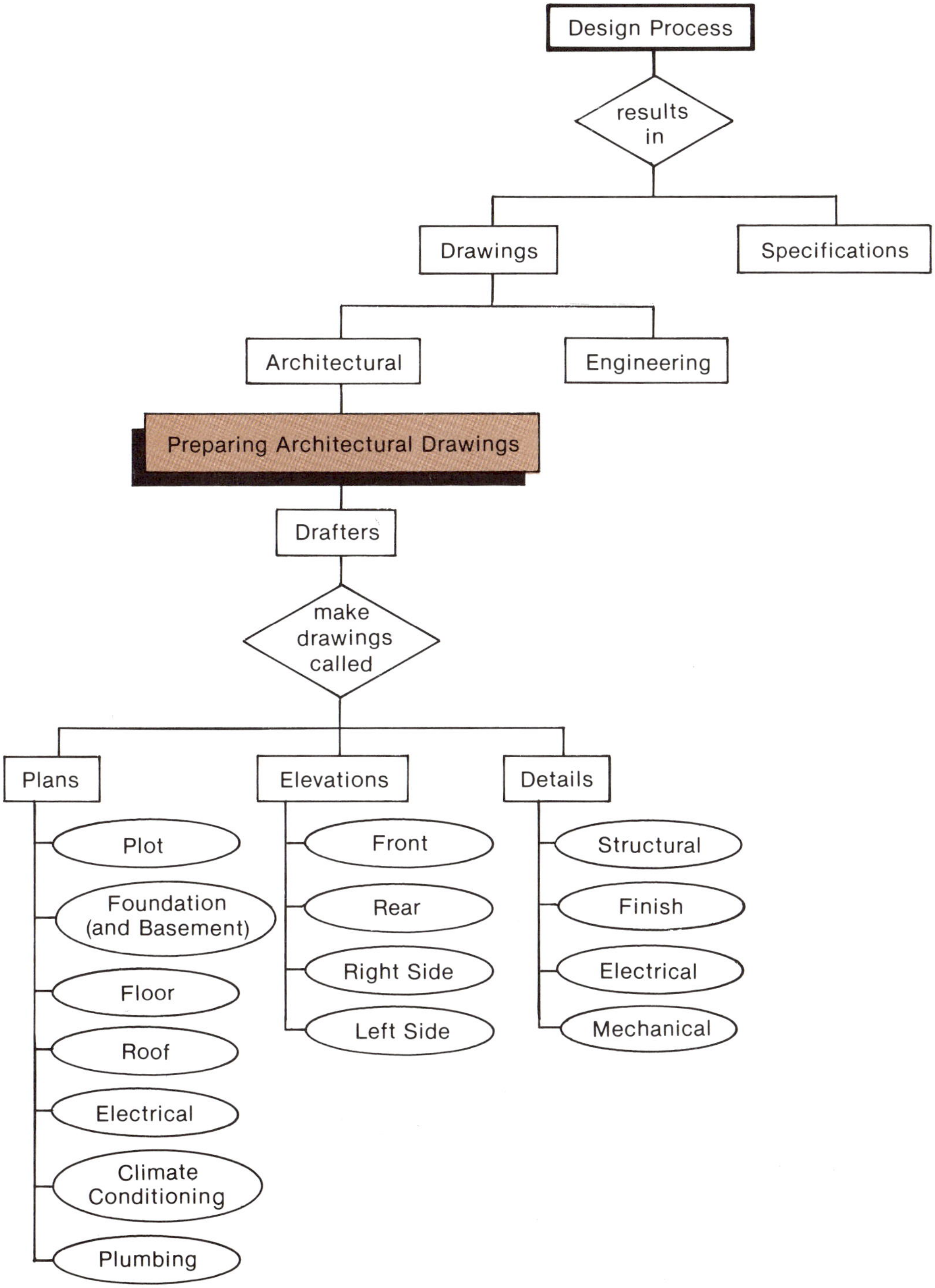
Design Process
results in
Drawings
Specifications
Architectural
Engineering
Preparing Architectural Drawings
Drafters
make drawings called
Plans
Elevations
Details
Plot
Foundation (and Basement)
Floor
Roof
Electrical
Climate Conditioning
Plumbing
Front
Rear
Right Side
Left Side
Structural
Finish
Electrical
Mechanical

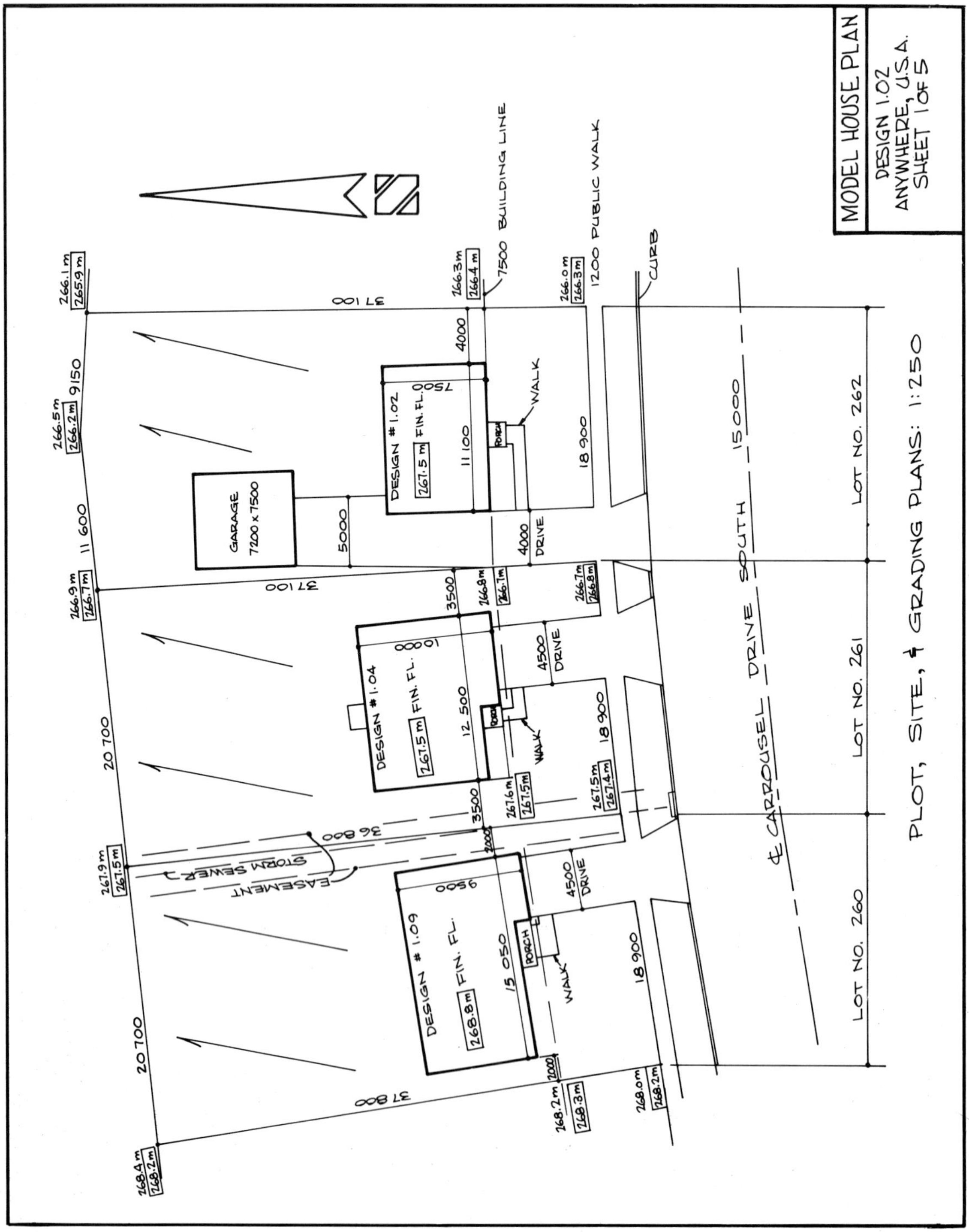

Fig. 17-1
Plot, site, and grading plans.

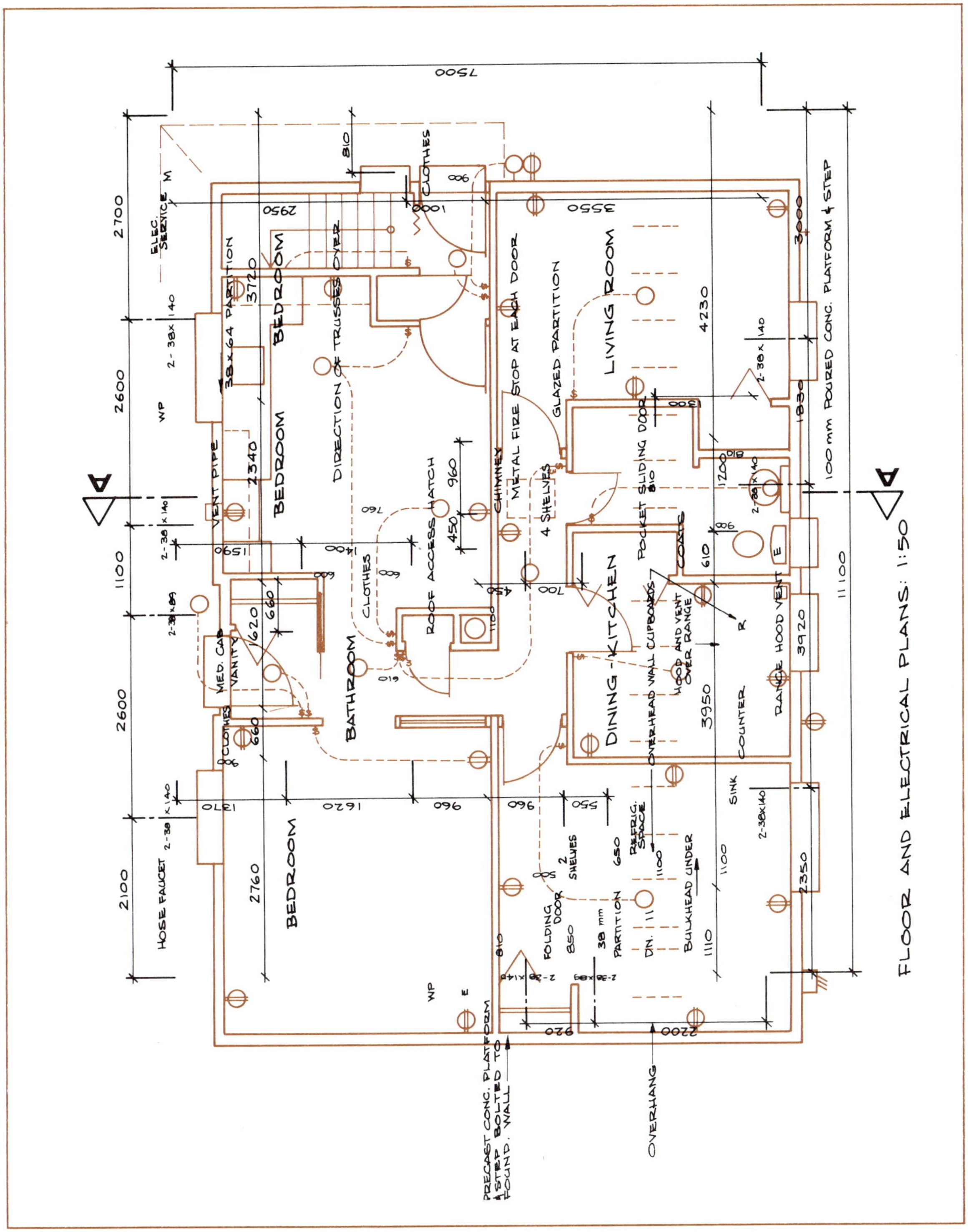

Fig. 17-2
Floor and electrical plans.

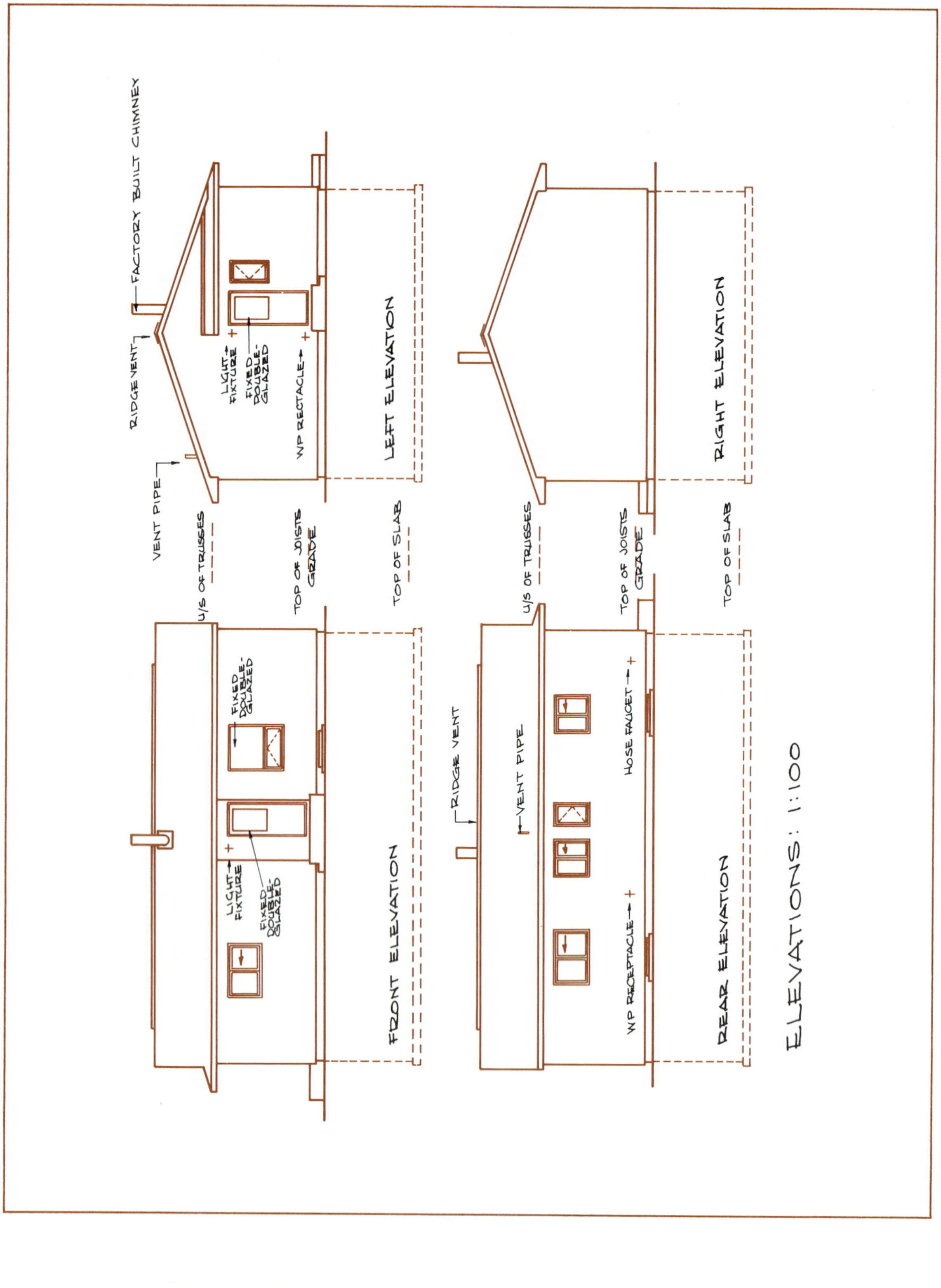
FACTORY BUILT CHIMNEY
RIDGE VENT
VENT PIPE
LIGHT FIXTURE
FIXED DOUBLE-GLAZED
WP RECTACLE
LEFT ELEVATION
U/S OF TRUSSES
TOP OF JOISTS
GRADE
TOP OF SLAB
RIGHT ELEVATION
FIXED DOUBLE-GLAZED
LIGHT FIXTURE
FIXED DOUBLE-GLAZED
FRONT ELEVATION
RIDGE VENT
VENT PIPE
HOSE FAUCET
WP RECEPTACLE
REAR ELEVATION
ELEVATIONS: 1:100

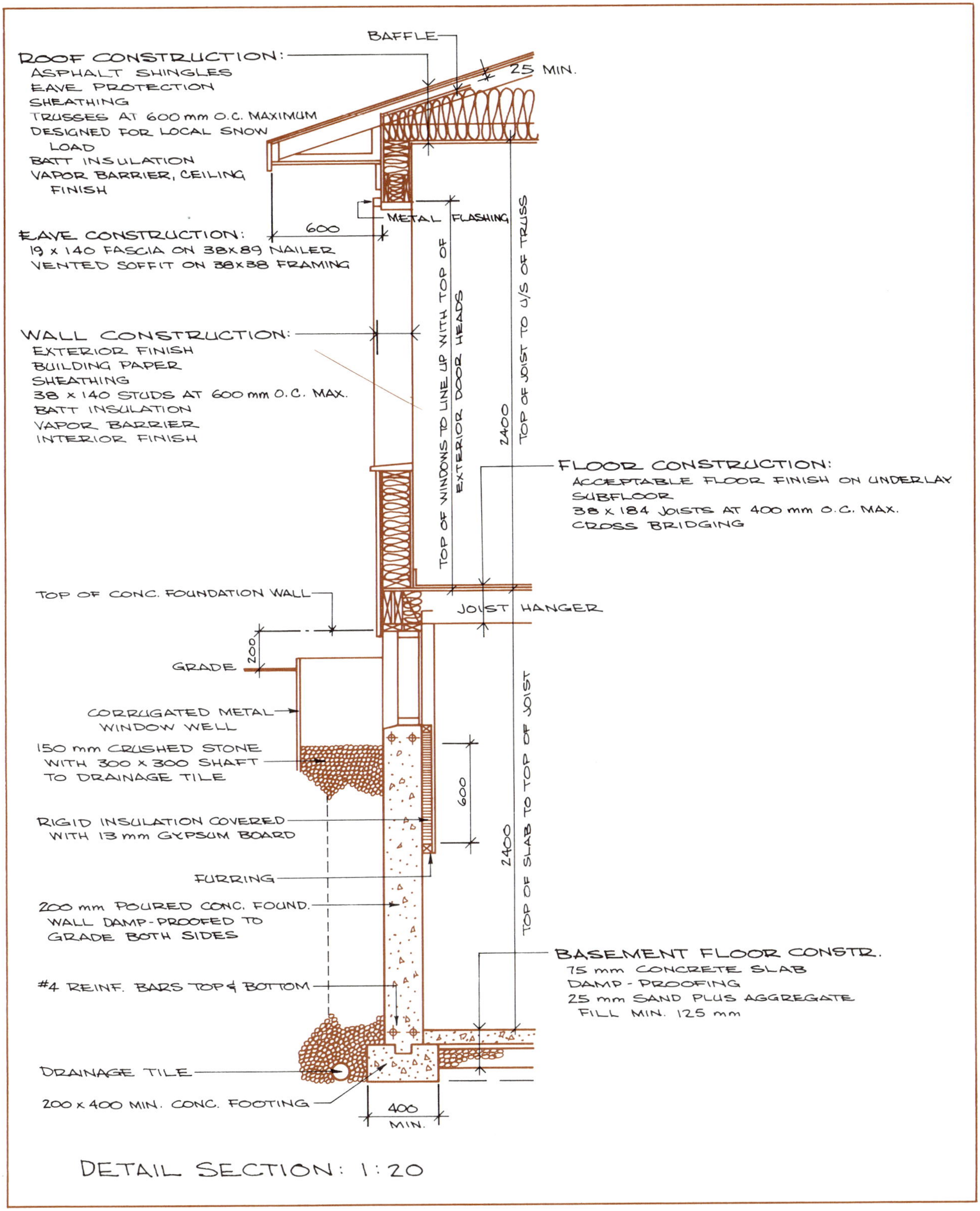

Fig. 17-4
Wall sections.

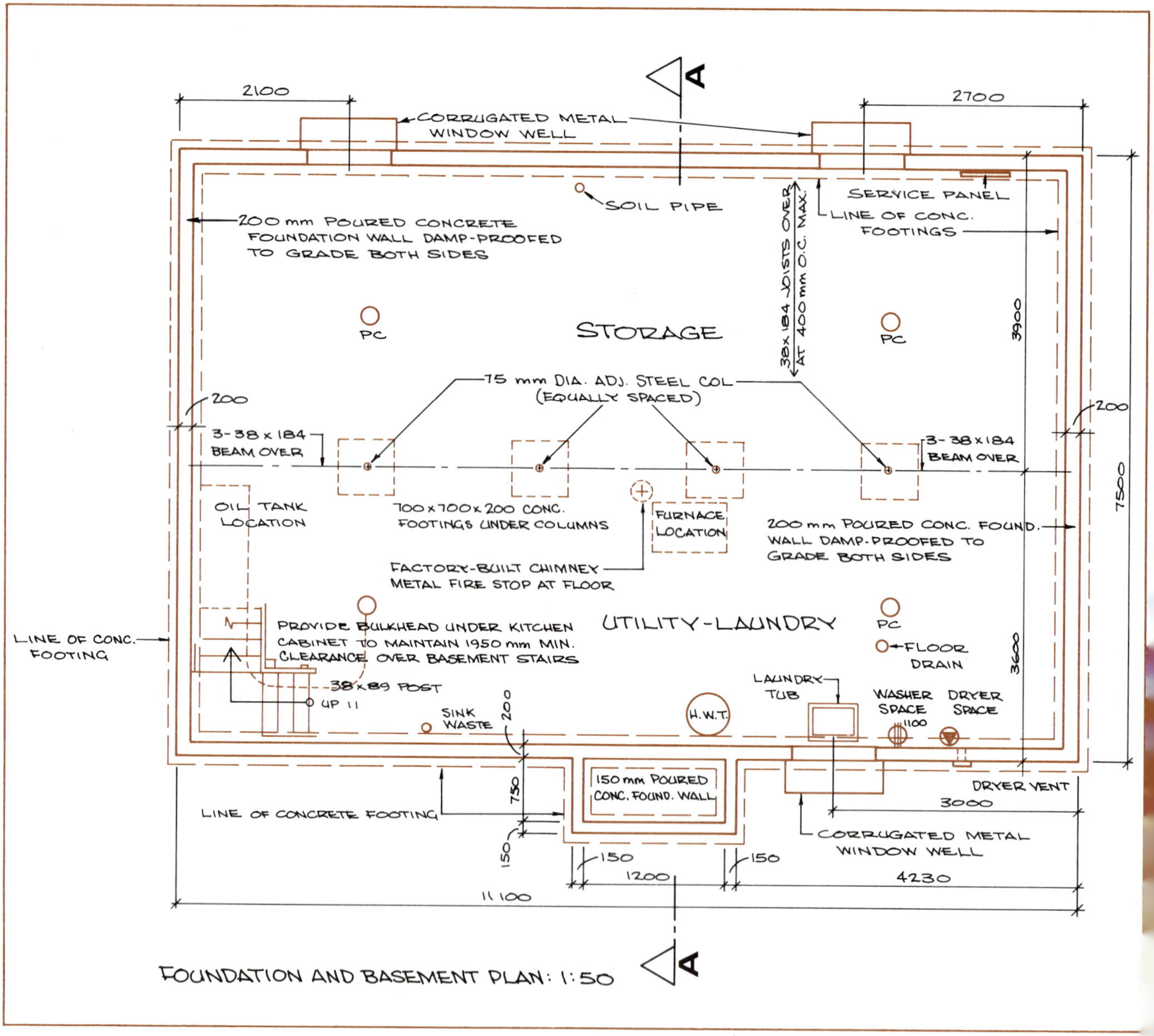

Fig. 17-5
Foundation and basement plan.

Preparing Engineering Drawings

Final Drawings

The designer's plan for constructing the project is communicated in the final drawings. These include all the views, dimensions, and details needed to build the project. Making these drawings is an important step in the designing and engineering process. Both architects and engineers are involved in preparing drawings.

Typical drawings prepared by architects are described in another reading. Architects may consult engineers about details of architectural drawings. For example, engineers may provide details for a complex electrical system.

Architectural firms and engineering firms often work together. Specialized design firms employ both architects and engineers. The architects plan the general designs of large buildings. But architects seldom play a major role in designing *heavy construction* projects. These are projects such as highways, industrial buildings, and dams. Most of this work is done by engineers.

Building heavy construction projects requires many engineering drawings. See Fig. 18-1. These drawings contain many kinds of information. They explain the project. This reading is about engineering drawings.

Fig. 18-1
A complex project such as this is built from many engineering drawings.

Suburban Propane Gas Corporation

Types of Engineering Drawings

There are three major types of engineering drawings. These are *civil*, *structural*, and *electrical-mechanical drawings*.

Civil drawings relate to the land on which a structure is built. They show how the earth's surface must be altered (changed) to suit a structure's needs. Plans for highways and interchanges are civil drawings. See Fig. 18-2.

Structural drawings show the location of the parts of a structure. They show how parts are to be fastened. The beams of a large building are one example. Bridge parts are another. See Fig. 18-3.

Fig. 18-2
Civil drawings of highways are very complex.

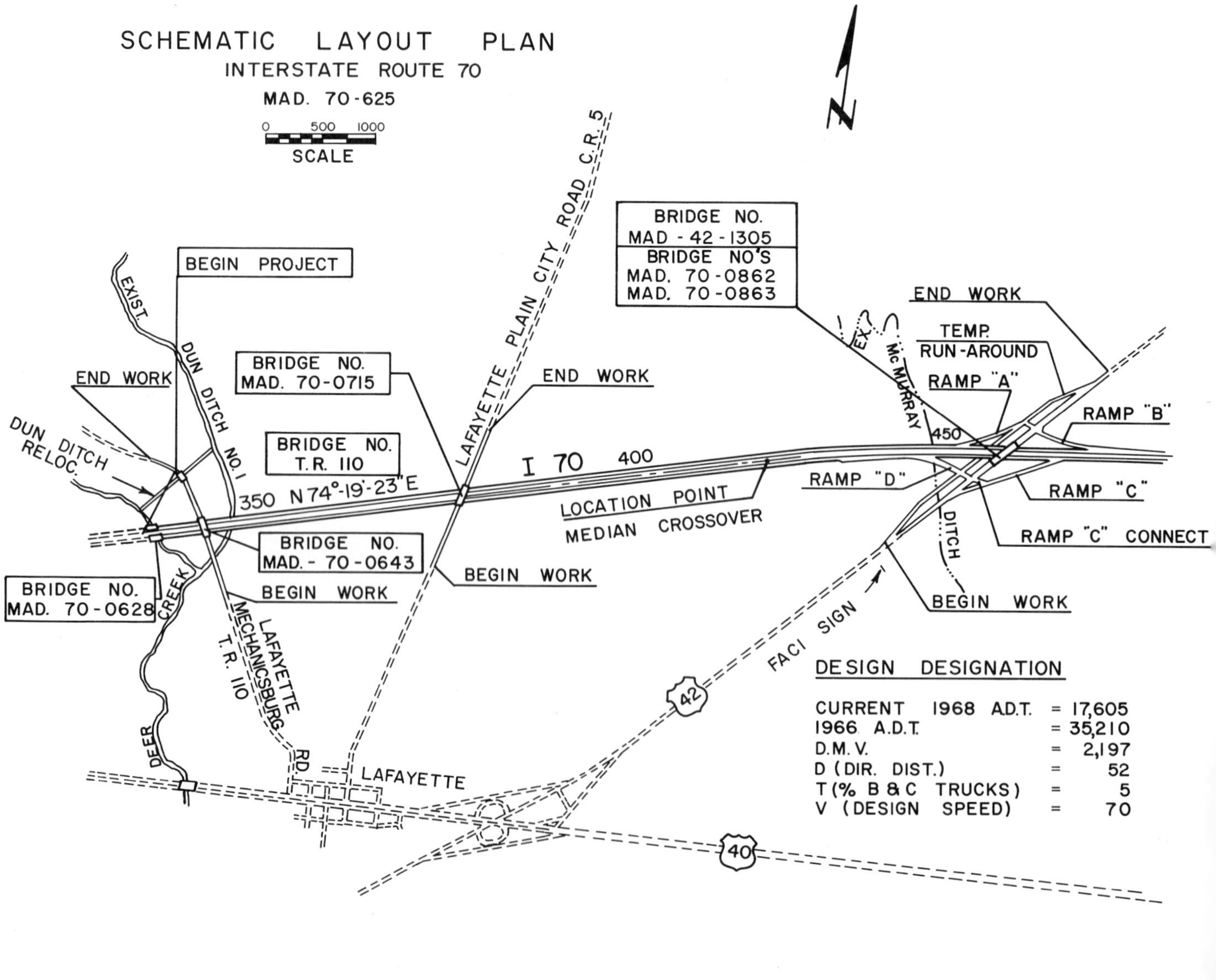

Fig. 18-3

This structural drawing shows details of one part of a bridge.

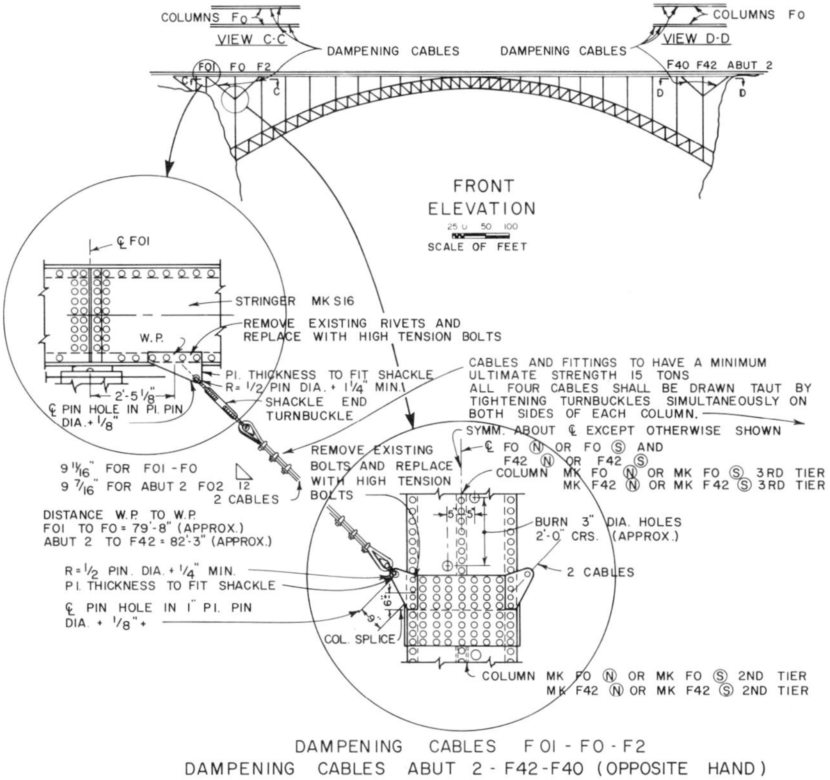

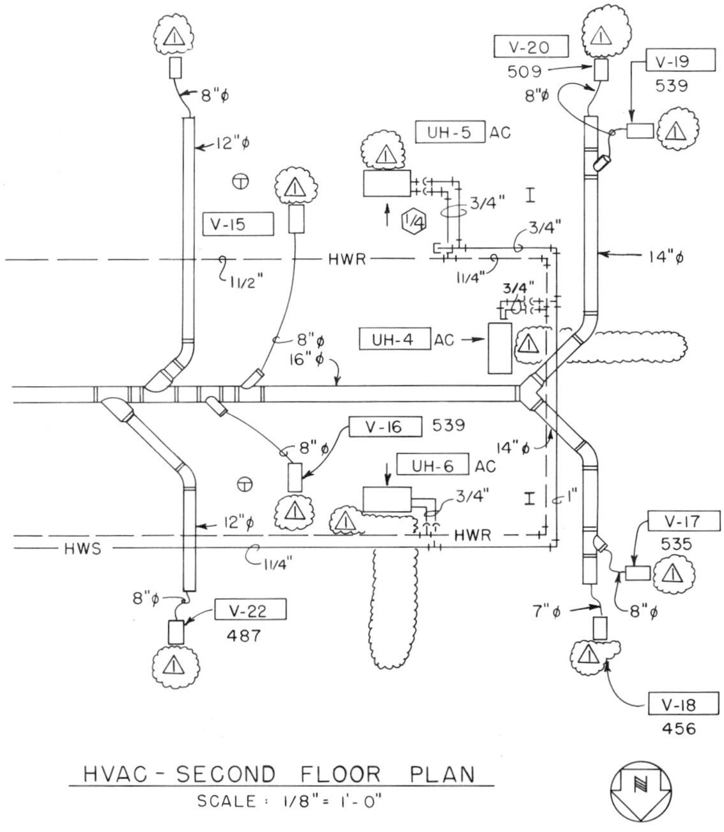

Fig. 18-4

A mechanical drawing showing a heating system.

Electrical systems are shown on electrical drawings. They show how electrical power will be supplied for lights, furnaces, elevators, and other electrical features. Mechanical features like plumbing and piping are shown on mechanical drawings. See Fig. 18-4. Often electrical-mechanical features appear together on one drawing.

Contract Drawings

Civil, structural, and electrical-mechanical drawings become parts of contract *documents.* These drawings are called *contract drawings.* They describe what a private company or a public group, such as a city council, wants built. The designing work is done by a team of designers. Engineers for the project supervise this work.

Contract drawings are rather detailed. They must describe completely what is to be built. They are detailed enough to describe all the information needed for making bids. Bids are offers made by contractors to build the project at a certain cost. To determine this cost, bidders must be able to tell the shape and size of the structure. They must know what materials and *standards* (types and qualities expected) are required. Several contractors all try to offer the lowest cost. This is *competitive bidding.*

Working Drawings

To build the project, contractors and construction crews need drawings that contain more details than many contract drawings. See Fig. 18-5. Drawings made for construction are *working drawings.* The contractor who is awarded the contract to build must develop detailed working drawings to guide the work. Sometimes these drawings are called *shop drawings.* Contract drawings and shop draw-

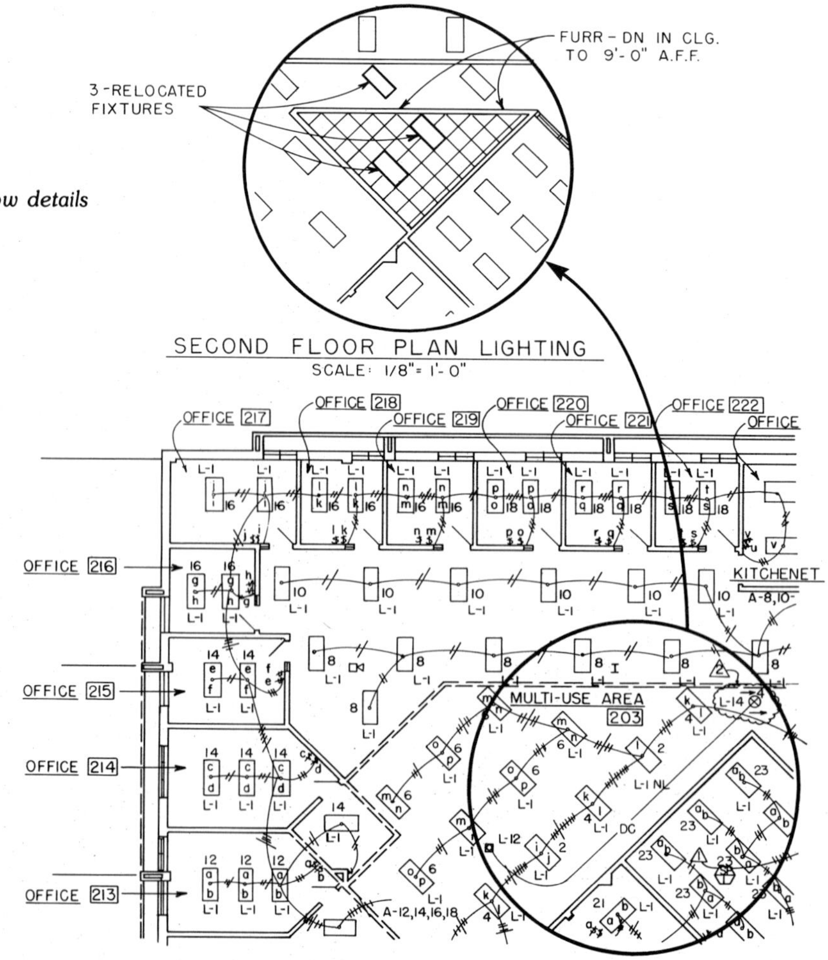

Fig. 18-5

Working drawings show details of construction.

A

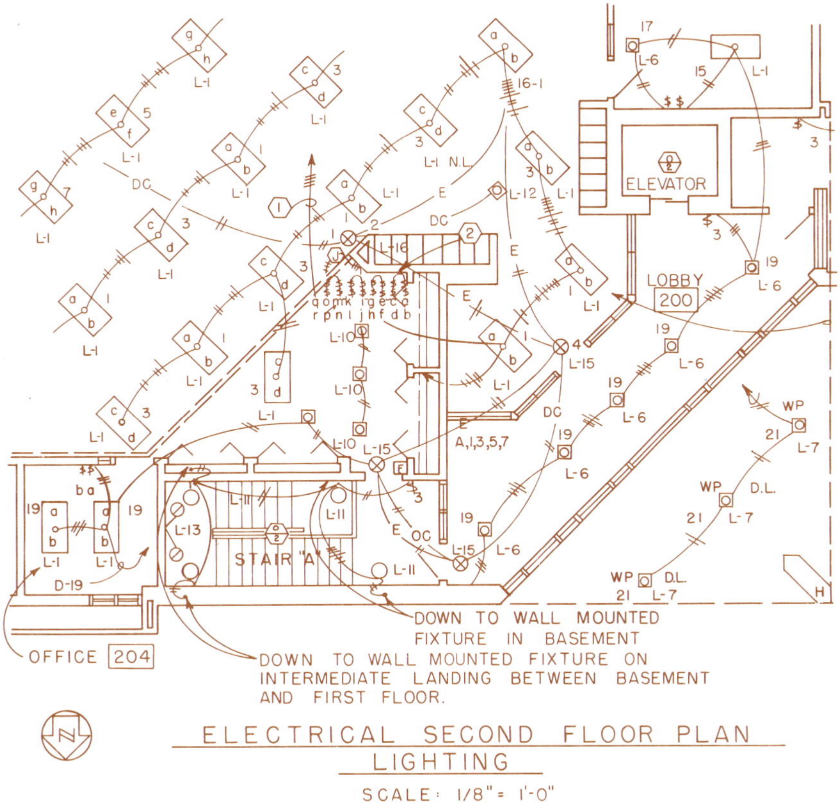

Fig. 18-6
Two types of electrical drawings are lighting (A) and power (B). The same area of an office building is shown in the two drawings.

B

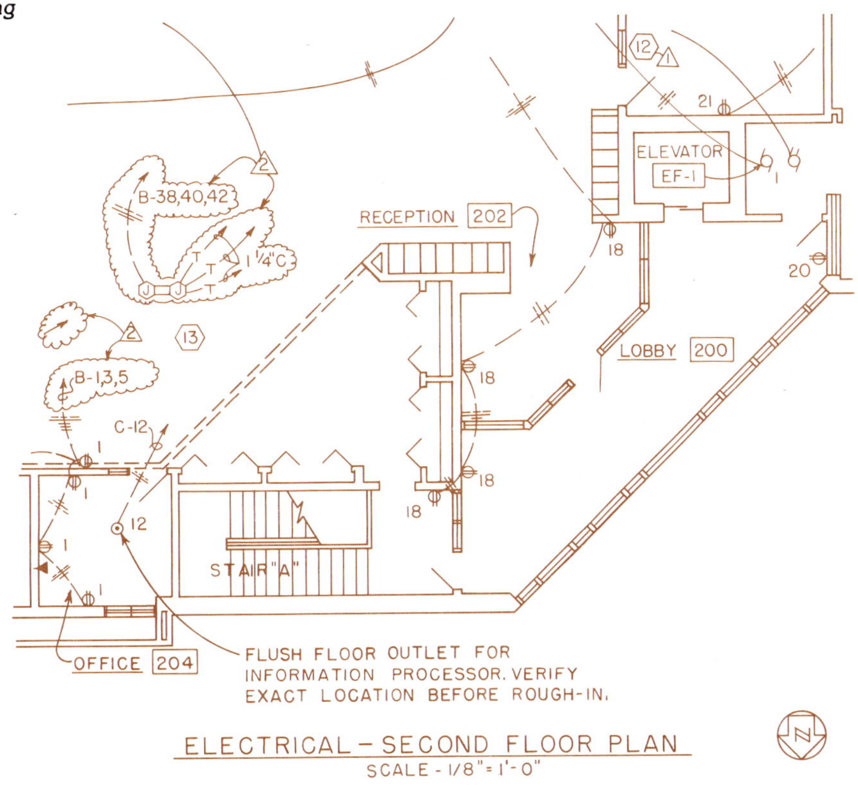

ings must be accurate and complete. Then the project can be constructed successfully.

Drawings for a large project are made in *sets*. There are four sets of drawings.

- Architectural drawings.
- Civil drawings.
- Structural drawings.
- Electrical-mechanical drawings.

These sets of drawings may be divided into more sets. For example, mechanical drawings might be divided into sets. One set might be plumbing drawings. Another might be drawings showing heating details. A complete set of drawings for a complex project may be made up of hundreds of separate drawings. See Fig. 18-6.

Drawings are made to *scale* and *dimensioned*. Each part of an object is dimensioned. Accuracy is very important. When the dimensions of all the parts are added together, they must equal the overall (total) dimensions. All parts on one drawing must relate accurately to all other drawings. Several drawings may have common parts. For example, an elevator shaft might be one part. This part must appear on each floor plan. The building might have forty floors. The shaft would be in the same place on each drawing.

Drawings are *reproduced* many times. That is, many copies are made. Sometimes the paper shrinks or stretches in the air. That is why the dimensions and scale *notations* are very important. A worker must never try to find a true size by measuring the drawing.

Summary

Engineers work mostly on heavy construction. They prepare civil, structural, and electrical-mechanical drawings.

Drawings are prepared to allow competitive bidding on projects. Such drawings are called contract drawings. They are detailed enough to allow for bidding.

The contractor who wins the bid must develop more detailed drawings. These are called working drawings or shop drawings. Each drawing in a set must be accurate. It must be drawn to scale and dimensioned. The overall dimensions must equal the sum of all the parts. There can be many drawings in a set. They provide every detail the work crew needs to build the structure.

Terms to Know

civil drawings	notations
competitive bidding	reproduced
contract drawings	scale
dimensioned	sets
documents	shop drawings
electrical-mechanical drawings	standards
heavy construction	structural drawings
	working drawings

Think About It!

1. What types of working drawings would you need to explain the construction details for a neighborhood playground?
2. Explain how architects and engineers would work together to design a large building.

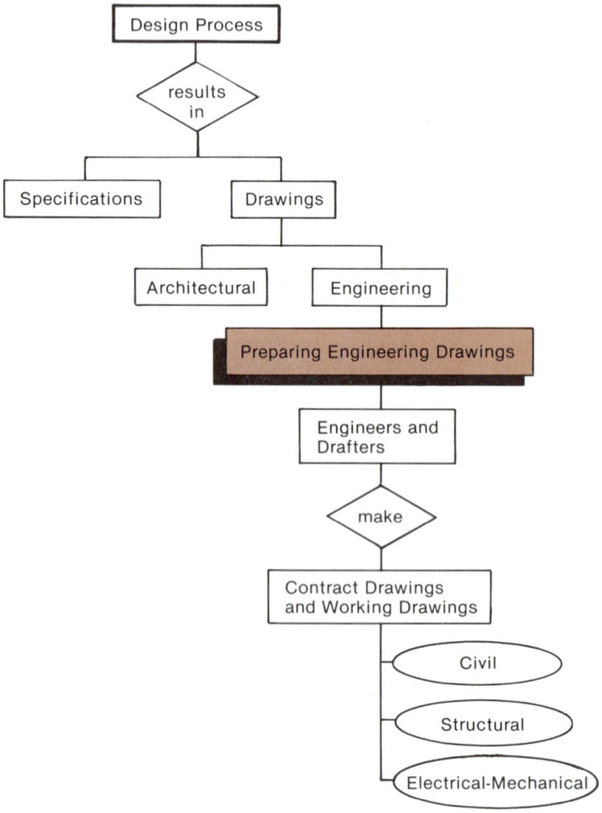

The Engineering Cycle

Esmark, Incorporated

Fig. 19-1
Designing an offshore oil drilling rig requires a thorough knowledge of the engineering process.

Engineering

Most construction projects are designed by architects or engineers. Often they work together as a team. For each structure, they consider

- appearance,
- function,
- strength,
- materials, and
- costs.

Architects are very interested in how a structure looks. Engineers are most concerned with its function and strength.

Heavy construction projects are designed by engineers. Industrial buildings, roads, and bridges are examples. Designing structures like these involves solving very difficult problems. See Fig. 19-1. Engineers must study all details of the structure's design. For example, suppose engineers are to design a bridge. They must make sure it can hold against the force of high winds. It must withstand the weight of thousands of cars over many years. It must be anchored to the earth so it will not fall down. Designing such a structure requires knowledge of the strengths of materials. And it often involves finding new solutions to design problems. This is *engineering*.

The design process used by engineers is much like that used by architects. The difference is in the type of design problems to be solved. The designing and engineering process is often called the *engineering cycle*.

The Designing and Engineering Process

The designing and engineering process has six steps. The work involved in these steps is often complex.

Problem Identification

The first step is *problem identification*. Suppose there is the problem of floods. Constructing a dam may be the solution. Or perhaps there is a lack of electricity. New power stations may have to be built. Sometimes the problem is traffic jams. The solution may be to construct freeways. But before beginning to design, the engineer must clearly understand the problem. Certain facts must be learned. For example, why is the project to be built? What purpose is it to serve? What safety measures are needed? The information is collected. Then it must be evaluated (judged).

Preliminary Ideas

The second step is the process of developing *preliminary* (beginning) *ideas*.

The engineer considers many ideas for solving the design problem. At this stage it is important to find many different solutions. Ideas are recorded as sketches. Notes are added to make ideas clearer and to record thoughts. All ideas are studied and restudied.

Refinement

Step three involves *refinement* (improvement) of the best ideas. Early in the design process, the engineer selects several ideas and refines them. Each idea is developed in more detail. The engineer determines lengths, sizes, shapes, and angles. Drawings are made of several of the best ideas.

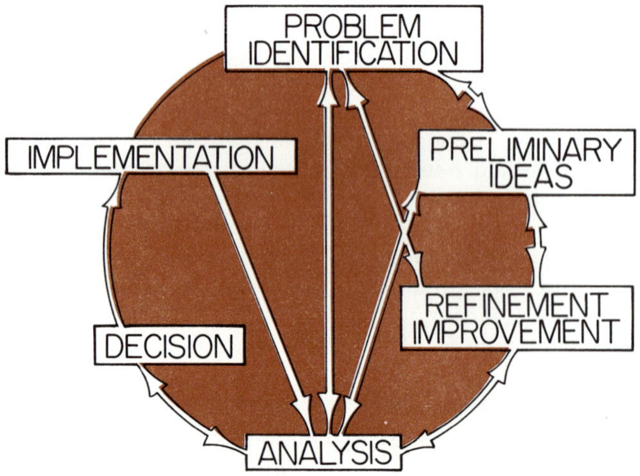

Fig. 19-2
Solving a design problem is complicated. Ideas must be questioned continually. Often ideas do not meet requirements. This leads designers back to other ideas.

Analysis

During step four, the ideas are *analyzed*. Engineers look at the best features of each design idea. To do this, they study four factors.

- Site. Special features of the land have to be considered. They may affect the design.
- Structure. The structure must remain useful under many conditions. A model of the design idea may be made and studied. This helps identify any faults in the design.
- Function. The design must show that the project will serve its intended purpose.
- Costs. Construction costs are considered. One design solution may be more economical than another.

Decision

Step five is the *decision* step. The best engineered design is chosen. The engineer presents the best solutions to the owners or clients. Graphs, charts, and diagrams may be prepared. These help explain the design. Together the designer and client decide which design solution will be used for construction.

Implementation

The last stage of the designing and engineering process is *implementation*. Detailed drawings and specifications are prepared. They explain the details of construction to the builder. The engineer works closely with the builder during actual construction. At this stage the design is not changed very much. The structure is usually built as planned.

Cycle of Activities

The six steps help us understand that the designing and engineering process is not simple. And the six steps are not necessarily followed in the order just described. See Fig. 19-2. Engineers must think through detailed ideas. Ideas may start at any point. For example, a new idea may grow out of testing a working model. Then the engineer starts at the beginning again.

Design ideas may not be taken through all six steps. Some ideas do not survive. They may cost too much to build. Or, they may not meet safety needs. Ideas can be dropped at any step. New ideas may be added at any time.

Engineering is not done by chance. It is very systematic. New ideas develop. Old ideas are

The Continental Group, Incorporated

Fig. 19-3
Civil engineers may supervise the construction of buildings as well as many other types of structures.

discarded. During the engineering process, plans change. That is why the process is called a *cycle*.

Engineers in Construction

Engineers need very special training. A college degree is very important for them. They must learn a great deal about science and mathematics. Engineers must study and learn from experience. They must develop ways to use the materials and forces of nature. Their work benefits us all.

There are about twenty-five *branches of engineering*. Many kinds of engineers are involved in construction. But only the engineers most common in construction are mentioned here.

Civil engineers design and supervise construction. See Fig. 19-3. They work on roads, harbors, airfields, and bridges. They work on water supply and sewage systems. They work on buildings and most other types of structures. Civil engineering is a very broad field. It includes structural, sanitary, hydraulic (water), and transportation engineering.

Electrical engineers design and supervise the electrical parts of construction. They design facilities for generating and distributing electrical power. They work with motors and communications equipment. They design electronic installations such as computer systems. When construction involves any kind of electrical system, electrical engineers are involved. See Fig. 19-4.

Mechanical engineers design machines and machine systems. They work with the production, transmission, and use of power. They design heating and cooling equipment for structures. Mechanical equipment such as an elevator is designed by mechanical engineers. During actual

Western Electric Company, Incorporated

Fig. 19-4
Electrical engineers design and supervise the construction of electrical systems.

Sun Company

Fig. 19-5
Mechanical engineers design and supervise the construction of mechanical systems.

construction, they supervise the installation of such equipment. See Fig. 19-5.

Other kinds of engineers work in construction, too. *Mining* and *chemical engineers* are examples. In fact, most kinds of engineers become involved in some way in designing construction.

Summary

A series of six steps is used in engineering construction. Engineers use this step-by-step process to find solutions to design problems. It is a systematic process. Often it is called the engineering cycle. The process can start or stop at any step. The cycle may be repeated.

Engineers and architects design construction. When engineers design, the process is called engineering. Engineers work on heavy construction projects. They design and supervise the important construction details.

Terms to Know

- analyzed
- branches of engineering
- chemical engineers
- civil engineers
- cycle
- decision
- electrical engineers
- engineering
- engineering cycle
- implementation
- mechanical engineers
- mining engineers
- preliminary ideas
- problem identification
- refinement

Think About It!

1. Suppose you are asked to design the stage sets for a school play. What questions might you ask to help identify the design problem?
2. After you sketch and note many preliminary ideas, you will refine some of them.
   a. What dimensions will you need before you start making drawings? (Think of the curtains, the ways of entering and leaving the stage, and all the other size limits.)
   b. Why will you need to know what the actors will do on the stage?

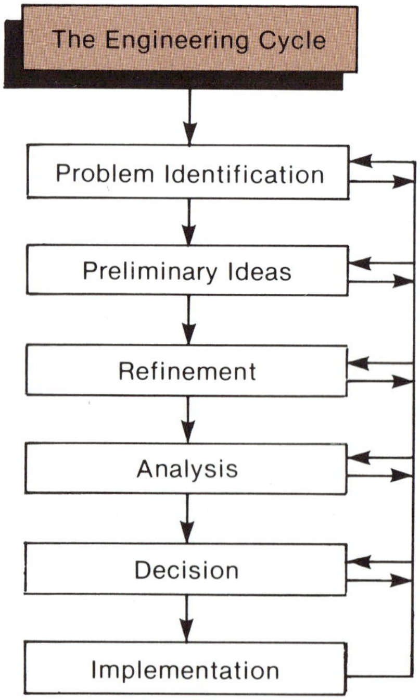

Reading 20

Contracting

The Contract to Build

When project plans and specifications are completed, the owner selects a contractor. To find the best one, the owner invites contractors' bids for the *contract* to build. See Fig. 20-1. The contractor with the best work record and bid will construct the project.

There are several methods of bidding and awarding contracts. The method described here is typical. It tells of a fair way to select a construction contractor.

Advertising

The first step in selecting a contractor is to advertise. See Fig. 20-2. The advertisement is an *invitation to bid* for the contract. Public contracts are usually advertised. This gives every contractor a chance to bid. Private contracts are sometimes advertised, too. The competition created by advertising usually results in lower costs.

Fig. 20-1
Most of the construction projects around us were built by contractors who bid for the contracts.

BIDDER'S COLUMN

Bids: March 30, 1982

Franklin and Castle Townships
North Shore County Wastewater System
Board of Public Works
Shore County, Washington
Contract No. 5
Lift Stations and Metering Stations
C421689 06

Sealed proposals for construction of — 2 built-in-place lift stations and 3 metering stations will be received by — Shore County Board of Public Works, Room 619, Luther Building, Vancouver and Sixth Street, Shore City, Washington 99908, until — 2:00 P.M. (Local Time), Tuesday, March 30, 1982, at which time and place proposals will be publicly opened and read aloud.

The Contract Documents may be examined at the offices of — Shore County Board of Public Works, Franklin Township.

Copies of the Contract Documents may be obtained from — Fitch Co., 418 Cascade Drive, Seattle — on payment of a fee of $25.00 per set, which will not be refunded.

Any Contract or Contracts awarded under this Advertisement are expected to be funded in part by a grant from the United States Environmental Protection Agency. Neither the United States nor any of its departments, agencies, or employees is or will be a party to this Advertisement or any resulting Contract. This procurement will be subject to regulations contained in 40 CFR 38.651, 38.653, and 38.654. The requirements for Bidders and Contractors under this order are explained in the Specifications.

As part of the procedure for determining the ability of Contractors to comply with the equal opportunity clause, prospective bidders may be requested to attend a meeting scheduled by the Environmental Protection Agency prior to opening of the bids where they will be instructed in the equal employment opportunity requirements of the agency.

A certified check, bank draft, or bid bond in the sum of 5% of the amount of the proposal will be required with each bid.

Fig. 20-2
Trade magazines may have special columns to advertise projects on which bids may be made.

An invitation to bid is a brief notice. It states three things.

- The type of work and the location of the job.
- Where and how to get plans and specifications.
- The time and place of the bid opening.

The *bid opening* is the time when the contractors can submit their proposals to the owner.

Invitations are sometimes mailed to some contractors. They also are sent to contractors' associations, such as a local chapter of the Associated General Contractors of America (AGC). There are also specialty contractor groups. National Electrical Contractors of America (NECA) is an example. Also, there are companies that publish reports of jobs being bid.

Great effort is made to notify all possible bidders. This is especially important for public projects.

Qualifications of Contractors

The owner must be satisfied with a contractor's work record. Sometimes this is determined before bidding starts. Only contractors considered capable are invited to bid. This is called *prequalification*. At other times, anyone may bid. Then the qualifications are determined after the bids are opened.

The contract is awarded to the responsible contractor whose bid is lowest. Often, the architect or engineer for the project assists in selecting the contractor.

Fig. 20-3
The bid opening is a public meeting. Individual proposals can be examined.

Bidding Documents

A contractor wanting to bid is given several documents. Among these are the instructions to each bidder. These tell the contractor how to submit a bid. They tell how the contract will be awarded. The contractor also receives project plans and specifications. The documents include the conditions of the contract, such as standards that must be met.

In addition to these documents, the contractor is given blank forms. These are to be used for posting bonds, reporting qualifications, and submitting the bid. All bidding documents must be completed.

Subcontractors and suppliers also obtain plans and specifications. Then they can bid their parts of the job to the general contractor.

The Bid Opening

Contractors may be given from ten to ninety days to prepare their bids. Usually, bids are opened thirty days after advertising starts.

Strict bidding rules are followed. This assures fair treatment of all bidders. Bids must be submitted in sealed envelopes. No bid may be submitted or changed after the opening. If it is a public opening, each contractor's proposal is read aloud. Anyone may attend the bid opening. See Fig. 20-3. Anyone may examine any contractor's proposal.

The Award

Each bid is carefully examined. The owner looks for the lowest *responsive bid*. A bid is responsive if it matches all conditions contained in the bidding documents. For example, suppose the owner specifies maple flooring. If a contractor's bid says pine will be used, the bid is not responsive.

The qualifications of the low (lowest) bidder are checked. There are several sources of information about contractors. One is the bidder's statement of qualifications. Another source is owners who have hired the bidder for other jobs. Banks and other businesses also have information about contractors.

If the low bidder is a qualified contractor, an award is made. The owner sends a *notice of award* to the successful bidder. A notice of the contract award is often listed in trade journals. See Fig. 20-4.

Kinds of Contracts

In construction, there are four general kinds of contracts.

GENERAL BUILDING

Notification of Contract Awards

NORTH STATE — Bids Nov. 12: addition to West Village Senior High School, Carson County, S. of existing school at West Village, to Lakeland E. MacKinley Inc., 620 W. Pond Rd., Oakville, $1,984,510; heating, to the Ryerson-Andris Co., Oakville, $314,865; plumbing, to Regal Plumbing & Heating Inc., Oakville, $139,500; ventilating, to L. and R. Mechanical Contractors Inc., King Park, $74, 650.

NORTON — Auditorium for senior citizens at Macklin Complex, to Craig-Smith Corp., $299,865. (Bids Nov. 29).

SPRINGVILLE — Replace fire station No. 6, related work, to J. Morton Construction Co., 215 Wheeler Dr., $258,700. (Bids Oct. 22).

OCTAVIA — Market Plaza 12,895 sq. ft., space for up to 7 tenants, adjoining Tony's Discount Supermarket, to Lin & Co., 450 Bow Ave., Libby, $380,550 estimated cost.

Fig. 20-4
Contract awards are commonly listed in special reports or in trade publications.

- Fixed price contracts.
- Cost-plus-a-fixed-fee contracts.
- Cost-plus-a-percentage-of-cost contracts.
- Incentive contracts.

Fixed Price Contracts

The *fixed price contract* is often used. In it, a contractor agrees to build a project. A price is set for the completed work. For example, a painting contractor agrees to paint a house for $600. The $600 is called the fixed price. It is a lump (single) sum paid for the work.

A *unit price contract* is a kind of fixed price contract. A unit price contract breaks down project costs. For example, a unit price contract might look like this.

100 000 cubic meters of common excavation at $.50 per cubic meter	$ 50,000
12 700 square meters of asphalt pavement at $10.00 per square meter	127,000
Total	$177,000

In the contract, the quantity of materials may have been estimated. Payment is made on the actual quantity. Suppose the contract said that 90 000 cubic meters were to be excavated. The contractor would receive 90 000 (m^3) × .50 (cost per m^3), or $45,000. Extra work would also be paid.

Cost-Plus-a-Fixed-Fee Contracts

Sometimes the *cost-plus-a-fixed-fee* (CPFF) *contract* is used. The owner agrees to pay the contractor the costs of construction plus a fixed sum. For example, suppose a contractor agrees to build a factory. The contractor and owner contract for cost plus a fee of $100,000. This fee goes to the contractor. It covers the contractor's *overhead* (operating costs) and *profit* (money left over after expenses).

This type of contract is used in many cases. It is most often used when work must begin quickly. Building can start before plans and specifications are completed.

Cost-Plus-a-Percentage-of-Cost Contracts

Cost-plus-a-percentage-of-cost (CPPC) *contracts* also are used. Contractors are paid a percentage of project costs. For example, a contract may be given to build a factory. The contract would provide for the actual cost, plus a percentage of the cost as a fee.* The higher the cost, the higher the contractor's fee. This kind of contract is used only in private industry. In fact, on most public work these contracts are illegal. This is because inefficient contractors receive higher fees than efficient ones. Costs for unneeded work still earn a fee.

Incentive Contracts

Finally, there are *incentive contracts*. In these, target estimates are made. If the structure is built for less than the estimated cost, the contractor may keep a part of any saving. If the actual cost is more than the estimate, the fee is reduced. Thus, contractors have incentives (good reasons) to keep down costs.

Negotiated Contracts

Sometimes owners and contractors talk together to agree on contract terms. The price is always one

*Suppose the actual cost is $1,500,000 and the fee is eight percent. To find this fee, the owner would multiply $1,500,000 (actual cost) by .08 (8%). The fee would be $120,000. The total contract price would be $1,500,000 + $120,000, or $1,620,000.

of the terms that are discussed. Sometimes the owner negotiates with more than one contractor. This helps the owner compare prices.

Contracts that result from these talks are called *negotiated contracts*. Negotiated contracts are only available to qualified bidders. Limiting the bidding saves time. It allows the construction work to begin sooner. CPFF, CPPC, and incentive contracts require some negotiation, even if they are advertised.

Bonds

Project owners often want protection against any failure of the contractor. They want to be assured that work will be done as agreed. They want to know that all the contractor's labor costs will be paid. For this reason, a performance bond and a payment bond may be required at the time a contract is signed.

Performance bonds guarantee that contractors will follow the plans and specifications. See Fig. 20-5. If they do not, the bonding company must fulfill the contract. The bonding company will probably hire another contractor.

Payment bonds guarantee that contractors will pay their subcontractors and suppliers. (Sometimes the bonds even guarantee that contractors will pay their employees.) If they do not make the payments, the bonding company must do so.

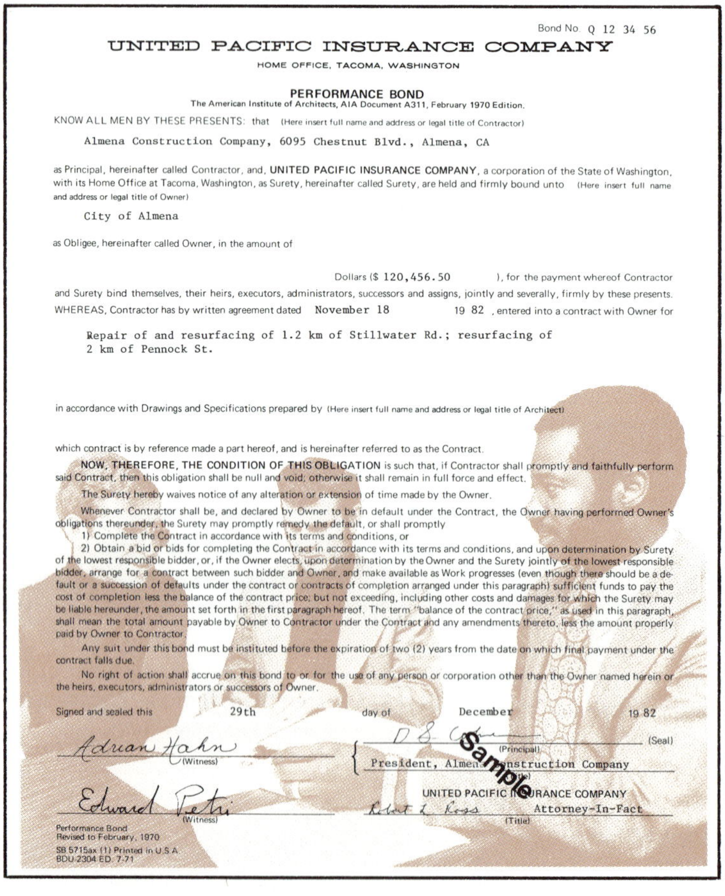

Bond No. Q 12 34 56

UNITED PACIFIC INSURANCE COMPANY

HOME OFFICE, TACOMA, WASHINGTON

PERFORMANCE BOND

The American Institute of Architects, AIA Document A311, February 1970 Edition.

KNOW ALL MEN BY THESE PRESENTS: that (Here insert full name and address or legal title of Contractor)

Almena Construction Company, 6095 Chestnut Blvd., Almena, CA

as Principal, hereinafter called Contractor, and, UNITED PACIFIC INSURANCE COMPANY, a corporation of the State of Washington, with its Home Office at Tacoma, Washington, as Surety, hereinafter called Surety, are held and firmly bound unto (Here insert full name and address or legal title of Owner)

City of Almena

as Obligee, hereinafter called Owner, in the amount of

Dollars ($ 120,456.50), for the payment whereof Contractor and Surety bind themselves, their heirs, executors, administrators, successors and assigns, jointly and severally, firmly by these presents.

WHEREAS, Contractor has by written agreement dated November 18 19 82, entered into a contract with Owner for

Repair of and resurfacing of 1.2 km of Stillwater Rd.; resurfacing of 2 km of Pennock St.

in accordance with Drawings and Specifications prepared by (Here insert full name and address or legal title of Architect)

which contract is by reference made a part hereof, and is hereinafter referred to as the Contract.

NOW, THEREFORE, THE CONDITION OF THIS OBLIGATION is such that, if Contractor shall promptly and faithfully perform said Contract, then this obligation shall be null and void; otherwise it shall remain in full force and effect.

The Surety hereby waives notice of any alteration or extension of time made by the Owner.

Whenever Contractor shall be, and declared by Owner to be in default under the Contract, the Owner having performed Owner's obligations thereunder, the Surety may promptly remedy the default, or shall promptly

1) Complete the Contract in accordance with its terms and conditions, or

2) Obtain a bid or bids for completing the Contract in accordance with its terms and conditions, and upon determination by Surety of the lowest responsible bidder, or, if the Owner elects, upon determination by the Owner and the Surety jointly of the lowest responsible bidder, arrange for a contract between such bidder and Owner, and make available as Work progresses (even though there should be a default or a succession of defaults under the contract or contracts of completion arranged under this paragraph) sufficient funds to pay the cost of completion less the balance of the contract price; but not exceeding, including other costs and damages for which the Surety may be liable hereunder, the amount set forth in the first paragraph hereof. The term "balance of the contract price," as used in this paragraph, shall mean the total amount payable by Owner to Contractor under the Contract and any amendments thereto, less the amount properly paid by Owner to Contractor.

Any suit under this bond must be instituted before the expiration of two (2) years from the date on which final payment under the contract falls due.

No right of action shall accrue on this bond to or for the use of any person or corporation other than the Owner named herein or the heirs, executors, administrators or successors of Owner.

Signed and sealed this 29th day of December 19 82

(Witness) (Seal) (Principal)

President, Almena Construction Company (Title)

(Witness) UNITED PACIFIC INSURANCE COMPANY

Attorney-In-Fact (Title)

Performance Bond
Revised to February, 1970
SB 5715ax (1) Printed in U.S.A.
BDU-2304 ED. 7-71

Fig. 20-5
The contractor must assure the owner that the project will be completed according to the plans and specifications. A performance bond makes this guarantee.

Not long ago, the cost of bonds was rather small. It was about one percent of the contract amount. Today, this cost is rising. But it is quite rare for contractors not to do what is required. Therefore, bonding companies do not often take over jobs.

The Agreement

The *agreement* is a statement of the work to be done under the contract. The price is also stated. The agreement refers to the plans, specifications, instructions to bidders, general and special conditions, and the bonds. It is signed by the owner and the contractor.

Notice to Proceed

It usually takes several days for contractors to arrange for bonds. The contractor signs the agreement and forwards it to the owner. The bonds are sent along. Then the owner signs the agreement and issues the *notice to proceed*. This notice gives the contractor permission to begin the work.

Summary

There are many steps involved in contracting. First the owner advertises with an invitation to bid. Contractors complete bidding documents. Then they submit bids to the owner.

The owner considers the bids of qualified contractors. The contractor with the lowest responsive bid is awarded the contract.

The four general types of contracts are fixed price, cost-plus-a-fixed-fee, cost-plus-a-percentage-of-cost, and incentive contracts. Cost-plus-a-fixed-fee contracts are very common. Cost-plus-a-percentage-of cost contracts are used only in private industry.

Sometimes, owners and contractors negotiate contracts. They decide on the terms and the price.

After a contract is awarded, the agreement is signed. The contractor signs it first and sends it to the owner to sign. Bonds are also sent with the contract. These assure the owner that the work will be completed as promised.

Terms to Know

- agreement
- bid opening
- contract
- cost-plus-a-fixed-fee contract
- cost-plus-a-percentage-of-cost contracts
- fixed price contract
- incentive contracts
- invitation to bid
- negotiated contracts
- notice of award
- notice to proceed
- overhead
- payment bonds
- performance bonds
- prequalification
- profit
- responsive bid
- unit price contract

Think About It!

1. If you were an owner, what information would you want to know about a contractor before awarding a contract?
2. A contract is usually awarded to the contractor who submits the lowest responsive bid.
   a. What is a responsive bid?
   b. What might happen if bids did not have to be responsive? Give examples.

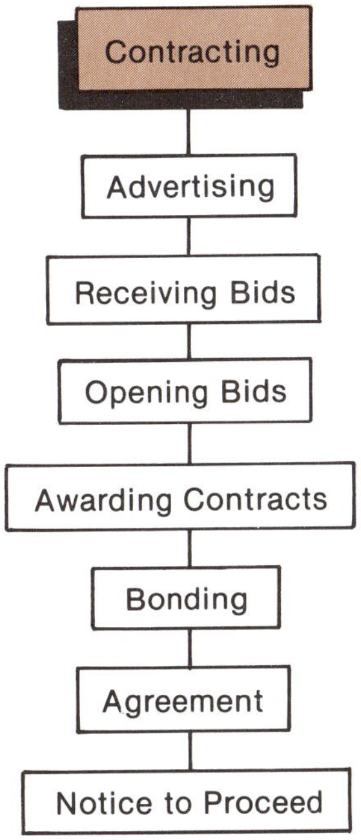

Reading 21

Estimating and Bidding

Deciding Whether to Bid

There are two sides to the contracting process. You have read how contracts are awarded. This reading will tell how the contractor bids for a contract.

Suppose a contractor receives an invitation to bid on a contract. A decision must be made. Should a bid be submitted? If so, a *bid estimate* (proposal) must be prepared. This requires planning. Bids are prepared in many different ways. The following is a typical procedure.

To decide whether or not to bid, the contractor must answer certain questions. See Fig. 21-1. "Is this the type of work I know how to do? Do I have a chance of being the low bidder? Can I make a profit? Is the job risky? Do I want to work in the area of the job site? Can I hire the needed workers? Do I have the money to support a job of this size? Do I have too much to do elsewhere? Do I have or can I rent equipment? Is this job too small? too big? the right size? Do I want to work with the owner? with the architect or engineer? Will the bonding company give me the necessary bonds?" These and many other questions must be answered. Then the contractor will reach a decision.

Fig. 21-1

The contractor must visualize and think through the construction of the structure before arriving at a decision to bid.

Fig. 21-2
Every detail of construction must be considered when preparing a bid estimate.

Fig. 21-3
Estimators compute the cost of materials, labor, equipment, and overhead. To do this, they must understand the technical information found in contract drawings and specifications.

The Bid Estimate

Once the decision is made to bid, the contractor makes an estimate. An estimate is a careful calculation of what the job will cost. See Fig. 21-2. The contractor bases the final bid on the estimate.

Most small contractors do their own estimating. Medium-sized and large contracting firms usually hire people called *estimators*. Estimators must understand the construction process. They must be able to read contract drawings. They must understand specifications. Estimators must be able to answer many questions. See Fig. 21-3. What are the current costs of materials? What are the hourly rates for the workers needed? How much work can these workers do in a day? How can equipment be used most efficiently? (The less time it takes to use equipment, the lower the cost.)

The estimator is a key person. The success or failure of a contractor depends largely on the estimator. If estimates are too high, bids will be lost. If estimates are too low, no profit will be made. In fact, money may be lost. To estimate as accurately as possible, the estimator should visit the site.

Approximate Estimates

Shotgun (approximate) *estimates* are made early. These are made to get a rough idea of what the cost will be. Suppose the size of a building is obtained from the drawings. A shotgun estimate can be made on the total number of square meters (m^2) or cubic meters (m^3). Suppose the office building requires 30 000 m^2 of floor space. Also suppose that a typical cost for an office building is \$300 per m^2. The estimator multiplies 30 000 m^2 (size) by \$300 (cost per m^2). The shotgun estimate is nine million dollars. Approximate estimates such as this are rarely used for final bids.

The Final Bid Proposal

A final bid proposal is based upon careful bid estimates. The costs of materials, labor, and equipment are figured. Overhead and profit are added to these costs. This gives the final price.

Materials

Material costs are determined by what is called a *takeoff* or quantity survey. Estimators study the plans and specifications. They count how many times a certain item appears on the plans. For example, an estimator may count 200 windows on the plan. The specifications indicate their size and kind. To determine the cost of the windows, the estimator may

- get the price from a catalog.
- call the company's purchasing department.
- call a company that sells windows.
- look in files to find out how much windows cost the last time they were used.

Suppose the windows cost $250 each, including delivery. The total cost of windows is 200 (windows) × $250 (cost), or $50,000.

Labor

Labor costs must be calculated by the estimator. See Fig. 21-4. Suppose 100 doors need to be installed in a building. The estimator determines the work time needed to hang doors. *Labor productivity* rates must be obtained. Suppose it takes a carpenter one hour to hang a door. In addition, two hours of laborer time are required.

The estimator must know the hourly wage of these persons. Assume a carpenter earns $10 an hour. Labor costs for carpenters would be 1 (hour) × 100 (doors) × $10 (labor cost), or $1000. Assume the laborers earn $6 per hour. Costs for laborers would be 2 (hours) × 100 (doors) × $6 (labor cost), or $1200. Installing the 100 doors would cost $2200. This does not include the cost of the doors themselves or other costs for workers.

Inland Steel Company

Fig. 21-5
The cost of renting or buying equipment must be estimated.

Fig. 21-4
The estimator must figure out how much it will cost to pay the workers.

One such cost is Social Security insurance. Another cost is worker's compensation (accident) insurance. Still another cost is unemployment insurance.

All labor costs for the project must be added together. This gives the total labor estimate.

Equipment

Most construction jobs require equipment. *Equipment costs* are figured on an hourly basis. The contractor often rents some equipment. This involves an hourly rental cost, plus costs of fuel and repairs. The cost of moving the equipment to the site and back to the owner is figured, too. See Fig. 21-5.

Many contractors own their equipment. The cost of owning and using it is also figured on an hourly basis. The total cost is obtained by adding all costs involved. These include *depreciation*. This is the loss of value from year to year. Costs also include insurance against theft and vandalism. Fuel and repair costs are added. Many other costs must be added to find the total cost. Then, hourly costs are figured on the total cost.

Contractors often use *unit prices*. These prices combine labor costs and equipment costs. For example, a trenching machine may cost $20 per hour. This includes the cost of moving it to and from the site. Suppose the operator is paid $8 per hour. The estimator knows how the trenching machine will perform. It will dig a trench 0.5 m wide, 1 m deep, and 50 m long in one hour. The cost for a 50 m length of trench is $20 + $8, or $28. The unit price is $28/50 m, or $.56 per meter. Suppose the contract calls for a trench 1000 m long. The estimator multiplies 1000 m by $.56 (cost per m). The estimate is $560.

Overhead

Overhead costs are indirect costs. They are not directly related to any one part of a project. There are two types of overhead costs. One is home-office costs. See Fig. 21-6. The other is job-overhead costs. Examples of *home-office costs* are office rent and utilities. Others are travel expenses and the salaries of executives and office personnel. *Job-overhead costs* might include field work such as surveying and soil testing. They include the salaries of superintendents and supervisors. Other job-overhead costs might be for barricades, building permits, and storage.

All overhead costs must be calculated. They are added to the total cost estimate.

Fig. 21-6
Office personnel, rent, and utilities are home-office overhead costs.

Profit

The contractor must make a *profit*. Otherwise the company will go out of business. The contractor spends money to build the project. The owner pays the contractor for the project. The difference is the contractor's profit or loss.

Sometimes the contractor's profit is called a *markup*. It is usually expressed as a percentage. This percentage is not set by the estimator. It is a decision of the top *management* of the contracting firm.

Management considers several factors to determine the percentage of profit. One is the size of the job. Another is the extent of risks involved. Another is the eagerness of the contractor to get the job.

Suppose the total cost for materials, labor, equipment, and overhead is $1,000,000. The contractor might set a markup of seven percent. The bid would be $1,000,000 + $70,000, or $1,070,000.

Subcontract Work

A contractor often obtains the services of *subcontractors*. They do specialized kinds of work. Earthmoving is an example of subcontract work. See Fig. 21-7. Subcontractors prepare their estimates in the same way general contractors do. They estimate the costs of materials, labor, and equipment. To this cost they add their overhead and profit. Then they quote a price to the general contractor. This price is the cost of the subcontract work.

Fig. 21-7
Sometimes clearing a site calls for very special equipment or skills. A subcontractor is hired to do this work.

Esco Corporation

Fig. 21-8

J & J Construction Company
Summary Sheet — Bid Estimate
Gymnasium — Camp Swampy

Item	Materials	Labor	Equipment	Subcontract	Total Cost
Clearing		300	700		1 000
Excavation, borrow, and compaction	3 000	4 000	5 000		12 000
Concrete and forms	50 000	23 000	2 000		75 000
Masonry	70 000	61 000	1 000		132 000
Carpentry and millwork	19 500	11 000	500		31 000
Steel				120 000	120 000
Roofing and flashing				112 000	112 000
Insulation	10 000	3 700			13 700
Lath, plaster and interior partitions	21 100	33 600			54 700
Wood flooring	42 200	20 900			63 100
Tile flooring	19 100	5 400			24 500
Painting				33 300	33 300
Hardware	19 800	2 100			21 900
Glass and glazing				14 700	14 700
Electrical				98 300	98 300
Plumbing, heating, and air conditioning				212 000	212 000
Parking lot				23 400	23 400
Cleanup		4 300	500		4 800
	254 700	169 300	9 700	613 700	1 047 400
Taxes and insurance	10 300	23 800	200	24 500	58 800
					1 106 200
Overhead 7%					77 400
					1 183 600
Profit 6.5%					77 000
					1 260 600
Bond 1%					12 600
Total cost					$1 273 200

Summary Table and Bid Proposals

After all estimated costs are known, the estimator totals them. First, the costs of materials, labor, and equipment are totaled. The cost of subcontract work is added. Then all items are placed in a summary table. See Fig. 21-8. This is the estimator's best estimate. It is given to top management (the contractor).

The contractor may adjust some items. For example, the profit percentage may be increased or decreased. This may be done for many reasons. For example:

- The contractor may expect much competition from other contractors, and wants to be low bidder.
- There may be an unusually high or low risk involved.
- The contractor may find a more economical way of doing the job.

Management may change other estimates. Even the overhead may be adjusted. Careful consideration is given to all factors. Often the estimator is consulted again before the bid proposal is made final. Then the final decision is made. The bid proposal is ready to be submitted.

Summary

Estimators or contractors prepare estimates of what the work will cost. Estimates include costs of materials, labor, equipment, overhead, and profit. They are always carefully checked. The success or failure of a contractor depends very much upon the ability of the estimator.

From the estimate, the contractor prepares a final bid proposal. The bid proposal may be adjusted before it becomes final. When it is final, it is submitted to the project owner.

Terms to Know

bid estimate
depreciation
equipment costs
estimators
home-office costs
job-overhead costs
labor costs
labor productivity
management
markup
material costs
overhead
profit
shotgun estimates
subcontractors
takeoff
unit prices

Think About It!

1. In preparing a bid for a construction project, where does an estimator get the following kinds of information?
   a. Sizes and kinds of materials needed.
   b. Labor costs.
   c. Equipment costs.
   d. Overhead costs.
2. What might cause a contractor to raise or lower the markup included in the bid?

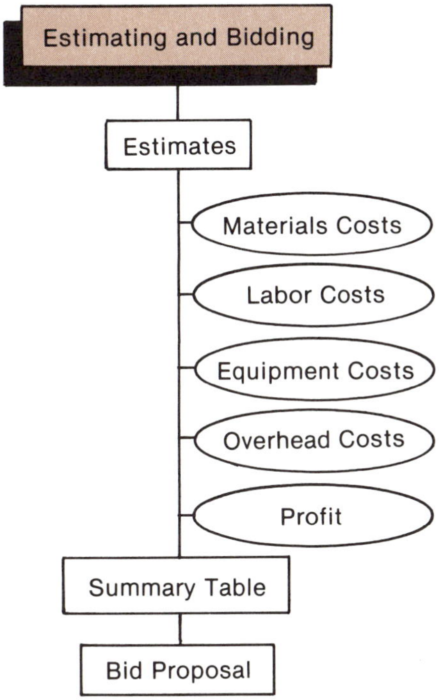

Reading 22

Scheduling

Fig. 22-1
Scheduling is important. What would happen if this load of concrete arrived at the site before the forms were built?

Fig. 22-2
List of Work to Be Done, Cost, and Time

Item	Cost	% of Job	Time*
Order and deliver pipe	$ 400	40%	3 days
Assemble pipe and fittings	100	10	1
Dig hole	100	10	2
Erect pole	200	20	1
Order concrete			1
Place and harden concrete	100	10	2
Make painting subcontract			3
Paint	100	10	1
Total	$1000	100%	

*Note: Because some items are done at the same time, times are not totaled.

The Work Schedule

When the bid for a contract is accepted, the contractor receives a *notice of award*. Next, a *schedule* must be prepared. This is the contractor's plan for the order in which work will be done. See Fig. 22-1. A work schedule is required by supervising architects and engineers. The specifications often set a deadline for receipt of the schedule.

Analyzing the Job

Scheduling begins with *analyzing* the job in detail. The contractor must plan the best use of people, materials, equipment, and money. Many decisions have to be made in a short time.

Suppose the job is installing a flagpole. The total cost of the job is $1000. A subcontractor will do the painting. All other work will be done by the contractor's work crew. The contractor makes a list of the work to be done. Cost estimates are listed in detail. See Fig. 22-2. (This example does not include the contractor's profit.)

Preparing a Work Schedule

Contractors commonly use two methods of scheduling work. These methods are the *Bar Chart Procedure* and the *Critical Path Method (CPM)*. These are tools that management uses to help achieve efficiency. Although they are different in form, each method permits the contractor to plan the order of work. Each can be used to monitor (check) progress. The Bar Chart Procedure will be used here to show how work is scheduled for the flagpole job.

The contractor develops a bar chart using the list of work to be done. (See again Fig. 22-2.) The availability of equipment, materials, and transportation is checked. From experience, the contractor knows which tasks must be done first. The pipe should be ordered and received first. It should be put together by the time the hole is dug. There is no point in digging the hole for the flagpole until the pipe (to make

Fig. 22-3
T. & L. Construction Co.
Installation of Flagpole — Camp Swampy
Bar Chart

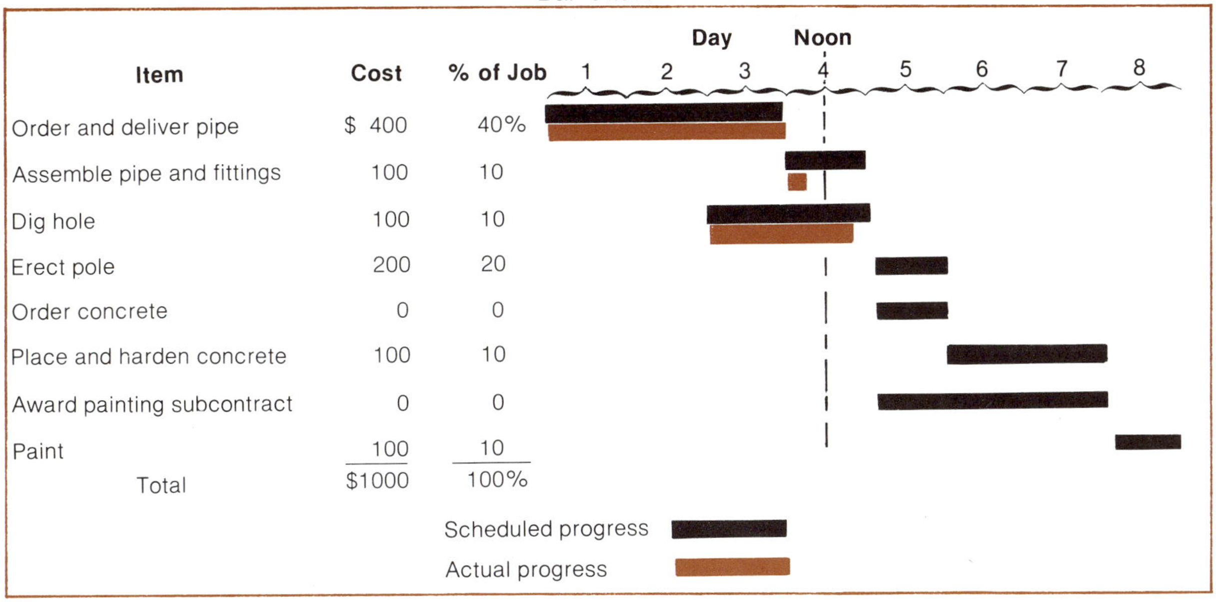

the pole) is ready. If it rained, the sides of the hole might cave in. When the hole is dug, the pipe can be erected (put up). After the pole is erected, concrete can be placed to help hold it. When the concrete has hardened, painting may begin. But before the painting can begin, the painting subcontract must be signed.

Using this knowledge and experience, the contractor makes a bar chart. See Fig. 22-3. Items, their costs, and the percentage each task is of the total job are listed on the bar chart. The work days are listed across the top. The items of work (tasks) are listed down the left side.

The black bars in Fig. 22-3 show *scheduled* (planned) *progress*. Notice that the first three days are for obtaining the pipe. By the end of Day 4, the pipe is to be put together. If the hole has been dug, the pole can be erected on Day 5. Two days are allowed for digging the hole. Digging preparations should start on Day 3. During the fifth day, the pole will be erected. Also on Day 5, the supplier of ready-mix concrete will be contacted. The concrete will be placed and set up during Days 6 and 7. The painting subcontractor should be contacted as early as Day 5. Making a deal with the painters may take three days. They must buy the paint and make preparations. Painting can be done on Day 8. Then the job should be completed.

This work schedule is given to the supervising architect or engineer. This is required by the *conditions* of the specifications.

Monitoring Job Progress

Once the job has begun, the contractor checks or observes the progress. This is called *monitoring*. The contractor uses the bar chart to keep track of work done.

Comparing Scheduled and Actual Progress

Study Fig. 22-3. The black bars are the scheduled progress. The colored bars represent work completed. This is called *actual progress*. The contractor examined the job at noon of Day 4. It was found that the pipe had been delivered. But the assembly was only one-fourth done. It should have been half finished by noon. Also, the contractor found that the hole was seven-eighths completed. By noon it had only needed to be three-fourths done. Pipe assembly was one-fourth of a day behind schedule. The hole was one-eighth of a day ahead of schedule.

Reassigning Tasks

After making this discovery, the contractor *reassigned* the crew. More workers and tools were put to work on pipe assembly. Fewer persons worked

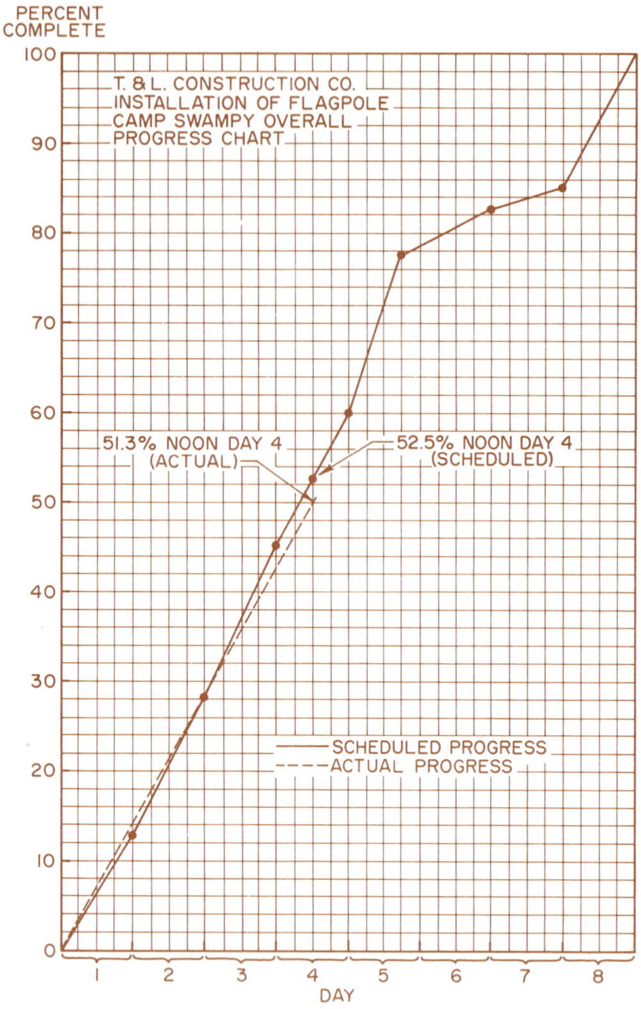

Fig. 22-4
A dotted line connects the points of actual progress. When the dotted line is above the solid line, the job is ahead of schedule. When the dotted line is below the solid line, the job is behind schedule. Look at Day 3½ (noon Day 4). The point on the dotted line is at 51.3%. The point on the solid line is at 52.5%. Scheduled progress (52.5%) minus actual progress (51.3%) is 1.2%. The job is 1.2% behind schedule.

on digging the hole. The pipe needed to be fully assembled by the end of Day 4. Then it could be erected on Day 5.

Monitoring Costs

A contractor monitors job costs by using an *Overall Progress Chart.* See Fig. 22-4. Often this chart is placed at the top of the bar chart. Progress is determined by the number of dollars that have been spent at a given point on the job. This quantity is found by multiplying the amount of time allowed for each scheduled item by its percentage of the job. For example, three days are allowed for pipe ordering and delivery. It requires forty percent of the cost of the total job. Look again at Fig. 22-3. One-third of forty percent of the job is 13.3 percent. To locate this on the Overall Progress Chart, Fig. 22-4, begin at zero. Move across the bottom to the end of Day 1. Move up to 13.3%. This point is marked on the chart. It shows that 13.3 percent of the project money should have been spent by the end of Day 1. The progress expected each day is figured and marked. These points are connected. See the "scheduled progress" line in Fig. 22-4.

Work schedules are good estimates of time needs. A certain amount of time is allowed for completing each part of a job. As work progresses, some of the contractor's estimates prove to be correct. Others are wrong. Bad weather or unexpected events may cause delays. By monitoring the job, the contractor knows when and where changes must be made.

Reporting Job Progress

On very big jobs, meetings must be held to check progress. See Fig. 22-5. Reports are given to tell

Fig. 22-5
The contractor meets with others involved in the project to decide whether the job is proceeding on schedule.
General Electric Company

how the job is going. The contractor is always at the meetings. The owner, the architect or engineer, and the suppliers usually attend also. Subcontractors may be present. Meetings may be held weekly, monthly, or any time they are needed. Work is checked against the schedule. The charts are carefully studied. If the job is behind schedule, ways to improve progress are discussed.

Revising Schedules

At times, schedules are revised (changed). This may occur when nothing can be done to stay on time. Maybe the weather is bad for a long time. Or perhaps work stops because of a fire or strike. Delays are avoided as much as possible. No one likes to have a job done late.

Sometimes the project is to be changed in some way. The owner, architect, or engineer may want to alter the plans. This may call for a new schedule. The new schedule is made in the same way as the first one.

Summary

After a contract is awarded, a schedule is developed. The contractor analyzes the job and marks work schedules on charts. Schedules are planned to use people, materials, equipment, and money in the best way. The owner or the architect or engineer receives a copy of the schedule.

The contractor monitors the progress of work. If all goes well, the work is done as scheduled. Sometimes changes must be made in the way work is done. Schedules may be revised. The goal is to complete the job on or before the scheduled date.

Terms to Know

actual progress
analyzing
Bar Chart Procedure
conditions
Critical Path Method (CPM)
monitoring
notice of award
Overall Progress Chart
reassigned
schedule
scheduled progress

Think About It!

1. What might happen if a contractor made these errors in scheduling?
   a. The inside walls were plastered and painted before an electrician arrived to install wiring.
   b. The kitchen walls and floor were finished before the plumbing was done.
   c. A truck arrived with ready-mixed concrete, but the forms to hold the concrete were not built.
2. Suggest some other scheduling errors that might be very expensive mistakes.

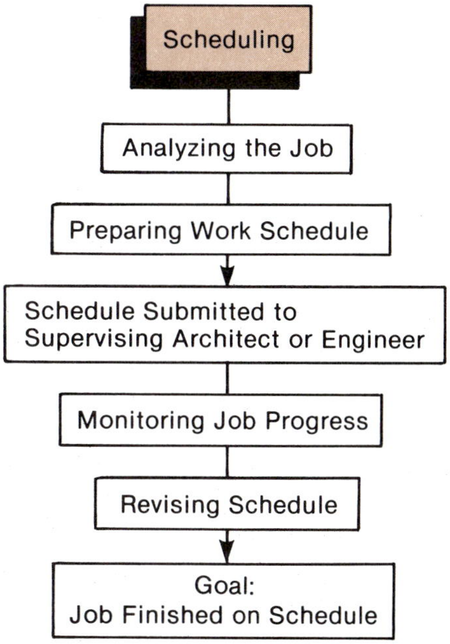

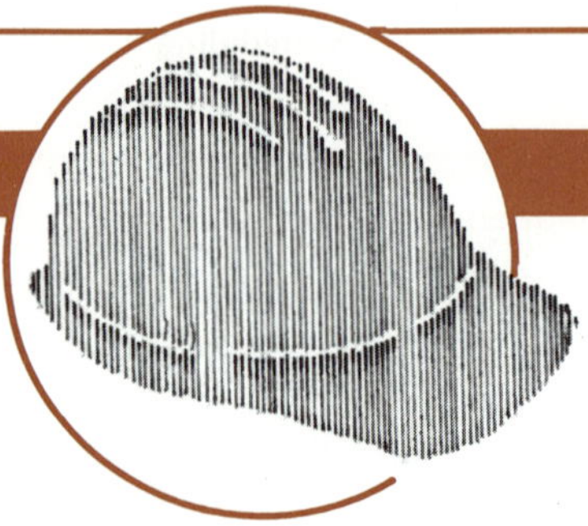

Reading 23

Working as a Contractor

Responsibilities of the Contractor

The *contractor* is in charge of project construction. This involves many responsibilities (duties). It also involves working with the project owner, architect and engineer, subcontractors, and suppliers. This reading discusses these duties and relationships. It also tells about the contractor's *business organization*.

You will find additional information about the contractor in other readings. Reading 20 describes how the contractor is selected. Other readings explain how the contractor bids for a contract. They also explain how work is scheduled.

The contractor has many different kinds of responsibilities. See Fig. 23-1. All the labor and materials for building a project are furnished (supplied) by the contractor. Work must be done exactly as agreed with the project owner. And it is important for workers to follow the *plans* for construction exactly. The contractor makes certain that the materials and methods used are those set by the *specifications*.

The contractor must make sure a *superintendent* is always on the job. The superintendent oversees all the workers. This includes those who work for subcontractors as well as those who work for the contractor.

Construction is regulated (controlled) by laws. Contractors must obey all of them. Some of these laws relate to labor relations and licensing. Others relate to safety, traffic, and sanitation.

The contractor protects the owner by providing the necessary *insurance* and *bonds*. These assure the owner that work will be completed. Furthermore, all payments will be made for labor and

Ceco Corporation

Fig. 23-1
Even on complicated projects, a contractor must make sure that all work is done as agreed in the contract.

materials. The protection required is stated in the contract (agreement) between owner and contractor.

The contractor must supply samples of materials as required in the contract. Shop drawings (detailed drawings for construction) must be furnished. Workers, suppliers, and subcontractors must be paid on time.

In other words, the contractor is responsible for completing the project. All work must be done as agreed in the contract.

Usually, the quality of materials and work must be *guaranteed*. The guarantee lasts for at least one year after the job is done.

Responsibilities of the Owner to the Contractor

The owner has responsibilities, too. They include supplying the *real estate* (land) for the project and paying the contractor as agreed in the contract.

Sometimes the owner wants to make changes in the project. If the project is underway, plenty of time must be allowed for changes. The owner should not delay the contractor.

Some projects require more than one contractor. On these projects, the owner must arrange for someone to organize the work. This is usually an architect or engineer.

Responsibilities of the Architect or Engineer

Architects and engineers work on the sites for large projects. This reading refers to them simply as engineers.

The work done by the contractor is supervised by the engineer. He or she sees that the contractor follows the plans and specifications.

The engineer also tests and approves all materials. See Fig. 23-2. For example, the contractor may provide a sample brick. The engineer may examine it for color and test it for strength. The brick may also be tested for its ability to withstand extreme weather conditions.

The engineer examines and approves

- shop drawings.
- equipment specifications.
- the contractor's schedule.
- general methods of operation.
- the superintendent on the job.
- the subcontractors hired.

The engineer also is responsible for explaining all the *contract documents*. For example, suppose the specifications recommend a Toledo No. 422 lighting fixture. The contractor wants to use a Sandusky No. 148 instead. The engineer would rule (decide) whether this is equal to the fixture specified. If it is, the No. 148 fixture can be used. Or, suppose the contractor finds rock where the drawings show soft sand. The engineer would rule whether the contractor should be paid extra to correct the problem.

In short, the engineer is the agent of the owner. She or he acts in place of or for the owner. See Fig. 23-3.

Fig. 23-2
Engineers examine construction materials to be sure that they meet the specifications for the project.

Fig. 23-3
Engineers represent the owner on the construction site. They check work being done and often make decisions about changes.

Fig. 23-4

Typical Organization of a Construction Contracting Company

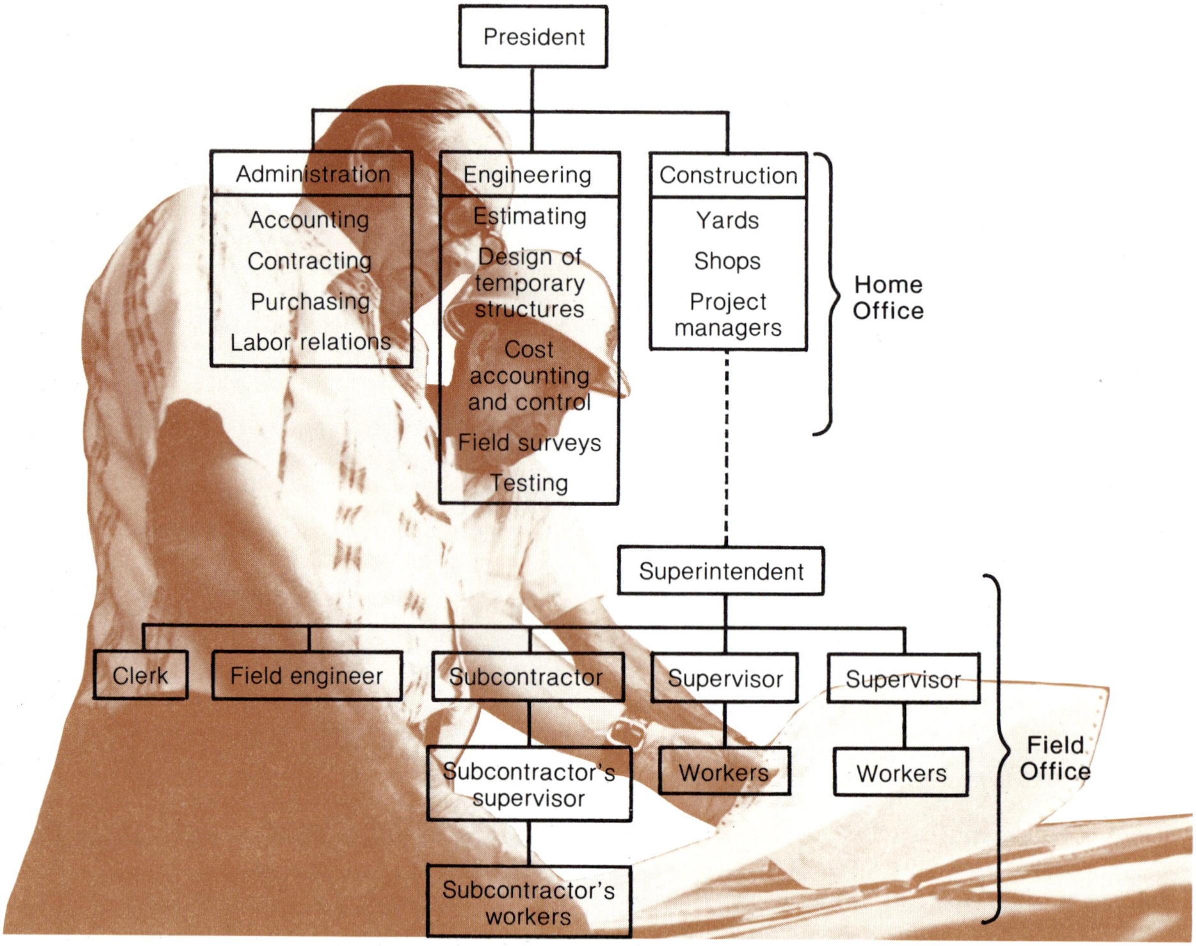

Fig. 23-4

Contractors organize their employees. Each person reports to someone in a higher position.

Organization of the Contracting Firm

Contractors manage people, materials, tools, and equipment. All of these are important. But people are the key to successful contracting. Each person is assigned a certain job. In this way, the contractor organizes the workers. Organizing employees may be done in various ways. The method used depends on four things.

The amount of business the contractor does.
The types of jobs the contractor does.
The various abilities of the workers.
The desires of the contractor.

Figure 23-4 shows how a typical construction company may be organized. However, another contractor may organize differently.

Home Office

At the head of the organization there is usually an owner or president. This person is in charge of the construction company. This *chief executive officer* has the overall responsibility and authority for all projects. The president's office is usually at the *home office*. This is the central business office. It may be a long way from any job site.

At the home office, there are usually three groups in authority next to the president. They are administrative, engineering, and construction groups.

The groups are shown separately on the organization chart. But the people in them work closely with each other. For example, the superintendent in the *construction* group may need supplies. The order is sent to the purchasing section of the *administrative* group. Or, suppose there is a question about the type of material needed. The purchasing group may ask the *engineering* group for an answer.

Administrative Group

The administrative group may include several sections. Accounting, contracting, purchasing, and labor relations are handled by this group. Workers in the *accounting* section of the group handle money matters. They keep the books (money records). They work with the payrolls. That is, they pay the contractor's employees, subcontractors, and suppliers. They also prepare bills to send the owner. They often do the paperwork for the hiring of personnel.

The *contracting* section prepares contracts. Agreements are made with owners, subcontractors, and suppliers. People in this section also prepare contract changes. They arrange for bonds and insurance.

The *purchasing* section has many duties. Workers make sure materials are available on the job. They buy materials and equipment and keep a file of catalogs and price lists up-to-date. Workers in purchasing handle the rental of equipment. They also arrange for transportation of materials to the job.

The *labor relations* section handles labor- management matters. These people work with individuals or unions to agree on labor contracts. They represent the contractor if labor problems arise.

Engineering Group

Temporary structures may be needed at a site. For example, before work can begin, materials and equipment must be hauled to the site. A temporary bridge and road may be needed. People in the engineering group design such structures. They keep records of construction costs. Costs are controlled. This keeps then from becoming too high. The engineering group also conducts needed tests and field surveys. Another job of this group is preparing bid estimates. These are proposed costs of building a project.

Fig. 23-5
The superintendent is in charge of the construction project. The field office is the superintendent's headquarters on the site.

Construction Group

The construction group actually builds the project. People in this group run yards, shops, and field offices. Look again at Fig. 23-4. *Yards* and *shops* are at the home office. Large equipment and lumber are stored in the yards. Parts of a project may be built in a shop. The *field office* is a temporary office set up at the construction site.

Often there is a *project manager*. This person works with the construction group at the home office. A project manager also supervises the activities of the field office. This person acts as the home office representative of the job site superintendent. For example, suppose that project work may be delayed because a roller is needed. The roller has to be rented and delivered to the site. The superintendent calls the project manager and states the problem. The manager asks the purchasing section to speed rental and delivery.

The Field Office

Usually a small field office is located on the site. This office is the superintendent's headquarters.

The Superintendent

The superintendent is in charge of all work done at the site. This includes the work of the contractor. It also includes work done by subcontractors. See Fig. 23-5. There may be many types of workers at the site. Or, there may be only a few workers. The number of people needed depends largely on the size of the job. It also depends on how complicated the project is.

The superintendent may have a clerk. This person takes care of paperwork in the office. For example, the clerk prepares the payroll and receives reports from timekeepers or supervisors.

Other People Who Manage Work at the Site

On large jobs, there may also be a *field engineer*. The field engineer works at the site, solving engineering problems that arise. This engineer also sees that the project layout is carried out properly. Layout is the measuring and locating of elements on the site. On large projects, the field engineer may have a survey crew.

Various *supervisors* work under the superintendent. Usually each trade group has a supervisor. For example, there may be eight carpenters working on the site. Usually a carpenter supervisor is in charge. There also may be supervisors in charge of steel, concrete, or earthmoving work.

Subcontractors also have supervisors who direct specialized work on the job. On very large projects, a subcontractor may even have a superintendent and several supervisors. This may occur, for example, when electrical work is done on a skyscraper.

Summary

The main responsibility of the contractor is to get the job done. The project owner's job is to furnish the site and the money. The engineer represents the owner. This person sees that the plans and specifications are followed.

Employees, subcontractors, and suppliers help the contractor get the job done. To work effectively, the contractor's employees are organized. Each performs a certain duty. Each reports to someone with more authority. The superintendent for the general contractor supervises all the work. This includes subcontract work as well as the work done by the contractor's own employees.

Terms to Know

accounting
administrative
bonds
business organization
chief executive officer
construction
contract documents
contracting
contractor
engineering
field engineer
field office
guaranteed
home office
insurance
labor relations
plans
project manager
purchasing
real estate
shops
specifications
superintendent
supervisors
yards

Think About It!

1. A contractor must be sure that all the correct materials are used. What other duties (responsibilities) does the contractor have?
2. If you were going to work in one of the groups in a contractor's office (administrative, engineering, construction), which would you choose? Why?

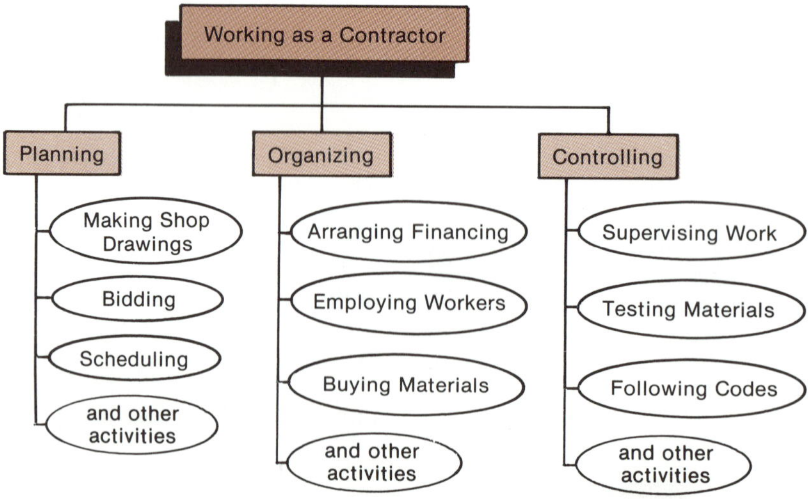

Reading 24

Training and Educating for Construction

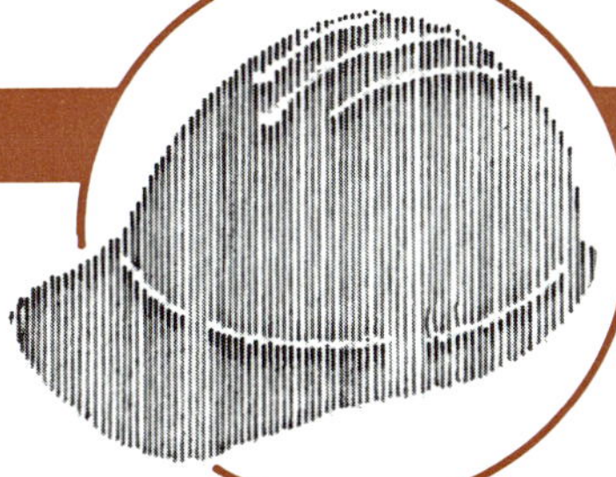

Preparing for Jobs

It takes hard work to become a professional baseball player. You must learn to hit, field, and run bases. And you must practice. Finally, you become part of a team. It is the same for persons who work in construction. They must learn work skills. And they must become experienced workers on a construction crew.

Workers go through training to learn their jobs. And they need further training to advance to better jobs. Suppose a carpenter wishes to become a supervisor. First, he or she must learn how to supervise the work of others.

Groups Requiring Training

There are three major groups of construction workers. Some workers perform *management tasks*. Others do *production tasks*. Still others perform *office tasks*. Each group has different training needs. See Fig. 24-1.

Managers plan, organize, and control production systems. There are many kinds and levels of construction managers. Examples are community planners, estimators, contractors, supervisors, architects, and engineers. Most of these people complete construction education programs after high school. They may attend a college, university, or technical institute.

Production workers carry out the plans of management. They do the on-site work.

Production workers have many different specialties. For example, there are bricklayers, electricians, plumbers, and carpenters. All must complete high school. Then they may enter apprenticeship programs.

Construction workers who perform office tasks also do important work. Their work is much like that done in any office. *Office workers* include receptionists, secretaries, and clerks.

Educating Architects, Engineers, and Construction Managers

Architects and engineers design construction projects. Civil engineers, sanitary engineers, and architects work mainly in construction. Electrical and mechanical engineers work in both construction and manufacturing.

Architects and engineers must attend college. Their college studies usually last four or five years. During this time they take general education courses such as English and history. They also study mathematics, science, and specialized subjects such as drafting.

Students in architecture and engineering must work hard. Their specialized courses are usually difficult. They must study and attend classes regularly. Their work must be average or above.

A good high school education improves the chances of being accepted into college. It also improves the chances of doing well. Architecture and engineering students need to have good high school grades. Mathematics and science grades are especially important.

Each engineering school sets its own entrance requirements. However, all require similar things. A typical engineering school requires

- three or four years of mathematics.
- three or four years of English.
- two or three years of science.
- one or two years of industrial arts.

In all these subjects, the higher number of years is preferred. Usually above average grades are necessary. For detailed information, check with specific colleges.

Many architects and engineers work as managers for contractors or as contractors. Many experienced office workers become office managers. Many skilled workers become contractors. Some people become managers by studying professional construction

MECHANICAL ENGINEER

Take responsibility for all aspects of improving existing lines of heating, ventilating, and air-conditioning products: marketing and technical analysis; design and testing; cost and manufacturing analysis; liter tion and education of sales pec product expert. Send resume o Heating and Air-Conditioning Fourth Ave., Springfield.

MANAGERS

ARCHITECT

Experienced architect with knowledge of estimating, purchasing, letting contracts to suppliers, and client contact, needed for well-paying position with long-time local building contractor. Work involves commercial, industrial, and public works design, building, and remodeling. Excellent opportunity. Send resume with salary requirements to Box 507-A, Oakville Dispatch.

CLERK/TYPIST

Good typist needed to type and file routine forms. Must be capable of 50-55 wpm accurately. Answer phones and open mail for building contractor. Call Ramirez Contracting, Personnel Department, 967-0347, 8-5 Mon.-Fri.

PUBLIC WORKS OPERATIONS SUPERVISOR
City of Verbena

Responsible for directing the construction and maintenance of the city's streets and s through management of a depart- l division. Duties include supervision, istration, planning and evaluation of uction activities. Requires gradua- rom a high school or vocational , 2 years college in construction logy preferred; minimum of 5 years construction and maintenance of -works type facilities such as streets and sewers, at least three of which must be at the supervisory level. Salary under review. Send resume to Personnel Division, City of Verbena, 921 Vine Ave., Verbena, by September 30. Equal Opportunity

OFFICE WORKERS

Construction Superintendent

Position for person with experience in volume production of Residential and Multi-Family Projects required. Need FHA inspection requirements and complete knowledge of the city. Send brief outline of experience and salary requirements to P.O. Box 640, Daily Herald.

Estimator

Take-off and estimating of sheet metal duct and fittings. Experience required. Some management responsibilities. Resume only. Send to P.O Box 45, Wood River.

ROOFERS (BUILT-UP)

Pay based on experience and ability.
Benefits included. 665-3030.

Piping Designer

Immediate opening for an experienced Piping Designer. Prefer person with working knowledge of the ANSI Power Piping Code. Opportunities for advancement. Send resume to Laura Davis, Personnel Manager, Calhoun Construction Design, Box 30-A.

APARTMENT MAINTENANCE PERSON. Must have complete knowledge of nditioning and heating, as well as phases of maintenance. Must own tools, vehicle. Must get along people. Good starting salary and its. Call Newport Community Ser- 334-9006, Mon.-Fri. 8-6.

Detailer

To prepare shop drawings of structural steel and miscellaneous metals. Samples of work will be required at time of interview Local established company with ex benefits. Send resume to Detailer, P. AA4, Waterbury, 00665.

PLUMBER, repair expert. Need state license and 10 yrs. experience in plumbing and heating. 976-4006, 331-5114.

Welders

Medium-sized company in North end needs arc or MIG welders. We offer medical, bonus, and paid vacations. We have steady work and no layoffs. Positions are for day shift, with wages based on experience. See L. Howard at Lakeland Steel Fabricators, 6597 E. Superior.

Drywall Finishers – Experienced only. For apartments and custom homes. A-1 Drywall Co., 403 State St., City.

PRODUCTION WORKERS

Fig. 24-1

Jobs in construction may require experience or special training. But they usually lead to further career opportunities.

management in college. They are not engineers. They are not skilled workers. They are specialists in managing construction.

Training Production Workers

The construction industry gets production workers from several sources. Persons may walk into a contractor's office and be hired. This usually happens when the industry is booming and *unskilled labor* is needed.

Young workers begin in construction trades as helpers or *apprentices*. An apprentice is a beginner who trains with a skilled worker.

There is usually a pool (available group) of unemployed skilled workers. They are already trained and qualified. But they may need to learn new skills and techniques.

Workers may enter the industry in many ways. But it is always necessary for them to become skilled workers. Being skilled helps them find permanent and rewarding (satisfying) work in construction.

Apprenticeship

Skilled workers go through a training period called an *apprenticeship*. Apprentices receive two kinds of training. See Fig. 24-2. They receive between 6000 and 10,000 hours of *on-the-job training*. They also get 144 or more hours of classroom instruction per year. Apprentices become skilled workers in about four years.

Labor unions and contractors work together to set up apprentice training programs. Apprentices accepted into these programs usually are between eighteen and twenty-five years old. They must be in good physical condition. Most often, it is necessary for them to have completed high school courses in mathematics and science. Mechanical aptitude is helpful. (This is the ability to use machines and understand how they work.) They should be handy with tools. Apprentices are graded on their progress. As they learn, they are advanced.

An apprenticeship program is often supervised by a *joint apprenticeship committee*. Committee members are representatives of local employers and unions. Members of the committee have many important tasks.

Deciding how many apprentices are needed.
Setting standards for apprentices' education, training, and experience.

- Determining which employers can provide good training.
- Settling disputes between employers and apprentices.

Illinois Laborers' and Contractors' Training School

Fig. 24-2
Apprentices in construction learn a trade in a classroom and on the job. Skilled workers teach them.

The committee also *certifies* the apprentices when they have completed the required courses. This means that the committee states officially that the person has met the set standards.

Apprenticeship programs are *registered* (listed) with the state apprenticeship agency. They are also listed with the United States Department of Labor. The Department of Labor has a Bureau of Apprenticeship and Training.

In some areas there is no joint apprenticeship committee. Apprentices may make agreements with employers or employers' representatives. This arrangement may be good. But it does involve some risks. There may be disputes or differences between the two parties. If so, the apprentice does not have the protection of the apprenticeship committee. And training and grading may be done by only one person. The skills or training abilities of this person may vary. If they are weak, the apprentice's skills may not be up to other contractors' standards.

A person interested in joining a building trade may shorten the apprenticeship period by taking *preapprenticeship* instruction. This instruction is given only for some trades. It is offered in high schools, trade schools, or junior colleges. A joint apprenticeship committee or employer may give credit for this training. To learn more about preapprenticeship and apprenticeship programs, ask your teacher or guidance counselor. See Fig. 24-3.

Training Office Workers

Office workers keep construction operations moving smoothly. They usually receive their initial (beginning) training outside the construction industry. For example, they may take business courses. These courses are offered in high schools and colleges. Or, they may attend general office training programs. Programs are run by businesses, industries, and the government.

Summary

Training is important in the construction industry. It enables workers to get jobs. It also helps them to advance to better jobs.

There are three main kinds of workers in the construction industry. These are managers, production workers, and office workers.

Managers plan, organize, and control systems. They are usually trained in a college, university, or technical institute.

Production workers do the actual project construction. Often, they are trained through apprenticeship programs. These programs take about four years. The training involves on-the-job experience as well as classroom instruction. Apprenticeship programs are usually operated by a joint apprenticeship committee. Committee members represent local employers and unions.

U.S. Department of Labor

Fig. 24-3
Are you interested in learning a building trade? Information is available at your school counseling office.

Office workers usually receive their training outside the construction industry. They may take business courses in high school or college. Or, they may attend office training programs.

Your teachers and counselors are good sources of information about construction education and training. People you know who work in construction can also advise you.

Terms to Know

apprentices	office tasks
apprenticeship	office workers
certifies	on-the-job training
joint apprenticeship committee	preapprenticeship
management tasks	production tasks
managers	production workers
mechanical aptitude	registered
	unskilled labor

Think About It!

1. Listed are the three groups of people who do construction work. Name several jobs in each group.
   a. Managers.
   b. Production workers.
   c. Office workers.
2. A skilled worker is trained in an apprenticeship program.
   a. Who sets up and runs this kind of training?
   b. What are the requirements to become an apprentice in a construction trade?
   c. What kinds of training are required to become a skilled worker? How much of each kind of training is required?

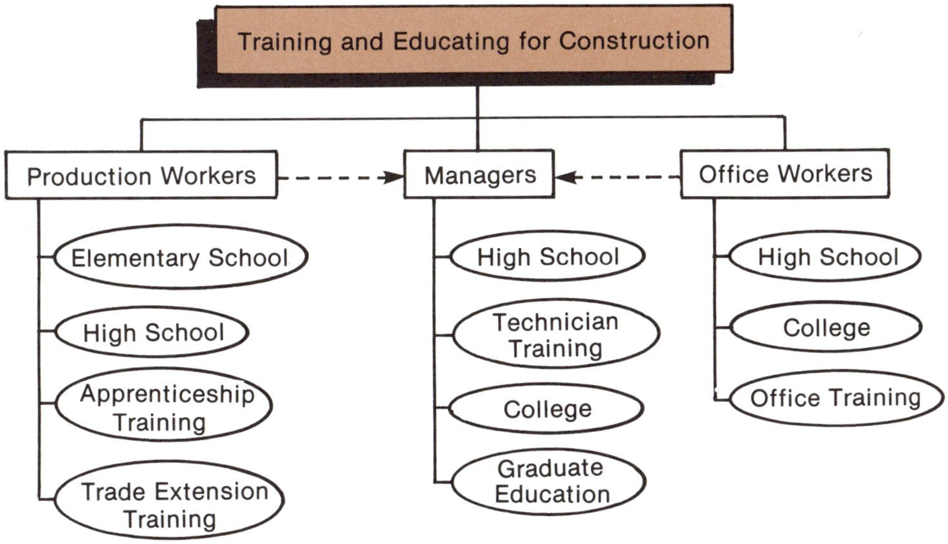

Reading 25

Hiring and Advancing Construction Personnel

Personnel Practices in Construction

Millions of people work in construction. Each one was hired for a job. Some have changed jobs by advancing. Someday, each of the workers will retire.

Hiring, advancing, and retiring are three major personnel practices. You will learn about them in this reading. Later, you will learn about training and working, the other major personnel practices.

Hiring

Every construction worker has been hired at least once. Many have been hired a number of times. They have worked on many different projects. See Fig. 25-1.

Hiring involves

recruiting,
selecting, and
inducting or assigning.

Recruiting

Recruiting is finding and attracting needed workers. Recruiting may be fairly simple. A contractor of small jobs may simply hire a person who is watching the work. There is even a name for this practice. It is called "hiring off the bank." (Here, "bank" means a mound of earth.)

For large contracting firms, recruiting may be difficult and expensive. This is especially true when specialists are needed. Engineers and superintendents are specialists. Supervisors of workers in each trade are specialists, too. These persons are needed on most large jobs. The right person must be found for each job. And the job must be attractive to that person.

The contractor must make the job attractive to recruit workers. People must want to work at the available jobs. Otherwise the jobs will remain unfilled.

Workers are recruited in several ways. Advertisements may be placed on the site. Or, they may be published in newspapers or trade magazines. Recruiters (employers or their representatives) may check with schools and colleges. Or, they may place *job orders* with employment agencies. Job orders describe the work to be done. They also tell what worker skills are needed.

Workers are often recruited through *labor unions.* See Fig. 25-2. Labor unions are organized groups of workers. Each group bargains with employers for certain working conditions, wages, and benefits. The agreement reached is called a *collective bargaining agreement.* Union contractors (employers) may recruit workers as described in this agreement (contract). Usually, they can call the union *hiring hall* and place an order for the number of workers needed. Then the union refers workers to the contractors.

Fig. 25-1
Construction workers are usually hired project by project. The same contractor may rehire them, or they may work for many different contractors.

The procedure is stated in the *referral* (or hiring hall) *provision* of the contract. This provision is a service to everyone involved. It frees contractors to accept any size job. They can be sure of a source of qualified workers.

The workers are helped, too. They receive help with job placement. And their wages and working conditions are set by an agreement.

Selecting

Selecting is a screening process. It is choosing the best qualified person from among those recruited.

Selecting may be easy or hard. This depends on the kind of job to be filled. The number of people who want the job is another factor. And the company's hiring policy also affects selecting.

The employer must know the facts about the job and describe them well. What skills and knowledge are required to do the job? Do *applicants* (persons applying for the job) meet these requirements? Which applicants meet them best? The employer must find out many facts about the applicants.

There are many ways to learn facts about applicants. First, there is the *application form*. See Figs. 25-3 and 25-4. On it, information is listed about applicants and the work they have done. This helps the employer learn whether the person meets the company's standards.

Some kind of application form is required for most jobs. In most cases, the applicant fills out the main part of the form. This should be done neatly and accurately.

In some cases, the employer fills out a form. The employer asks the applicant questions and records the answers. The applicant's answers are important. But the way the applicant answers questions is important, too. Both help the employer judge each applicant.

Testing may also be used in selecting. There are two major kinds of tests. There are proficiency tests and aptitude tests. *Proficiency tests* measure abilities, such as typing or bricklaying.

Aptitude tests measure the applicant's ability to learn a job. These tests are especially important for testing people who have not done the work before. Some people may not have the needed skills. Aptitude tests tell how well they can learn them.

Interviewing is another way of selecting workers. An interview is a talk between an employer and an applicant. Interviewing is generally thought to be the most important part of selecting.

Applicants should prepare carefully for interviews. It is very important for an applicant to make a good first impression. This impression may determine whether the person is offered the job. Applicants should know about the construction industry and about the employer. They should be able to state exactly what jobs they are interested in.

Applicants may be interviewed by more than one person. A supervisor, superintendent, or manager may do the final interviewing. See Fig. 25-5.

Fig. 25-2
Skilled workers who belong to a union can report at the union hall to find a job.

APPLICATION FOR EMPLOYMENT

(PLEASE PRINT PLAINLY)

PERSONAL Date ______

Name ______ (Last / First / Middle) Social Security No. ______

Present address ______ (No. / Street / City / State / Zip) Telephone No. ______

Position(s) applied for ______ Rate of pay expected $ ______ per week

Would you work Full-Time ______ Part-Time ______ Specify days and hours if part-time ______

Were you previously employed by us? ______ If yes, when? ______

List any friends or relatives working for us ______

What method of transportation will you use to get to work? ______

If your application is considered favorably, on what date will you be available for work? ______ 19 ______

RECORD OF EDUCATION

GIVE NUMBER OF YEARS ATTENDED		NAME OF SCHOOL	CITY	STATE
ELEMENTARY SCHOOL	YEARS			
HIGH SCHOOL	YEARS			
COLLEGE	YEARS			
OTHER	YEARS			

List diplomas, degrees, or certificates received ______

RECORD OF EMPLOYMENT

List your present and past employment, beginning with most recent.

FROM MONTH—YEAR	TO MONTH—YEAR	FIRM EMPLOYED BY	WHAT DID YOU DO?	EMPLOYER'S ADDRESS	NAME OF SUPERVISOR	REASON FOR LEAVING

PERSONAL REFERENCES (Not Former Employers or Relatives)

Name and Occupation	Address	Phone Number

Fig. 25-3
To place the right workers in the right jobs, employers need information about applicants.

FOR OFFICE USE ONLY	
Possible Work Locations	Possible Positions

FOR OFFICE USE ONLY	
Work Location ______	Rate ______
Position ______	Date ______

APPLICANT — DO NOT WRITE ON THIS PAGE
FOR INTERVIEWER'S USE

INTERVIEWER	DATE	COMMENTS

FOR TEST ADMINISTRATOR'S USE

TESTS ADMINISTERED	DATE	RAW SCORE	RATING	COMMENTS AND INTERPRETATION

REFERENCE CHECK

Position Number	RESULTS OF REFERENCE CHECK	Position Number	RESULTS OF REFERENCE CHECK
I		IV	
II		V	
III			

This *"Application for Employment"* is prepared for general usage throughout the United States. In the opinion of our legal counsel this form meets all Federal and State fair employment practice laws and complies with the Fair Credit Reporting Act. Furthermore, our attorneys have undertaken a monitoring service to advise us of any revisions to be made as a result of any future changes in the laws. However, we assume no responsibility for the inclusion in said *"Application for Employment"* of any questions which may violate local and/or State and/or Federal laws.

—2—

Fig. 25-4
On part of the application form, information about the applicant is recorded by the interviewer.

Checking references is another method used to select workers. A *reference* is a statement of an applicant's qualifications. (Qualifications are the experience and abilities that make a person able to do a job well.) Most application forms ask for the names of persons who know the applicant. These persons may be asked to write letters to recommend the applicant. See again Fig. 25-3.

Two kinds of references are usually needed. *Character references* are personal statements about the applicant. They are written by people who know the applicant well. *Former employer references* are statements by previous employers. They tell about the applicant's work at previous jobs. If applicants are young, their school records are almost always checked. The school is considered a "former employer."

People's values affect the decisions they make. No two employers look for exactly the same qualities in an applicant. And no two applicants want exactly the same things in an employer. Employers have different strengths and weaknesses. Workers, too, have different strengths and weaknesses. Good recruiting and selecting practices match the right employers and employees.

Inducting

Inducting is introducing new workers to their jobs. This helps them feel they belong to a team. It also helps them learn how to do their jobs. See Fig. 25-6.

Inducting tells new workers about the company. Information about procedures and work rules is given to workers. Benefits and services for workers should also be described.

Some construction personnel may go through formal induction programs. The programs may even involve classroom instruction.

Skilled construction workers are usually expected to know their jobs. See Fig. 25-7. For this reason, they may receive little formal introduction to their new assignments. When hiring skilled workers, contractors may simply have them report to the superintendent. This person assigns them to job supervisors. And supervisors tell them what work they are to do.

Fig. 25-5
Prepare carefully for an interview. How well you do may determine whether you get the job.

Fig. 25-6
The supervisor shows a new worker how to do the job correctly and efficiently. This is part of inducting.

Fig. 25-7
Skilled construction workers have spent much time learning their jobs. When they report for a job, they already know how to do the work.

Advancing

Opportunities for advancement affect which workers will be hired. Ambitious workers (workers who try to achieve) are usually interested in advancement. They may also be the best workers.

Advancing practices result in changes in the company's worker assignments. There are four major kinds of advancing practices.

- Promoting
- Demoting
- Transferring
- Separating

Promoting

Promoting is moving to a higher job level. Promoted workers are usually given more responsibility. Higher level jobs often require more knowledge or skill. Promotions are given for doing good work. They very often mean more pay.

Demoting

Demoting is moving to a lower job level. This is the opposite of promoting. Demoting usually relieves workers of some responsibilities. Lower level jobs may require less knowledge and skill. Demoting may or may not mean less pay. Often demoting is done on a temporary basis.

Sometimes demotions are given to discipline workers. Other times the work force has to be cut back. Or, maybe the nature of the work has changed. Some people may be demoted just because fewer high level workers are needed.

Transferring

Transferring is moving a worker to a different job. The new job is at the same level and pay.

Transferring may be done because the worker wants a change. Workers who are not happy in their jobs have low *morale*. They may feel better about themselves and their work if they are transferred.

Usually the employer has reasons for making a transfer. For example, transferring is often the best way to fill new job openings.

Separating

Separating may be discharging, relocating, or laying off workers. Retiring is also separating. But it is a special and permanent action.

Discharging means firing. Discharged employees are not usually rehired.

Relocating is changing a worker's site assignment. The new site may be far away. Or it may be close by.

Laying off is letting workers go temporarily. The job may be nearing completion. Or, the work may have slowed down. Sometimes bad weather causes layoffs. Workers who have been laid off may or may not be rehired.

Retiring

Retiring is a special form of separating. Retired workers who have qualified receive benefits. Usually, workers have to work a minimum number of years to qualify. Often, the years must be spent with a single employer. In construction, this poses special problems.

Sometimes a construction worker works with only one employer throughout his or her whole career. But many workers have a number of employers. Construction jobs are often only for the length of a contract. When the project is completed, the jobs end. Workers may then go with another employer to another site.

When workers change employers, they may lose retirement benefits. But many construction workers belong to labor unions. For many, the union is the only long-time contact. That is why unions usually offer *pension plans*. Funds for these plans may come from workers' union dues. Or, they may come from payments made by employers. They deduct (subtract) a certain amount from employees' pay.

All construction workers are covered by Social Security insurance. But most want additional retirement benefits.

The Advantages of Good Personnel Practices

Good personnel practices benefit a company. They attract workers. Fair treatment makes workers more loyal.

Good personnel practices benefit workers, too. Workers receive fair pay, good working conditions, and retirement pay. Everyone gains from fair and efficient personnel practices. See Fig. 25-9.

All personnel practices are important. They affect workers directly. They affect workers' morale and their work. Indirectly, they affect families of workers

ELECTRICAL ENGINEER

We have a position for a BSEE plus 4 years experience who will perform, direct, and coordinate electrical engineering efforts (power, control, lighting).

We offer salaries according to ability, plus an attractive company-paid benefit program including medical, dental, life insurance, pension, profit sharing, and continuing education. Twice a year we conduct salary and performance reviews to assure your success. Please send your resume or call our Personnel Department for an interview.

J.L.Peters, Inc.
Consulting Engineers,
Greenville, 696-4490

An Equal Opportunity Employer

Fig. 25-9
Good personnel practices help both the company and the worker.

and the entire community. In fact, personnel practices affect the total economy and the nation.

Much is known about construction production practices. American builders are respected around the world for their achievements. But, so far, construction personnel practices have not made as much progress. Increasing efforts have been made to improve them. Much has been done. Much remains to be done.

The construction industry faces difficult personnel problems. By nature, construction has periods of high and low activity. Contractors must move to where they are needed. The work cannot be brought to a central site. The products cannot be distributed to the sites where they are needed. And workers must go where the work is. Work cannot be brought to them.

New technology is developed daily. Change is rapid. Much understanding and cooperation are required to improve personnel practices. Then progress can continue.

Summary

Construction personnel practices include hiring, advancing, and retiring workers. Hiring includes recruiting, selecting, and inducting. Recruiting is finding and attracting job applicants. Choosing the right person for each job is selecting. Inducting is getting workers started in their jobs.

Advancing includes promoting, demoting, transferring, and separating. Retiring is a special kind of separating.

There are two other major personnel practices. These are training and working. They are discussed in other readings.

Personnel practices are very important construction practices. They affect workers' happiness and the company's success.

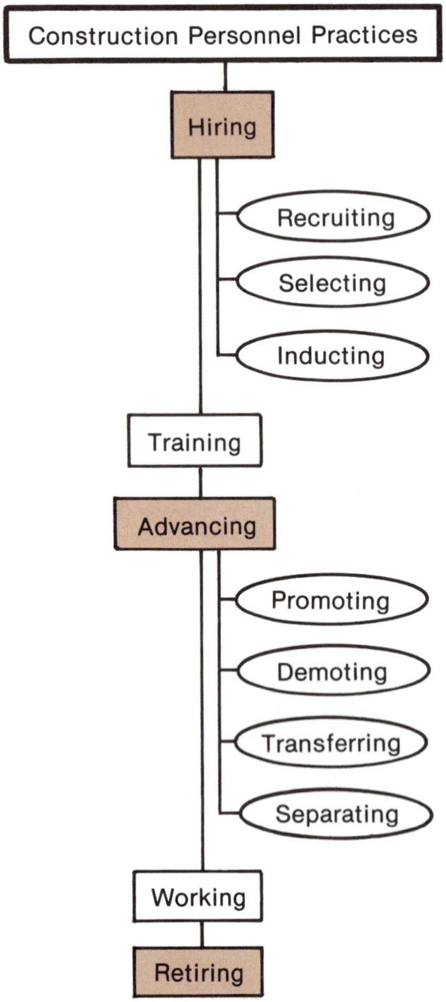

Terms to Know

advancing	job orders
applicants	labor unions
application form	laying off
aptitude tests	morale
character references	pension plans
collective bargaining agreement	proficiency tests
demoting	promoting
discharging	recruiting
former employer references	reference
hiring	referral provision
hiring hall	relocating
inducting	retiring
interviewing	selecting
job orders	separating
	transferring

Think About It!

1. To select the best person for a job, a contractor needs to know many things about the applicants.
   a. What questions may be asked on job application blanks?
   b. Why does the employer need to know the answers?
2. An induction program for new employees may be simple or complex.
   a. Why do you think a contractor would want a new worker to know something about the whole project?
   b. What kinds of rules and policies might the employee need to know about?
3. Compare promotion on a job with promotion to a higher grade level in school.
   a. How are ambition and hard work involved in both school and job advancement?
   b. How is promotion to a supervisor's job different from promotion to a higher grade? (Give several differences.)

Reading 26

Establishing Working Conditions

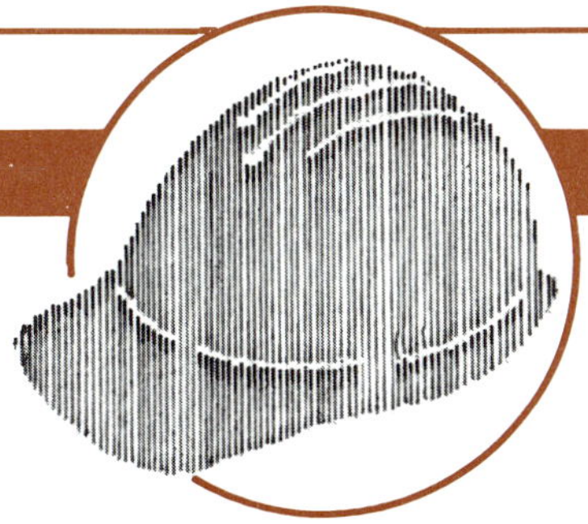

Rules for Working

Construction workers and an employer are much like a sports team and a coach. A coach needs a good team to win a game. And a team can't go very far without a good coach. The team and coach help each other to succeed. Like the players and coach, construction workers and employers must work together to build a successful project.

Every sport has rules that players and coaches must follow. In the same way, there are rules for construction workers and their employers. These rules cover three main areas of work.

- Physical working environment (where an employee works).
- Social working environment (relationships with other workers).
- Economic rewards (the worker's pay and benefits).

See Fig. 26-1.

Fig. 26-1
Working rules influence working environments. They affect work practices and safety. Rules also help set up good work relationships. And they establish the economic rewards workers receive.

A. Physical working environment

B. Social working environment

C. Economic rewards

Some things about the *physical working environment* are determined by the job. But there are often things that can be improved. Labor and management work together to make improvements. Safety equipment, lighting, and dust conditions may be improved, for example.

Certain conditions cannot be changed. Much construction work is done outdoors or in partly enclosed structures. Exposure to cold, hot, and rainy weather is common. However, many persons prefer construction work because it is done outdoors.

The *social working environment* includes many things.

- The way workers feel toward each other and toward their own jobs.
- The number of hours a worker must work each day.
- After-hours get-togethers (such as company bowling or a Christmas party).

Economic rewards include, of course, the money workers receive. But they are more than this. *Fringe benefits* are also economic rewards. Insurance programs and paid vacations are examples of fringe benefits.

Setting the Rules

The rules for work are set by an agreement between the employer and the employees. This may be done on an individual or group basis.

Workers may have only a spoken agreement with the employer. Or, they may work under an assumed agreement. An assumed agreement is not spoken or written. Suppose an employer does not say how long employees must work each day. The workers assume they are to work a standard workday (usually eight hours). They are working under an assumed agreement.

Most work agreements, however, are put into formal written contracts. A group of workers makes one agreement with an employer. Such an agreement is called a *collective bargaining agreement.*

Statements made in a collective bargaining agreement should cover all important details. The following is an example of part of a typical agreement. This particular passage covers the hours of work.

> Eight hours shall constitute a day of work and shall be performed between the hours of 7:00 A.M. and 5:00 P.M., local time (8 hours sometime within this 10-hour period) with one-half hour off for lunch between 11:00 A.M. and 12:30 P.M. (one-half hour sometime within this hour-and-a-half period), Monday through Friday.

The statements in the agreement must be very specific (exact). This will help avoid conflicts between labor and management. Each point should be carefully written into the agreement.

Most working conditions are set by *negotiation*. To negotiate means to discuss and decide. Workers and employers both need to be satisfied with the results of the negotiation.

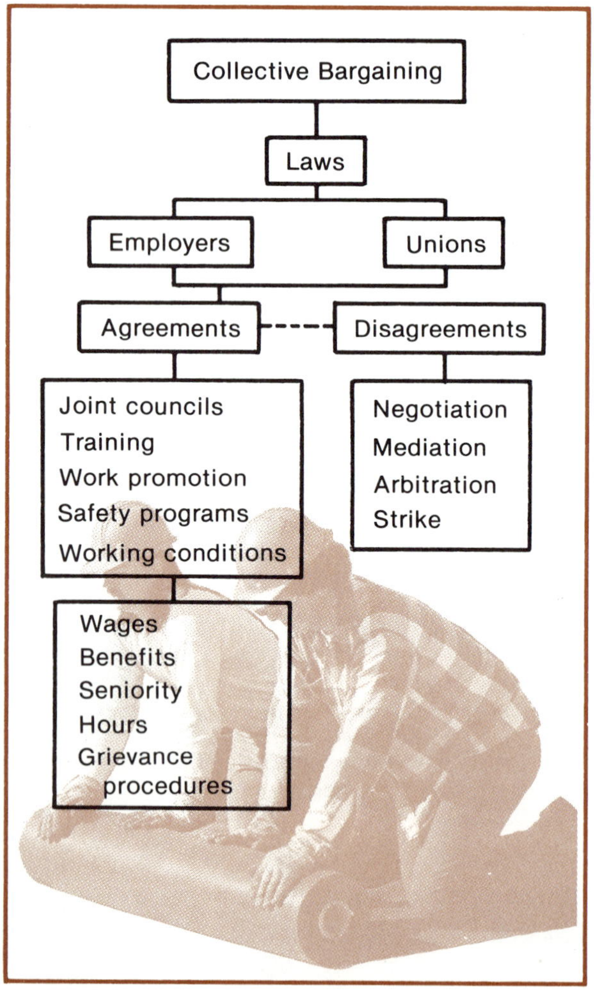

Fig. 26-2
When workers join together, they have more bargaining power. Workers and employers must cooperate.

Collective Bargaining

An individual may negotiate working conditions. But one person alone has little bargaining power. Often people with common interests join together. They may elect one person to speak for all of them. The elected leader has more authority than a single member. This person may seek benefits for the group. This is *collective bargaining.* See Fig. 26-2.

A *labor union* is a group of workers who have organized (joined together). They bargain collectively. Labor unions represent less than one-third of our labor force. Yet they play an important role in our economy.

Unions are important to everyone. Often the benefits obtained by local unions are shared by workers who do not belong to a union. Also, the actions of unions affect the price of goods we buy. Perhaps one day you will belong to a union. In the next part of this reading, you will learn more about labor unions. The building and construction trades unions are discussed.

The Building and Construction Trades Department of the AFL-CIO

Some unions are *independent unions.* Each works to accomplish its own purposes. Sometimes unions join together to make a larger union. In 1955, two unions joined together. One was the *American Federation of Labor* (AFL). The other was the *Congress of Industrial Organizations* (CIO). Together, they are called the *AFL-CIO.* More than 100 unions belong to the AFL-CIO. In January, 1980, the AFL-CIO had over 13,600,000 members.

The AFL-CIO is divided into departments. Each department is made up of unions of a certain category. One of the most important departments is the *Building and Construction Trades Department.* This department alone has over 4,000,000 members. In 1981, it had fifteen member unions. Just a few are Iron Workers, Boilermakers, Granite Cutters, and Carpenters unions. See Fig. 26-3.

Employees in other fields may work their entire careers with one company. However, construction employees often work with many different employers. And they work on many different sites. Therefore, construction employees have a special need for unions. The unions can protect their members' benefits as members move from one employer to another.

The Building and Construction Trades Department works for the good of the building trades. All the officers in this department are elected. Representatives of the department elect them at a special meeting. The department officers have several important duties.

- They work with the department's members.
- They work to organize members.
- They settle disagreements between members.
- They stay informed about local, state, and national labor policies.
- They tell people about the place of the building trades in our economy.

How Unions and Contractors Bargain

Unions have certain goals. They try to get as many benefits as possible for their members. People in management (the contracting firm, for example) have a goal, too. They seek to make a profit. Because the goals differ, conflicts sometimes arise.

Causes of Disagreements

There are four common causes of labor-management disagreements.

Fig. 26-3
Individual unions also band together. By belonging to a small local union this worker is also part of a large national union.

- Rate of wages.
- Working conditions or the environment in which the employees work.
- Hours of work.
- Fringe benefits or privileges, such as health care.

Unions favor improving wages and benefits. These things make life more pleasant for workers. But improving wages and benefits decreases profits. To make up the cost, the contractor will have to raise the bid amount. As a result, the contractor may even lose bids for projects. If this happens very often, the contractor will be out of work. There will be fewer jobs for workers. If everyone is to benefit, a fair balance must be found. This is done through collective bargaining.

Settling Differences

Through collective bargaining, workers and employers obtain agreements and contracts. In these, the rules and regulations for working together are written out.

The union may use several collective bargaining procedures to achieve its aims.

- Negotiation
- Mediation
- Arbitration
- Strikes

Negotiation

The union tries to settle differences by discussing or negotiating with management. Unions and management generally seek to settle differences peacefully.

Mediation

Sometimes labor and management do not reach an agreement. Then both parties (groups) may request an impartial mediator. This person does not favor one group over another. The mediator helps them reach an agreement. Both parties talk to the mediator about the disagreement. The mediator encourages discussion. By talking, the two groups may work out their differences. The mediator does not have the power to enforce a decision. *Mediation* is helpful. But it does not always work.

Arbitration

Sometimes national security or public welfare is threatened by the conflict. Then, *arbitration* may be used. An arbitrator is appointed. This person is a third party. This means the person is not involved in either side of the conflict.

The arbitrator's job is to arbitrate (settle) the disagreement. Arbitrators have the power to decide (enforce) a settlement. Neither labor nor management likes arbitration. Both believe that the parties directly involved should negotiate in good faith (trust). Arbitration is not used often in construction.

Strikes

The last resort is a *strike*. Strikes are voted by the union membership. Members of the union do not work during a strike. Strikes are used by the union to force management to bargain.

Workers are not paid by the employer for the days they are out on strike. Some unions have strike-benefit payments. This money is only a part of the amount wages would be. Strikes also hurt the employer. Profits and production are threatened. Overhead costs (expenses of operating) continue. Labor and management avoid strikes whenever possible.

Labor-Management Cooperation

Usually, labor and management have no conflicts. They work together on many programs. Examples are

- training
- work promotion, and
- safety programs.

Training

Training helps workers become skilled. Every union wants its members to be skilled. They will be hired by employers. Management wants skilled workers because they are more productive. They get more work done in less time.

Work Promotion

Work promotion is supporting new construction. Management supports work promotion. More construction means more contracts. More contracts mean greater profit. Unions want work promotion, too. Increased production results in greater employment for union members.

Safety Programs

Both labor and management have an interest in safety. Employers know that injuries cause the loss of valuable employees. Unions support safety programs because they protect their members. See Fig. 26-4.

Fig. 26-4
Safety programs may include classes on safety procedures and equipment.

People are needed to run work promotion, training, and safety programs. Problems that arise between workers and employers must be solved. Unions and management may set up joint councils or committees. Members of these councils or committees work together on matters involving both labor and management.

Summary

There are rules for work just as there are for sports. They cover physical and social working environments. They also cover economic rewards. Following the rules helps make construction projects successful.

Labor and management must agree on the conditions or rules. Sometimes they cannot agree. They try to negotiate an agreement.

A mediator or arbitrator may settle the problem for the two groups. Workers strike only as a last resort. Everyone involved loses from strikes.

Labor and management prefer to cooperate. They work together on work promotion, training, and safety programs. Workers and employers benefit from these programs.

Terms to Know

- AFL-CIO
- American Federation of Labor
- arbitration
- Building and Construction Trades Department
- collective bargaining
- collective bargaining agreement
- Congress of Industrial Organizations
- economic rewards
- fringe benefits
- independent unions
- labor union
- mediation
- negotiation
- physical working environment
- social working environment
- strike

Think About It!

1. What makes up your physical working environment? your social working environment? What "wages" (grades and scholarships, for example) and fringe benefits are available to you as a student? (For example, can you buy some things at student rates?)
2. Why do construction workers have a special need for unions?
3. Explain the differences between negotiating, mediating, arbitrating, and striking.

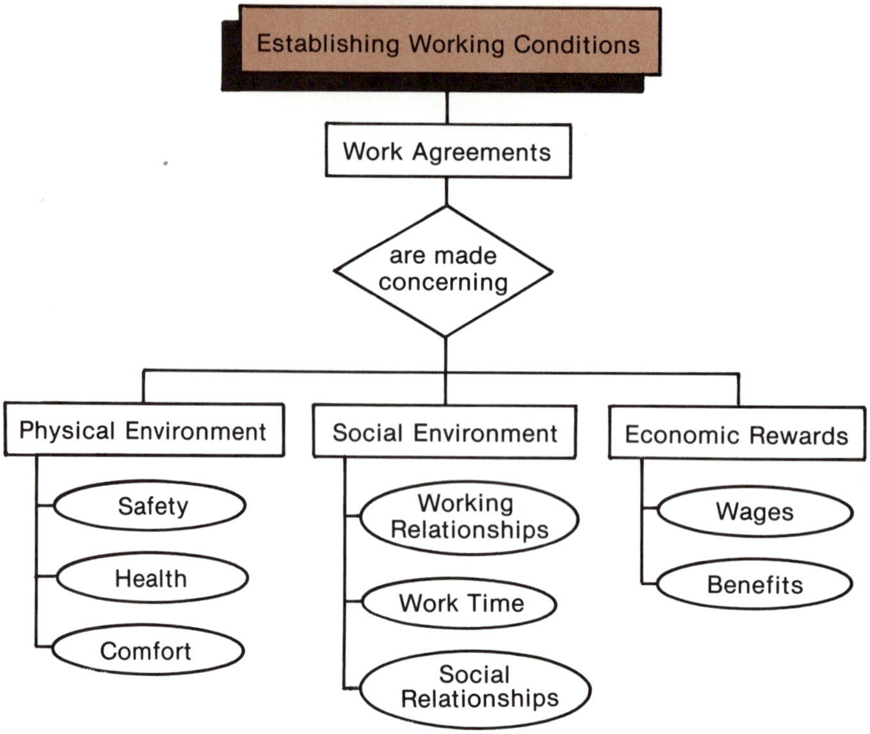
Establishing Working Conditions
Work Agreements
are made concerning
Physical Environment
Safety
Health
Comfort
Social Environment
Working Relationships
Work Time
Social Relationships
Economic Rewards
Wages
Benefits

Reading 27

Solving Labor-Management Problems

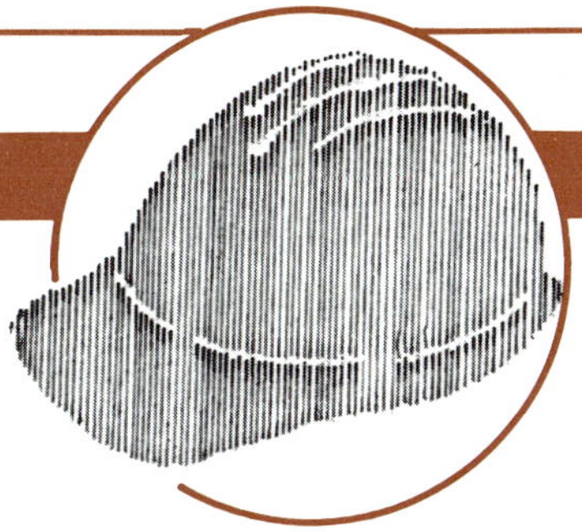

The Collective Bargaining Agreement

Workers and employers may *negotiate* a collective bargaining agreement. A collective bargaining agreement might be compared to the rules of baseball. Before you begin a neighborhood game, you agree on some rules. Then you play the game according to these rules. But problems may come up. These problems may not be able to be solved through the rules. Then the teams must discuss problems not covered in the accepted rules. They must try to reach an agreement.

It is the same in construction. The collective bargaining agreement lists the accepted rules for employers and employees in construction. They must work together to solve problems not covered by the agreement.

There are many parts to a collective bargaining agreement. One part is a statement of the rules for handling worker complaints. Complaints are referred to as *grievances*. The rules for handling them are called *grievance procedures*.

Grievance Procedures

Grievance procedures are designed to solve most problems between labor and management. They make it easy for people to discuss problems. They help each person to understand the way other people feel. And they are a means of letting workers know that their employers care about their problems.

No matter how understanding people are, problems still come up between them. It is important for employers to listen to worker grievances. Most grievances are about working conditions. Working conditions affect the worker on the job. See Fig. 27-1. The first step in solving problems is to talk them over. Employers can often solve small problems quickly. Solving small problems may help to avoid big problems. And doing this lets workers know that grievance procedures work. They will use these procedures again. Many problems can be solved. Many can be avoided. And workers will feel better about their employers and their jobs.

Grievance procedures may vary in different parts of the country. However, some things are important to all good grievance procedures.

- At any grievance discussion, there should be representatives from workers and management.
- Grievance procedures should always begin with the first-line level of management. This is usually the job supervisor.
- Progress should be orderly. Discussions should move from one level of authority to the next. This should continue until a solution is found.

Fig. 27-1
Employers must listen to complaints about working conditions. Solving a small problem quickly may help avoid a large problem later.

Each level of management should have the authority to settle the matter. For example, suppose a student has a grievance that involves school policy. The teacher should be able to rule on the matter. That ruling will probably be supported by the principal, superintendent, and board of education. Each level has the authority to settle the issue without going any further.

A set of grievance procedure steps may be complicated. Typical steps for a large company are shown in Fig. 27-2.

Mediation and Arbitration

Sometimes grievance procedures don't work. It is then necessary to use *mediation.* In extreme cases, *arbitration* may be used. Arbitration may be voluntary. Both sides may agree to it. Or it may be compulsory. The government may impose (force) arbitration in a major dispute (disagreement).

Sometimes mediation and arbitration are written into the collective bargaining agreement. They may be given as the final steps in the process.

People Who Mediate and Arbitrate

Qualifications (qualities and experience needed) are much the same for mediators and arbitrators. People chosen should be *neutral.* This means they should not favor either side. They must be fair. Knowledge of the construction industry is important. They must be known and respected in the industry or their community. They must know labor legislation and practices. Experience in labor negotiations is also helpful.

Mediators and arbitrators may be selected by the federal government. They usually come from the

Fig. 27-2

Typical Grievance Procedure in a Large Company

Step	Worker's Representative	Employer's Representative	Procedure
1	Union representative and employee	Job supervisor	These three parties will probably discuss the problem.
2	Union representative or business agent	Superintendent	These two parties may discuss the problem. They draw up a written statement of it.
3	Grievance committee	Local management head or representative of industrial relations department	These two parties will discuss the written statement. Then they may draw up a more complete statement.
4	International union officer	National corporation executive	These two parties may attempt to solve the problem. If no solution can be seen, they may begin steps toward having a neutral observer meet with the group.
5	**Neutral Party or Parties** A mediator or arbitrator may be called in to hear the problem. Representatives with the most authority, from both labor and management, meet with the neutral party or parties.		They attempt to reach a final solution.

Department of Labor, the Federal Mediation and Conciliation Service, or the National Mediation Board. Sometimes state governors appoint them. Or, workers and employers themselves agree on a mediator or arbitrator.

An arbitrator may come from the American Arbitration Association. This is a nonprofit organization. (It makes no extra money.) This association provides arbitrators for various types of disputes.

Mediation

The role of the mediator is to bring labor and management together. Efforts are made to keep both sides talking. See Fig. 27-3. The mediator tries to get them to settle the dispute.

A mediator may clarify or outline disagreements. This helps each side to understand the position of the other side. A mediator may make suggestions. Or, he or she may even propose a final solution. But the mediator has no power. The decisions of the mediator are not *binding*. That is, management and labor are not required to accept a mediator's decisions.

Fig. 27-3
Mediators meet with labor and management. They encourage discussion. Sometimes they suggest solutions to problems.

Mediation can be quite successful. But sometimes it is not. Arbitration may be needed.

Voluntary Arbitration

In *voluntary arbitration,* both sides agree on the arbitrator. Both sides agree to accept the arbitrator's decision. Voluntary arbitration is not often used in the construction industry. It may be used when settling grievances and *jurisdictional disputes.*

A jurisdictional dispute is a disagreement between people in different crafts. Members of various crafts are involved in construction. Plumbers, carpenters, and steam fitters are examples. They may be working on a job at the same time.

Disputes sometimes come up. Workers in different crafts may argue over who should perform a certain task. For example, suppose a steam boiler must be hooked up. It is needed to operate a pile driver. The steam fitters may decide that it is their job. It involves the piping of steam. The operating engineers may decide that it is their job. They operate the boiler and the pile driver.

Arbitration could be used to settle the issue. There are several contractor associations and labor unions that have agreed to arbitrate such disputes.

Compulsory Arbitration

Compulsory arbitration is imposed by the government in a major dispute. Workers and management do not like it. Workers do not want to be told they must work. Many workers consider this very close to "slave labor." And management does not want other people to decide how the business should be run.

In general, compulsory arbitration is not the law of the land. Members of the state and national legislatures (lawmaking bodies) avoid dealing with it. Compulsory arbitration is used in the most serious national emergencies. Few of these involve the construction industry.

The Arbitration Process

Issues to be settled may be complex. But the arbitration process is fairly simple. First an arbitrator is selected. Then both labor and management furnish (supply) this person with written statements. Each side explains its position in the dispute. Statements include lists of the issues or disagreements. Each

Fig. 27-4
The arbitrator holds a hearing to discuss the facts. Then an arbitration award is made.

Fig. 27-5
During a strike, production stops until the disagreement between labor and management is settled.

side offers evidence and arguments in favor of its position.

The arbitrator studies the two written statements carefully. Then a hearing (meeting) is held with representatives of both sides. See Fig. 27-4. Questions are asked. The case is discussed.

After the hearing, the arbitrator makes a decision. It is written as an *arbitration award.* The award must follow certain rules.

- Nothing should be said about issues not at stake.
- All issues under question must be in the award.
- The award must be enforceable. (There must be some way to make sure it is followed.)
- Facts, figures, and dates named in the award should be correct.
- The award must not violate (go against) the basic collective bargaining agreement.

Striking

A *strike* is a work stoppage. See Fig. 27-5. Strikes do not happen often. Labor and management avoid them whenever possible.

The main cause of strikes is the failure to reach a collective bargaining agreement. For example, sometimes important issues such as pay and working conditions cannot be settled. Then negotiations between labor and management break down. Strikes result. Strikes are labor's final effort to force a return to collective bargaining in good faith.

Strikes seldom happen if a collective bargaining agreement is in force. But these agreements have a time limit. When an agreement expires (time runs out), it must be renewed or a new agreement must be reached. Most agreements are reached or renewed without strikes.

How a Strike Is Conducted

Most strikes are organized by a union. If union leaders believe a strike is necessary, they call a union meeting. Members discuss the situation. See Fig. 27-6. Then they vote. If the members vote to strike, a time and date are set. On the set date, all members of that craft union leave their jobs. They are not quitting. They are only stopping work. They will return when an agreement is reached.

Members of only one craft may be on strike. But workers in other crafts may not be able to do their jobs. Then all construction work stops. Also, the striking union may assign workers to *picket.* Workers in other unions may decide to support the striking

Fig. 27-6
In a union meeting, workers have the opportunity to express their opinions.

Fig. 27-7
Management and labor usually reach agreements quickly. They work together to produce constructed products.

union. If so, they will not cross a picket line. They, too, will stay away from their jobs.

Settlement of Strikes

Everyone loses in a strike. Workers, contractors, and owners lose money. Even the general public may lose money. And good working relationships may be lost. Therefore, there is great pressure on both sides to settle strikes.

Most strikes are settled after a few days or weeks. See Fig. 27-7. But they may run for months or even years.

No one likes a strike. But in a free country, management and labor must be free to work out their problems.

Summary

Grievance procedures are ways of handling worker complaints. They are part of the collective bargaining agreement. Grievance procedures make it easy for employers and workers to talk things over. Talking may help resolve problems before they become serious.

Sometimes grievance procedures do not work. Then mediating, arbitrating, and even striking may be employed. It is rare for a neutral outsider to be called in to mediate or arbitrate. Most construction disputes are handled by negotiating or by following grievance procedures. If all else fails, strikes may result. This is harmful to both sides. Collective bargaining in good faith can settle differences and maintain good working relationships.

Terms to Know

arbitration	mediation
arbitration award	negotiate
binding	neutral
compulsory arbitration	picket
grievances	strike
grievance procedures	voluntary arbitration
jurisdictional disputes	

Think About It!

1. Suppose you want to watch a particular television program. Your parents or brothers and sisters want to watch another. Describe a way (grievance procedure) of settling these differences.
2. What do you think might happen on a construction project if no agreements were made about orderly procedures for handling complaints?
3. Has there been a strike in the construction industry near you lately? If so, what caused it? How was it settled?

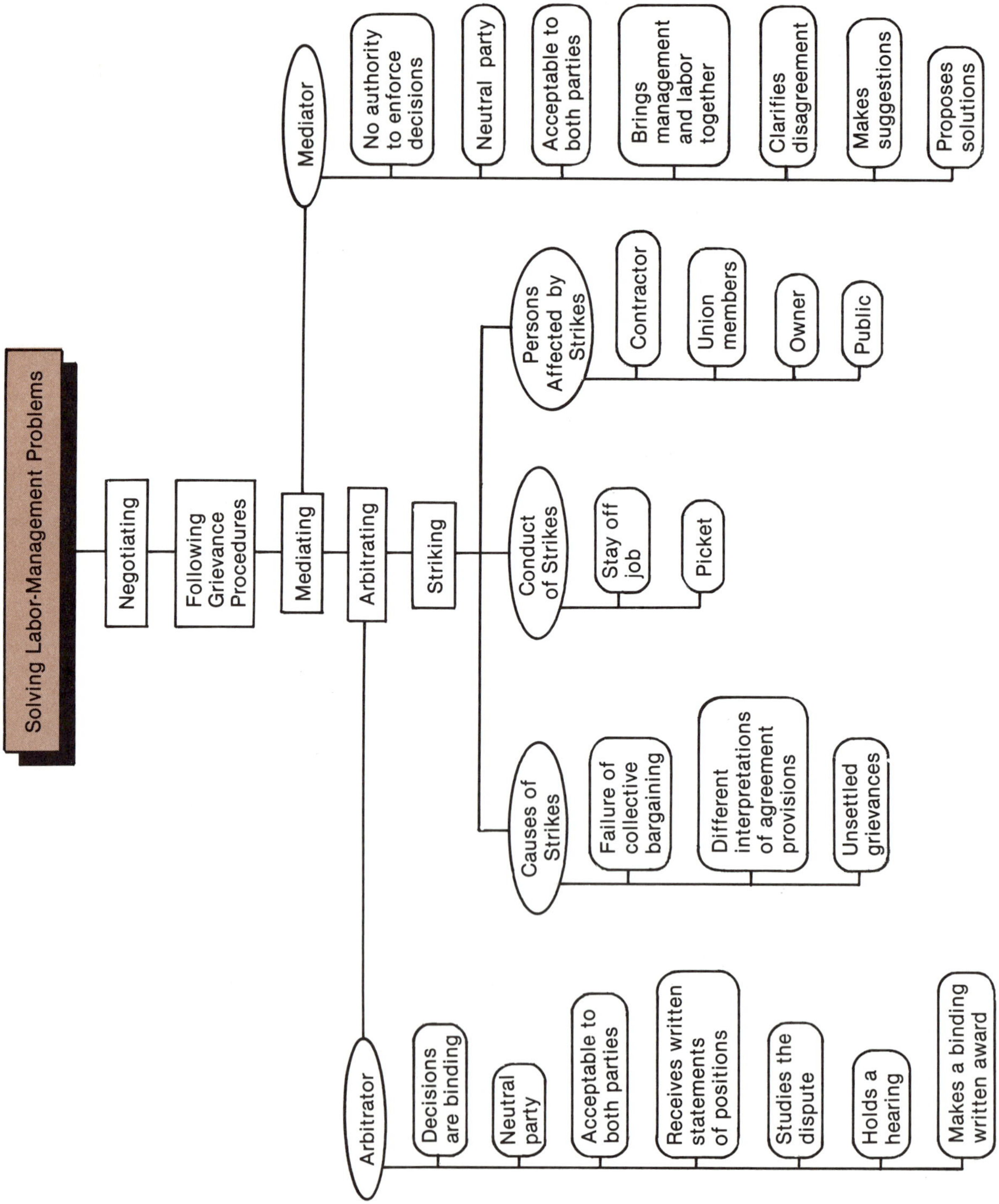
Solving Labor-Management Problems
Negotiating
Following Grievance Procedures
Mediating
Arbitrating
Striking
Mediator
No authority to enforce decisions
Neutral party
Acceptable to both parties
Brings management and labor together
Clarifies disagreement
Makes suggestions
Proposes solutions
Persons Affected by Strikes
Contractor
Union members
Owner
Public
Conduct of Strikes
Stay off job
Picket
Causes of Strikes
Failure of collective bargaining
Different interpretations of agreement provisions
Unsettled grievances
Arbitrator
Decisions are binding
Neutral party
Acceptable to both parties
Receives written statements of positions
Studies the dispute
Holds a hearing
Makes a binding written award

Reading 28

Getting Ready to Build

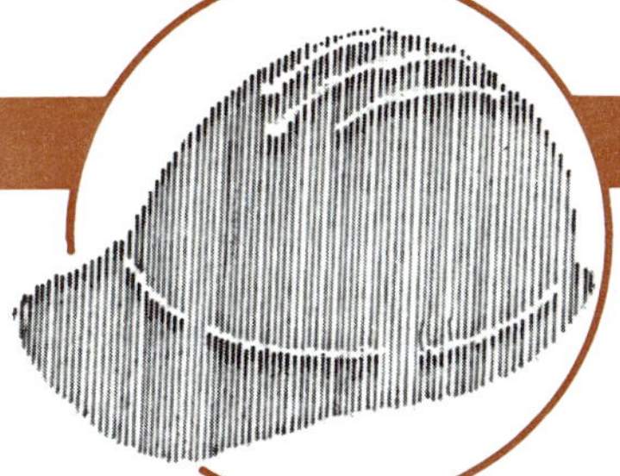

Before Construction Begins

The contractor has been awarded the contract to build. Now the site will be prepared and cleared. Many things must be done before construction can begin.

Following Laws

Contractors must know and follow many laws and other regulations. Laws are of many types. They may be national, state, county, township, or city laws.

National laws apply to the whole country. One national law, for example, requires all employers to keep many records on their employees.

State projects must follow state laws. For example, state laws may say what qualifications a contractor must have.

Cities and townships also have laws that apply to construction. For example, *zoning laws* limit what can be built in each area of a city.

Zoning

There are six major city zones.

- Single family housing areas.
- Multiple family housing (apartment) areas.
- General shopping areas.
- Commercial areas.
- Light industrial areas.
- Heavy industrial areas.

Zoning is one type of restriction. There may be others. For example, there may be a limit on how small a lot can be. The height of buildings may be limited, too. There also may be a minimum allowable floor space. These restrictions vary from one city to another.

Building Permits and Special Arrangements

City laws often require contractors to have a *building permit* (license) for any project. This permit is posted at the site. Authorities and the public must be able to see the permit. Building permits are needed even for making additions or alterations.

Sometimes large or heavy equipment must be moved to a site. Special plans have to be made with the highway department or police. Laws restrict the equipment that may be moved on streets and roads.

Have you seen bags covering parking meters at a construction site? This means that local police have given a contractor special permission to use these parking spaces to unload materials at the site.

Keeping Safe and Protected

In all construction, safety and protection are very important. The contractor is responsible for safety on and near the site. Workers must be safe. The tools, equipment, materials, and site must be safe and protected. Property near the site must also be protected. The work must not pose any danger to the public. See Fig. 28-1.

This responsibility is not taken lightly by a contractor. Most contractors have insurance against accidents and other problems. However, they must still take all possible steps to avoid safety problems.

Getting to the Site

In the city, streets are used to get to the site. In rural areas, roads are used. Sometimes, roads are built to get to the site.

At the project site, walkways are sometimes built. These are for workers and visitors to use. They help because almost all sites are bare earth. Without walkways, people would have to walk through mud when it rained.

Providing Temporary Facilities

Weather can cause serious problems in any construction project. Workers, materials, and equipment must be protected from weather. At most sites, temporary shelters are built for workers. Tents, canvas, or plastic covers are often used to protect equipment and materials.

A. Construction workers wear hard hats for protection.

B. Materials are stored and protected in a truck trailer.

C. Property near the site is protected by bracing and shoring excavation walls. Here high wooden fences are also used. They protect existing structures.

D. Trespassers could be injured. Therefore, many sites are fenced in.

E. A sidewalk canopy protects people walking near the site from falling materials.

Fig. 28-1

People and property on or near the construction site must be protected.

Often contractors and subcontractors need field offices. Many shelters used for offices are *prefabricated*. They are assembled at the site. Trailers are also very popular. They can be taken from site to site easily. See Fig. 28-2.

Storage areas are needed. These, too, may be prefabricated and put up at the site.

All temporary shelters must be planned. In a city, construction work may take most of the site space. The local government may be asked to give a special permit. Then parts of the street may be blocked and used for storage and offices. Some skilled workers may wait until the basement or first floor is in before they come to the job. Then they use one of these areas for offices and storage.

Providing Utilities

Electricity is needed at the site. It is used for lights and to run pumps and shop equipment. To get the needed electricity, power lines are run onto the site. Sometimes a portable generating plant is set up on the site to supply power. Usually, power lines are close to any construction area.

Communication lines are important at the site. See Fig. 28-3. Telephones or radios are needed at all projects. Portable two-way radios are a great help on some projects.

Water is needed at all project sites, for many functions. For example, it is used for sanitation, construction work, and drinking. It is needed for fire protection and water-cooled equipment. Cleaning and mixing also require water. Waterlines can be brought to the site from nearby water mains. When these are not near, water tanks on trucks can be used. In many remote areas, pumping stations are used. They are placed near rivers or lakes to bring water to the project. In other cases, wells are dug close to the site.

Fig. 28-3
Communications devices on the site save time and trouble. Instructions or warnings of dangers can be given instantly.

Fig. 28-2
A trailer makes a very good field office. When the project is completed, this "mobile headquarters" can be moved to another building site.

Site Clearing

Before temporary facilities can be placed and actual construction can begin, the site may need to be cleared. The site-clearing practices used vary with each site. Contractors must know all about a site. They must also know site-clearing practices.

Five site-clearing practices are most common. They are

- demolishing,
- salvaging,
- cutting,
- earthmoving, and
- disposing.

Demolishing

Demolishing means destroying. See Fig. 28-4. Demolishing by using high explosives is called *blasting*. When bulldozers or wrecking balls are used, it is called *wrecking*.

Fig. 28-4
Demolishing can be done by blasting or by wrecking.

A. Blasting Hercules Incorporated

B. Wrecking

E.I. DuPont DeNemours and Company

Fig. 28-5
Plastic explosives can be easily and safely loaded into a borehole.

Blasting is a popular method. Safer, plastic explosives were developed during World War II. These brought about changes in blasting techniques. Smaller, better-controlled blasts were made possible. See Fig. 28-5.

Wrecking can be done with many mechanical devices. A *crowbar* is a hand tool used for wrecking. A crawler tractor with a bulldozer blade is often used. For tall structures, a crane with a wrecking ball is used. This is a heavy ball hung on the end of a cable. It is swung against the structure that is to be wrecked. It acts as a battering ram.

Salvaging

Some items on a site may have some value. They should be saved for use on another project. Saving things from being demolished is called *salvaging*. See Fig. 28-6.

Fig. 28-6
A highway crossover will be built on this site. Construction workers salvage useful items. This sign will be put up in a different, safer place on the city hall lawn.

Daily Pantagraph

Fig. 28-7
Metal structures may be cut apart with a torch and then removed from the site.

When a building is to be demolished, many parts may be saved. Electrical switchboards, plumbing fixtures, doors, and windows are examples.

Cutting

Cutting has many uses. See Fig. 28-7. In wooded areas, cutting means bringing down trees and brush. Metal structures may also need to be cleared. For example, old steel towers or bridges are cut down. Cutting torches may be used to cut through metal.

Fig. 28-8
Extra dirt, brush, and other materials are cleared away by earthmoving.

Fig. 28-9
Unwanted materials may be loaded into a truck and hauled away. This is one method of disposing.

Earthmoving

Earthmoving is used to clear many sites. See Fig. 28-8. Earthmoving may also be a major part of the actual construction work. This is especially true with heavy construction projects, such as highways, tunnels, and earthen dams. Earthmoving will be studied further in a future reading.

Disposing

Disposing means removing materials that are not wanted. See Fig. 28-9. Disposing may be done by stockpiling, burying, spreading, or hauling away.

Clearing Water from the Site

Some construction, such as dam construction, is done in or around water. There are special problems involved with this type of construction. Water may have to be kept from the work area. *Caissons* or *cofferdams* are commonly used to do this. See Fig. 28-10.

Caissons are large tubes. They are placed vertically through water. They touch solid ground. Then they are emptied of water. This gives a dry place for working.

A cofferdam is a watertight wall. It may be a series of cells or units. Each unit sits on the bottom of a body of water. It is filled with soil and rock. Units are placed next to each other. They make a wall around the site. The area surrounded by the cofferdam is emptied of water. This makes a dry work area.

Fig. 28-10
A cofferdam is used to keep water away from the construction site.

Summary

Contractors consider many things before they begin to build. They are required to obey many laws. They also must follow many regulations. Contractors are responsible for providing safety and protection for the public and the workers. They must protect property close to the site, the site itself, and equipment and materials.

Often, temporary facilities are needed at construction sites. These may be prefabricated and put up on the site. They are used as offices, shelters, or for storing materials.

Utilities are also needed at the site. Power lines are run onto the site to provide electricity. Waterlines provide water needed on the site. They are run from nearby mains. Telephones and two-way radios provide communication lines on the site.

Often, sites must be cleared before construction can begin. Therefore, contractors must know their sites well. And they must know site-clearing practices.

Terms to Know

blasting
building permit
caissons
cofferdams
crowbar
cutting
demolishing
disposing
earthmoving
prefabricated
salvaging
wrecking
zoning laws

Think About It!

1. Zoning laws place limits on construction in each area of a city.
   a. Name the major kinds of zones.
   b. Describe some other restrictions or limits that may apply in some areas.
2. Safety and protection are very important in construction. At a construction site, who and what must be protected?
3. Look in your telephone directory. Write down the names of several companies that use site-clearing practices.

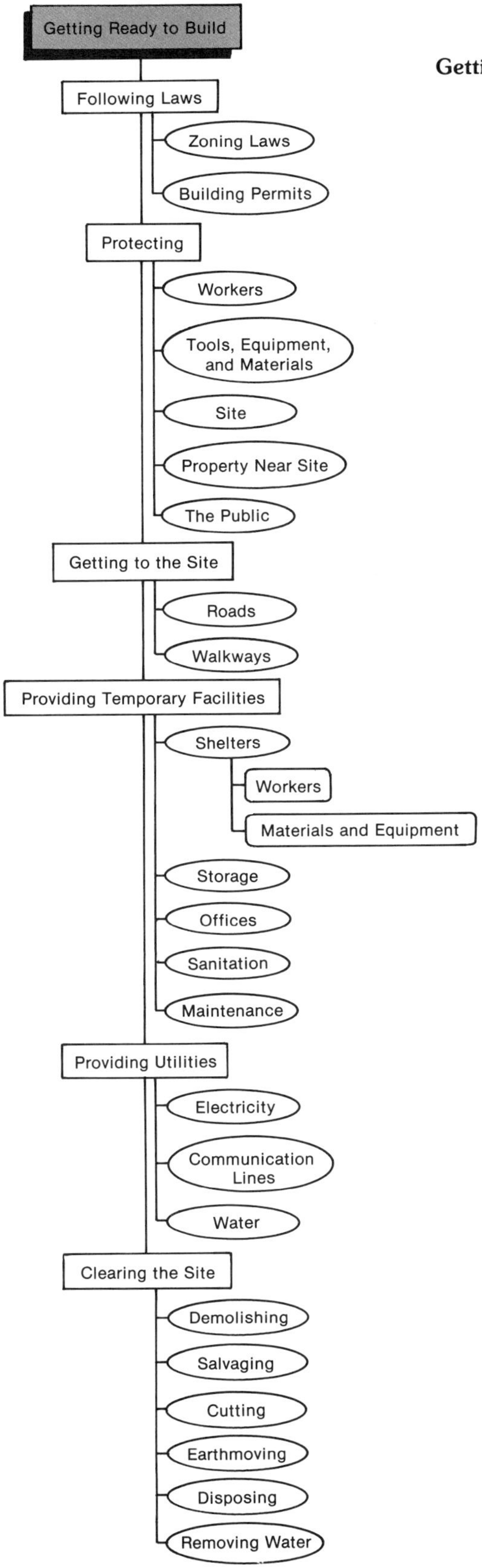
Getting Ready to Build
Following Laws
Zoning Laws
Building Permits
Protecting
Workers
Tools, Equipment, and Materials
Site
Property Near Site
The Public
Getting to the Site
Roads
Walkways
Providing Temporary Facilities
Shelters
Workers
Materials and Equipment
Storage
Offices
Sanitation
Maintenance
Providing Utilities
Electricity
Communication Lines
Water
Clearing the Site
Demolishing
Salvaging
Cutting
Earthmoving
Disposing
Removing Water

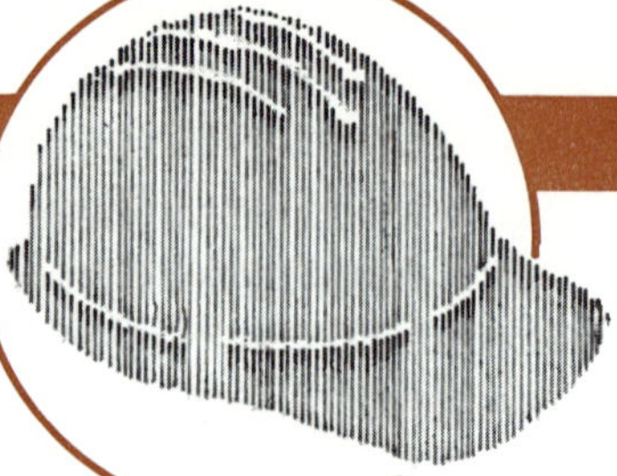

Reading 29

Locating Structures

Using Surveys

Before locations for structures are decided, construction sites are surveyed. The land is measured. Natural and artificial features of the site are located. All of this information is used to draw maps that accurately describe the site. A later survey sets the exact location for the structure.

In this reading, you will see how locations are determined by surveying. You will also learn about two major ways of locating structures. One way is used for locating structures such as highways. Another way is used for locating structures such as buildings and bridges.

Importance of Surveying

The site has been cleared. Temporary facilities have been set up. Drawings made earlier show the exact location and size of the structure. Now the location must be *laid out* on the site.

First, *surveyors* refer to a set of plans to find out the structure's dimensions. They set up special equipment to find the planned location for the structure. A *measuring tape* and a *transit* are used to measure angles and horizontal distances. (A horizontal distance is the distance from point to point on the ground.) An *engineer's level* and a *level rod* are used to measure vertical distances. (A vertical distance is one that is measured up from a point that has a known height.) See Fig. 29-1.

Surveyors mark the size and location of the structure by driving stakes into the ground. These stakes also guide workers in placing construction materials on the site. People looking at a finished structure are usually not aware of the job done by the surveyor. But the job is very important. It must be done carefully.

Surveying for Highways

There are several steps in laying out a highway. First, the correct direction must be found. This is the most important step. The surveyor finds land-

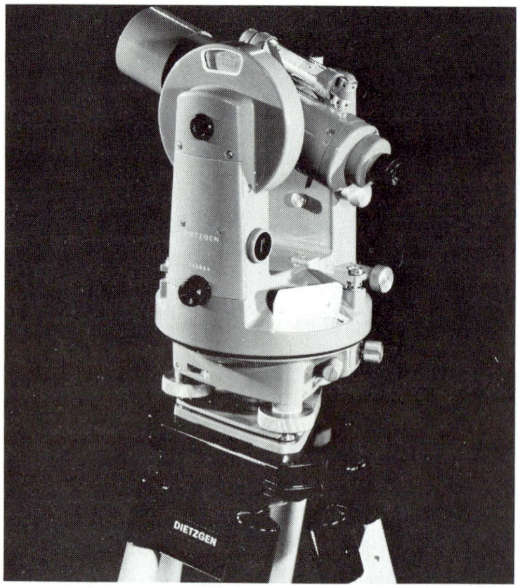

A. A transit is used with a measuring tape to measure horizontal distances.

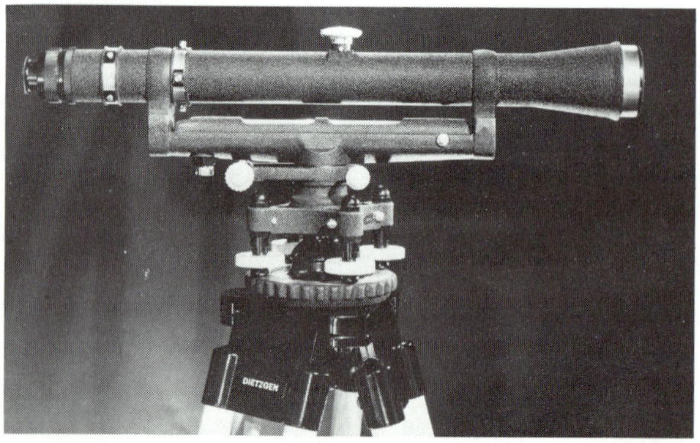

B. An engineer's level is used with a level rod to measure vertical distances.

Fig. 29-1
To locate a structure, both horizontal and vertical distances must be measured.

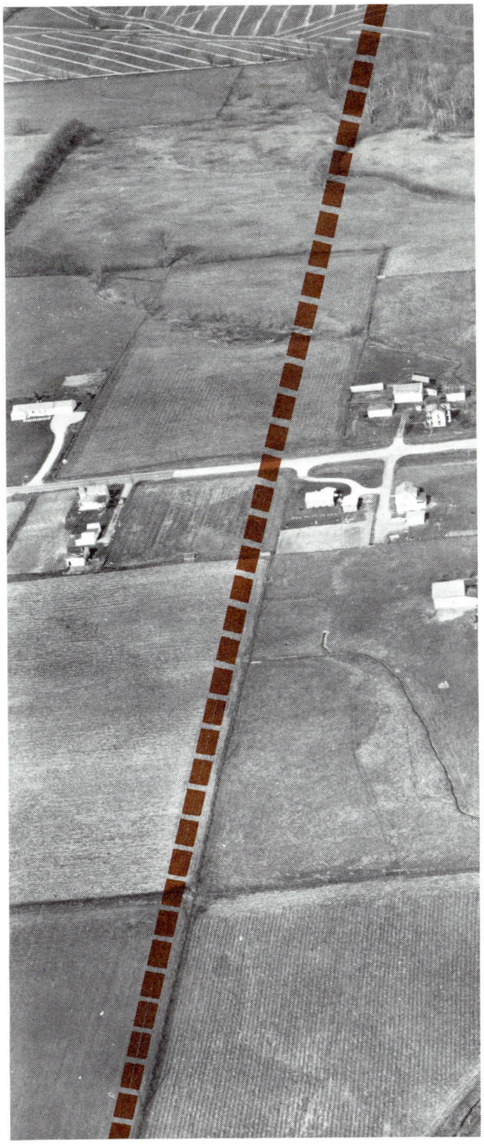

Fig. 29-2
The center line of a proposed highway is shown here by a dashed line.

marks, such as trees, on the plans. Plans often show the highway's location in relation to these landmarks. The surveyor uses the landmarks, a measuring tape, and a transit to find the *center line*. The center line is the highway's proposed route. See Fig. 29-2.

Control Points

Stakes are set up along the length of the center line. They are set about fifteen to thirty meters apart. The most important points along the center line are *control points*. They show where lines cross (such as for intersections). They also show where the center line turns. Control points are well marked with flags and stakes. The markers keep construction equipment from running over the control points.

To locate the edges of the pavement, the surveyor measures horizontally from the center line to each side. This completes the horizontal measurement needed for earthwork. All horizontal measurements are recorded for future use.

Next, vertical measurements must be made. These are made from control points called *bench marks*. Several bench marks are shown on the plans. They mark points of known elevation (height) above sea level. Surveyors refer to bench marks as they use the rod and level to find elevations along the center line. Elevations are measured at each center line stake.

Cut and Fill

Then the surveyor must find out how much cut or fill is needed. *Cut* is earth that must be removed. Earth is cut away to get the correct elevation. *Fill* is earth that must be added to get the correct elevation.

The surveyor compares the real elevation of each stake to the elevation shown on the plans. The difference between elevations is the amount of cut or fill needed. The amount at each point is marked on the center line stake. Construction workers will follow these markings to *grade* the ground. See Fig. 29-3.

Fig. 29-3
Stakes are marked to show the amount of fill or cut needed to grade the site correctly.

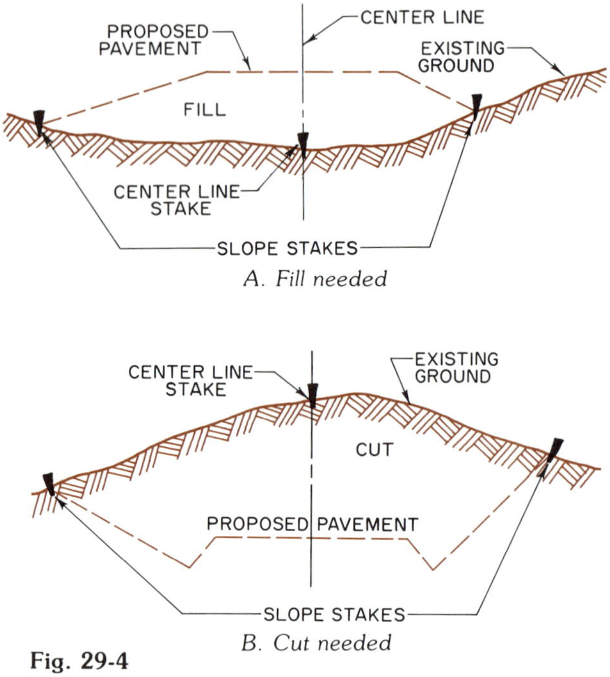

Fig. 29-4
Slope stakes are placed on each side of the center line of the proposed highway.

Fig. 29-5
Heavy equipment is used in grading a roadbed.

Stakes are also placed at the edge of the proposed (planned) pavement. The surveyor marks on the stakes the amount of cut or fill needed. The cut or fill should slope from each stake to a point on the ground. Stakes are set where the slope should be. These are called *slope stakes.* See Fig. 29-4. They show the slope of the ground to be constructed. All vertical measurements are recorded for future use.

Using Stakes in Highway Construction

When the earthwork is done, the surveyor puts up new stakes. Stakes are set on each side of the pavement to be constructed. The stakes begin at the control points. They are between eight and sixteen meters (m) apart. The stakes are about one meter outside the edge of the proposed pavement. The tops of the stakes are at the height the road will be. Workers follow these stakes to construct the *roadbed.* See Fig. 29-5. Later, other workers will follow the stakes to lay the pavement on the roadbed.

This procedure is followed for all earthwork and highway projects. Slope stakes are used for the earthwork. Then more stakes outline the structures

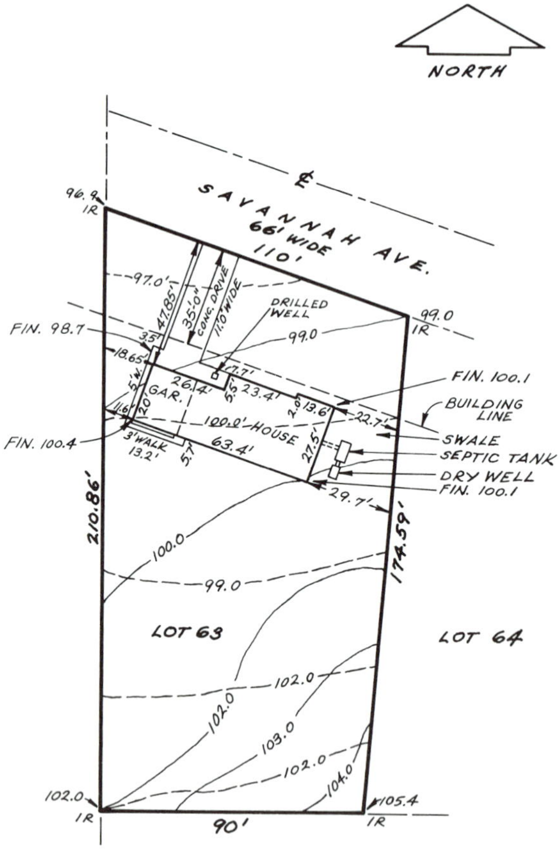

Fig. 29-6
The plot plan shows where the building is placed in relation to the property lines.

to be built. The procedure used for surveying buildings, dams, bridges, and other structures is different from that used for surveying highways.

Surveying for Buildings

To survey a building, the surveyor looks at the plans to find the building's location in relation to the *property lines*. See Fig. 29-6. The structure's location is marked on the site. The marks are used to set control points. Control points are placed wherever they are safe from construction equipment. Horizontal and vertical distances are measured from the control points.

Actual layout of the building is done using *batter boards*. Stakes are driven partway into the ground. Boards are fastened horizontally to the tops of the stakes.

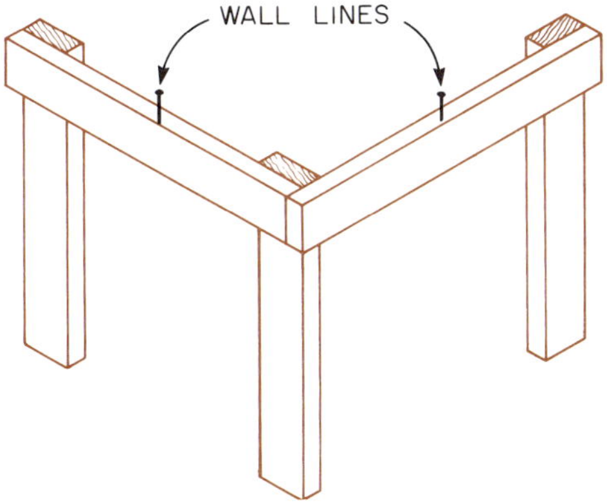

Fig. 29-7
Batter boards are built at each end of a planned building wall. These nails mark the locations of walls.

Horizontal Measurements

A wall is located using horizontal measurements. To stake a wall, the surveyor first locates it on the drawings. Then batter boards are built past each end of the planned wall. They are set back so equipment will not move or destroy them.

The location of the wall is marked exactly on batter boards. See Fig. 29-7. A saw cut or nail may be used as a marking. A wire or cord is attached to a marking on one batter board. Then it is stretched to a marking on the batter board at the opposite end of the wall. This locates the wall line. The surveyor does this for each wall of the building. See Fig. 29-8.

Corners are located where the wires or cords cross. Other markings are made on the batter boards to locate the foundation or columns. Sometimes separate batter boards are put up to mark each construction line. Markings on batter boards must be clearly identified. Otherwise, they may be confused with each other.

Vertical Measurements

Vertical measurements are needed. Batter boards are used in making vertical measurements, too. They are set in a special way. The top boards are set at a known elevation above ground.

A wire is stretched between the batter boards. Then, a *plumb bob* is hung from the wire. See Figs. 29-8 and 29-9. Vertical measurements can be made along the *plumb line*.

Fig. 29-8
Wires or cords are stretched between the batter board markers to show the locations of the exterior walls of a building.

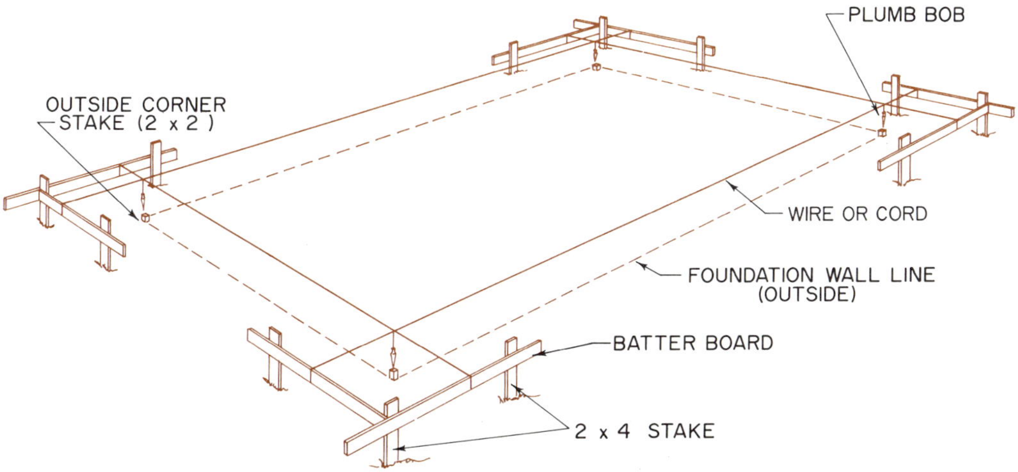

Fig. 29-9
A plumb bob.

Checking Accuracy in Building Construction

During construction, the surveyor sights through a transit to make sure that horizontal locations do not change. A level and rod are used to check vertical measurements. The structure's measurements are checked against the plans and bench marks. The surveyor makes this check many times during construction. It is especially important for buildings with many floors. Each floor must be the right height.

Structures may *settle* (sink) during or after construction. The surveyor makes checks to see if this is occurring. If it is, the surveyor records the amount of change. The amount is checked against surveying records to determine the amount of settling.

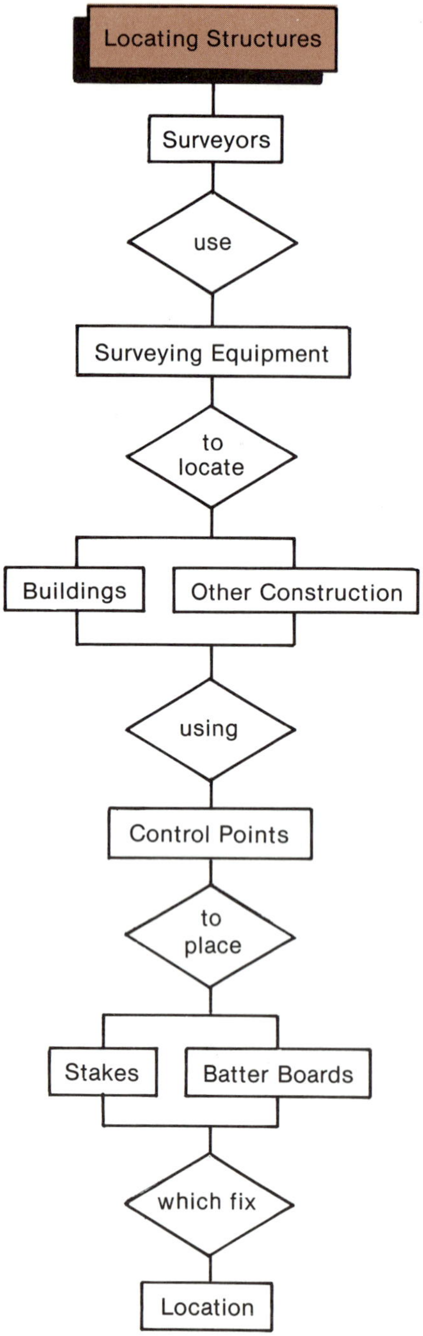

Summary

Surveying is done to locate structures on a construction site. Surveyors use special tools to measure exact horizontal and vertical distances.

There are two methods of surveying. One is used for highways. The other is used for buildings and other structures.

Stakes are used to locate a highway. They mark the center line, the width, and the slope of the proposed highway. Workers follow stakes to construct the roadbed and lay the pavement.

The location of a building or other structure is marked with batter boards. During construction, the surveyor checks to make sure locations do not change.

Terms to Know

batter boards
bench marks
center line
control points
cut
engineer's level
fill
grade
laid out
level rod
measuring tape
plumb bob
plumb line
property lines
roadbed
settle
slope stakes
surveyors
transit

Think About It!

1. What do you think would happen if a new home were incorrectly located and built on a portion of someone else's property? Ask your parents if they know of an example.
2. How are workers guided in constructing a highway? a building?
3. Have you ever seen surveyors working? Were you able to tell why they were surveying?

Reading 30

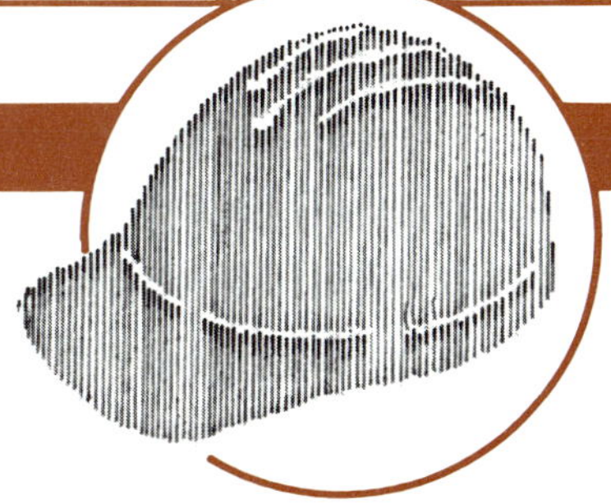

Earthmoving

The Importance of Earthmoving

Moving earth is an important activity in construction. Soil and rock are moved. This allows foundations to be placed below the surface. It also makes it possible for such things as pipes to be placed below grade. Earth must be moved for these and other reasons. The process is called *earthmoving*.

Earthmoving Equipment

Many years ago all earthmoving was done by hand. Only hand tools were used. Hand tools are still used today. See Fig. 30-1. But now large pieces of equipment have been developed. See Fig. 30-2. These machines are efficient. For larger jobs, they work faster and better than hand tools.

Large, heavy earthmoving equipment is expensive to operate and maintain. Some contractors specialize in earthmoving. They own their equipment. Other contractors own only basic equipment. Many contractors own no earthmoving equipment. They often lease or rent special machines. Charges for rented machines are made by the hour or day.

Fig. 30-1

Hand tools are particularly useful for small jobs.

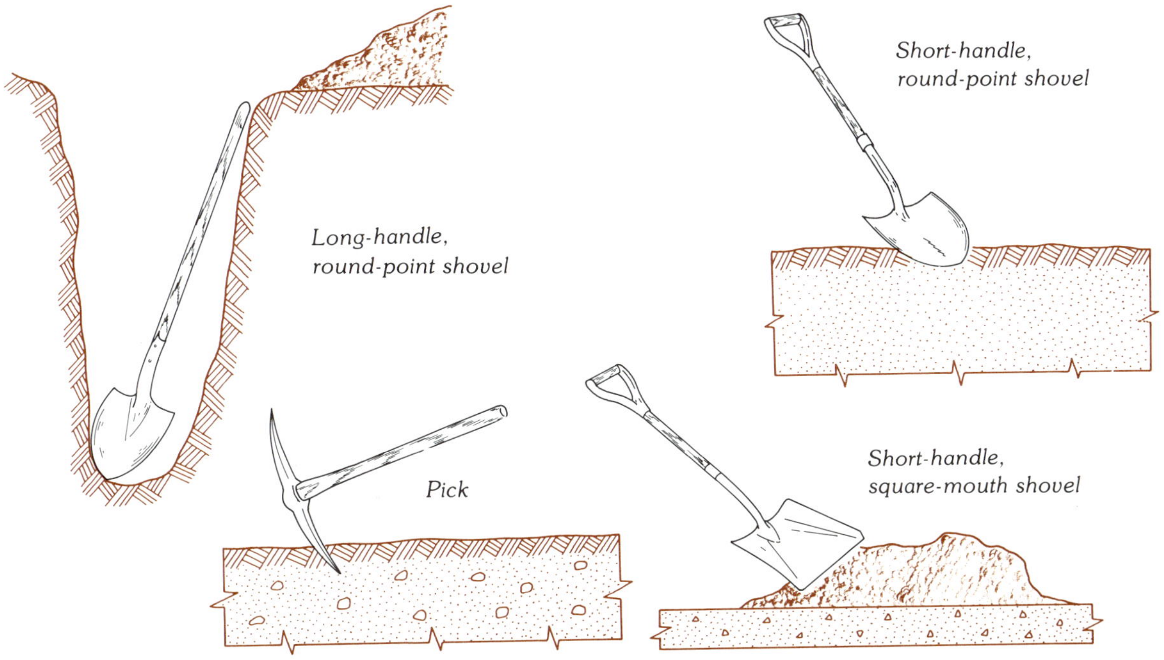

Bucyrus-Erie Company

Fig. 30-2
Large earthmoving equipment can get the big jobs done quickly.

Charges are also made for taking the unit to and from the site.

Many kinds of equipment are used to move earth. See Fig. 30-3. *Shovels* and *scrapers* such as the bulldozer are two kinds of equipment used. *Draglines, clamshells,* and *hoes* are other types of heavy earthmoving equipment. Tractors with special attachments are often used for smaller jobs.

Kinds of Earthmoving

Earthmoving involves digging and moving soil and rock. Digging is called *excavating*. There are two kinds of excavating. *Rough excavating* involves

Fig. 30-3 *Different kinds of earthmoving jobs require different earthmoving equipment.*

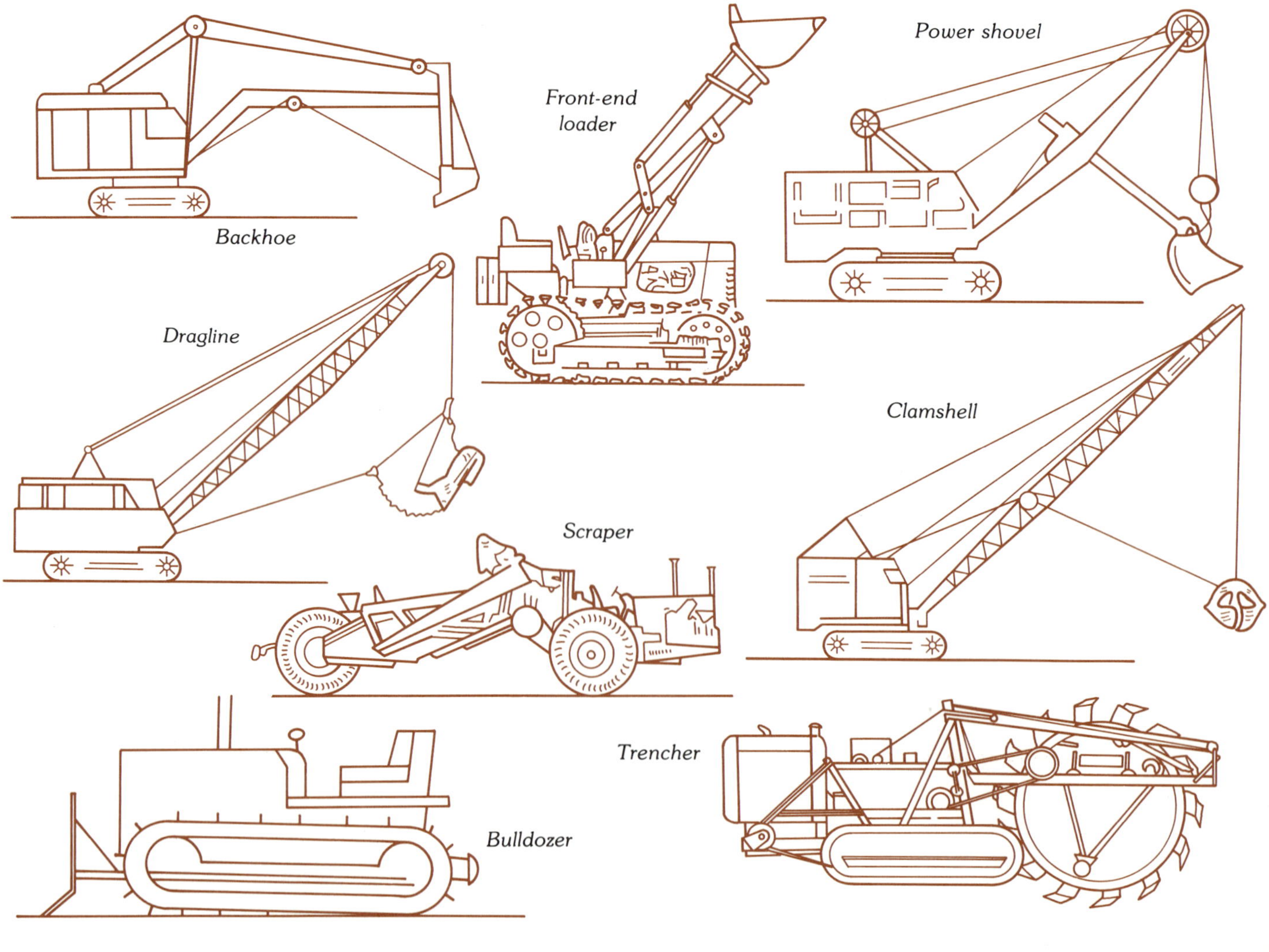

Fig. 30-4
A conveyor may be used to transfer materials.

digging up large amounts of earth. This leaves a crude pit, hole, or trench. The earth must be moved and used again in some way.

Finish excavating gives the final shape to the earth opening. Less material is moved in finishing an excavation.

Moving material from one area to another is called *transferring*. See Fig. 30-4. The moved earth is called *spoil*. Spoil is piled up, spread out, or used as fill material. These activities are called *disposing*. Excavating, transferring, and disposing are the main processes of earthmoving.

Rough Excavating

Excavating is done for four main reasons.

- To reach earth that will provide a good base for the foundations of structures.
- To build basements of houses and other buildings.
- To make cuts through hilly land for travel routes.
- To smooth uneven ground.

There are seven kinds of rough excavating. *Bulk pit excavating* involves digging a wide, deep area. Access to the area is very limited. Loose earth is hauled away from the site. This type of excavating is done to prepare the ground for a structure. For example, the basement of a downtown building is dug using the bulk pit method.

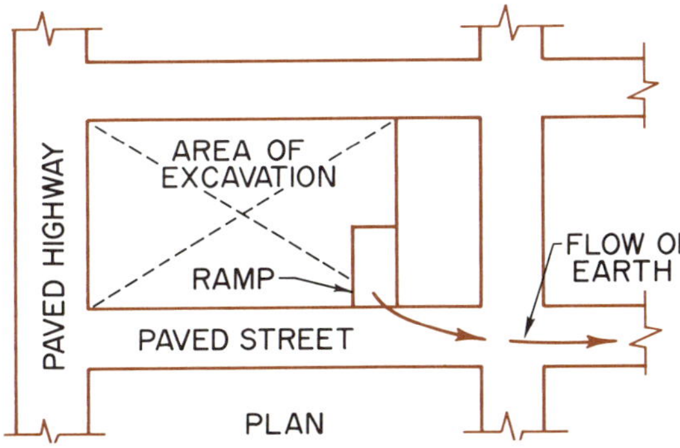

Bulk wide-area excavating involves wide, shallow digging. Usually there is easy access to the area. Grading for a highway overpass is an example.

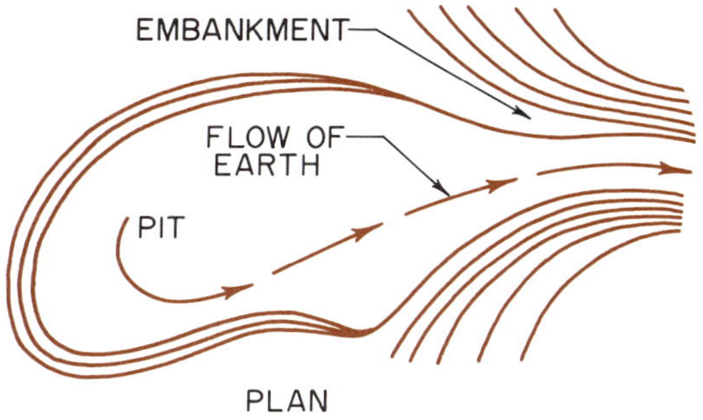

Loose-bulk excavating moves earth only a short distance. The material is piled close to the excavation. An example is moving soil to shape the land for a pond or lake.

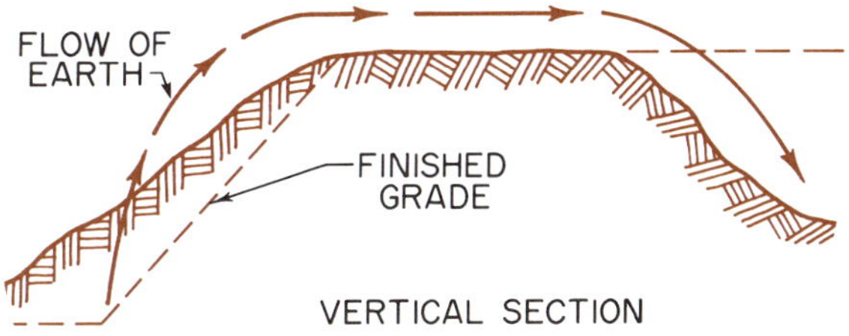

Limited-area vertical excavating involves lifting earth up and over vertical sides. The basement for a small home would be dug in this way.

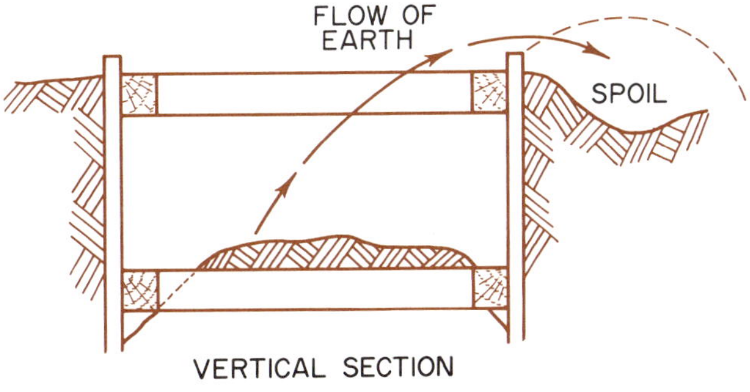

Trenching is digging long, narrow excavations called trenches. Trenches provide space for underground materials such as pipes and conduits (tubes for wires or cables). In trenching, the spoil is piled up on the surface. It is used later to refill the trench.

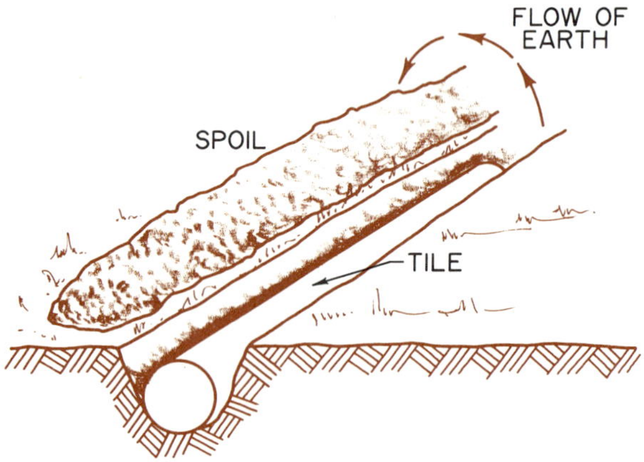

Dredging removes soil and other materials from under water. It is done to deepen river channels and harbors. Dredging is done in much the same way as loose-bulk excavating.

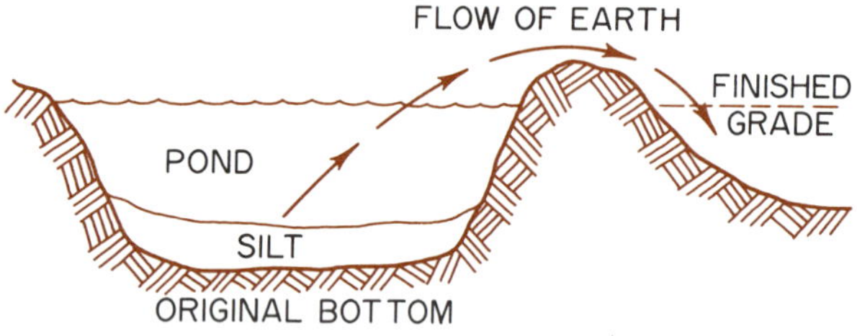

Tunnel excavating is done completely underground. Earth must be brought out of the tunnel. Roadways may be made through mountains or under rivers by tunneling.

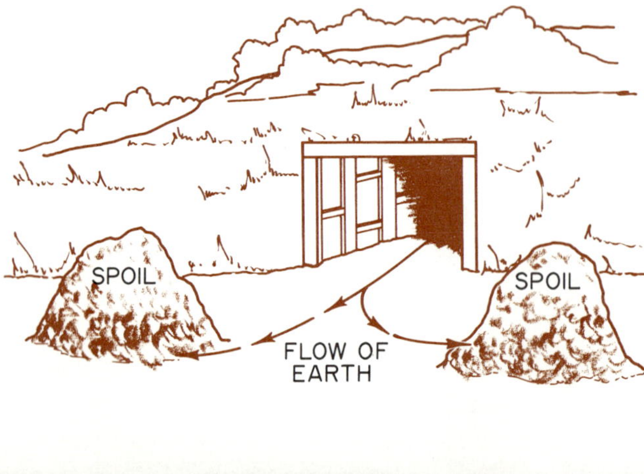

Processes of Loosening Earth

Sometimes earth must be loosened before it can be excavated. The earth may be too tough or dense. Common processes of loosening earth are

- blasting,
- breaking,
- scarifying, and
- ripping.

Blasting is done mostly in rock. Special explosives are set in holes drilled in rock. Blasting causes rock to break into pieces.

Breaking means applying strong blows to tear up hard materials. Breaking may be done with a tool called an air hammer. See Fig. 30-6.

Often tough or dense earth is loosened by large machines that have blades or teeth. Scarifying and ripping are two processes of doing this. They are like plowing. *Scarifying* is usually done in the upper 450 millimeters (mm) of soil. *Ripping* is done to depths of about 600 to 900 mm.

Fig. 30-6
Concrete may be loosened with an air hammer. This process is called breaking.

Transferring and Disposing

Workable or loosened earth is excavated. Then it is moved out of the way. This involves transferring and disposing.

Dump trucks are used to move excavated materials from one place to another. The excavated material may be stored for reuse. Or it may be disposed of. For example, the spoil may be used for fill or top dressing. (Top dressing is a thin layer of earth over an area.) Spoil may also be used to raise the general level of the construction site.

Sometimes water must be moved away from a site. Pumps are used to do this. Water is pumped away from the working area through pipes.

Finish Excavating

Finish excavating follows rough excavating. One main operation of finish excavating is trimming and shaping. The other is stabilizing.

Trimming and Shaping

A crude trench or pit is unfit for many foundations. Usually *trimming and shaping* must be done to the excavation bed and walls. This can be done by

- cleaning and washing,
- grading,
- sloping, or
- treating.

Fig. 30-7
Filling holes is part of grading. J.I. Case Company

Cleaning and washing removes mud from base rock. This is done so that the rock will make a good bond with the concrete. Water is used under high pressure to remove mud from rock surfaces. The mud is carried away through a drainage system.

Grading makes the bed and walls of the excavation level and firm. See Fig. 30-7. The high spots are cut away. Low, uneven spots are filled.

Sloping is another way of trimming and shaping earth. The sides of excavations are sloped to reduce the chance of slides.

Treating is done to make loose soil more solid. It is often done with water. Sometimes treating is done with a chemical solution.

Stabilizing

Excavating may weaken the support of structures next to the excavation. Earth and structures may need to be held in position. This is *stabilizing.*

There are five main ways of stabilizing the earth for construction. They are

- compacting,
- sheathing,
- bracing and shoring,
- piling, and
- cofferdamming.

Compacting makes earth firm. Soil is compressed (packed down). Large rollers pack the earth to make it firmer. See Fig. 30-8. For small areas a tamper is used to compact the surface.

Fig. 30-8
Compressing the earth is one way of stabilizing it for construction. This process is called compacting.

Sheathing forms walls to keep earth out of the excavated area. A framework holds up the sheathing. Common sheathing materials are wood planks and wood or metal panels.

Bracing and shoring are commonly used to support sheathing materials. Bracing is put horizontally between sheathing panels. It is used in narrow excavations. Shoring is used for wider excavations. It is placed diagonally against the excavation walls or panels. See Fig. 30-9. Shoring is held at the bottom by stakes. Bracing and shoring are used to support existing structures that are very near an excavation. *Underpinning* is a special kind of bracing and shoring. It is used to stabilize a structure that may be weakened by excavating. When necessary, earth under the structure is removed in sections. The earth is replaced by concrete or other rigid underpinning.

Fig. 30-9
Bracing and shoring support sheathing materials in excavations.

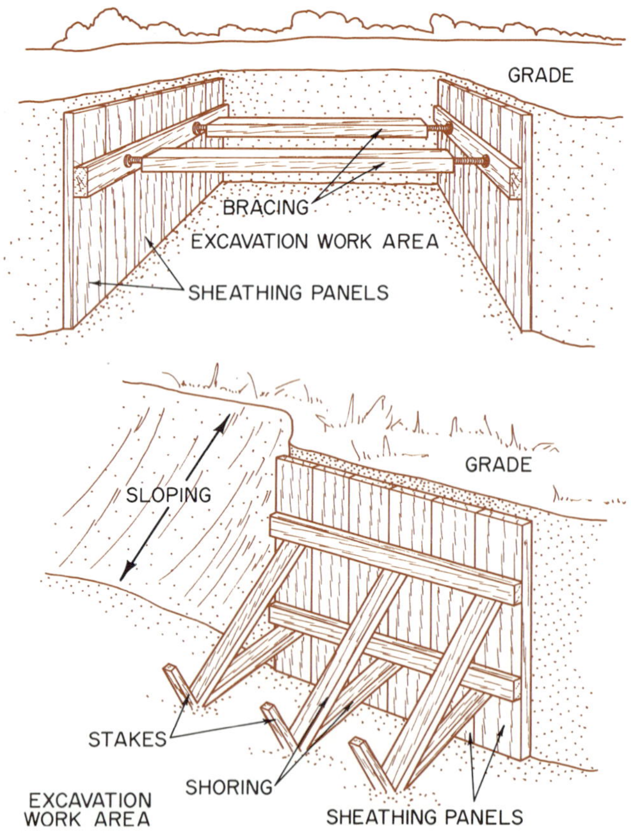

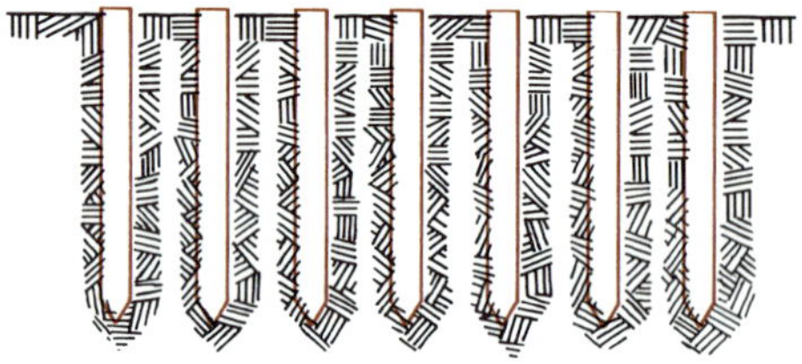

Fig. 30-10 Erico Products, Incorporated
Piles help support structures. They are driven into the ground.

Piling gives stability to a structure. Large concrete, steel, or wooden piles are driven into the ground. See Fig. 30-10. Piles improve the load-bearing capacity of the earth. They also help keep the structure from settling unevenly.

Cofferdamming is used in water-filled soil or when building is being done in water. Cofferdams keep water out of the working area. See Fig. 30-11. Sheets of wood, steel, or concrete are driven into the ground. They form a watertight wall around the work site. Shores and braces support the walls. Water and other materials are removed. Then, construction workers can work in a dry space.

Fig. 30-11
Structures such as dams and canals are built in waterways. Cofferdams are used to keep the working area dry.

Summary

Earthmoving is a process of moving soil and rock. The process allows substructures to be built on or below the surface. Earthmoving often involves large, expensive equipment.

Rough excavating moves large amounts of earth. This leaves a crude earth opening. The earth must be transferred from the excavation. It may be used as fill or for some other purpose.

Finish excavating is done to trim and shape the crude opening. It also involves stabilizing the earth.

Terms to Know

blasting
bracing and shoring
breaking
bulk pit excavating
bulk wide-area excavating
clamshells
cleaning and washing
cofferdamming
compacting
disposing
draglines
dredging
earthmoving
excavating
finish excavating
grading
hoes
limited-area vertical excavating
loose-bulk excavating
piling
ripping
rough excavating
scarifying
scrapers
sheathing
shovels
sloping
spoil
stabilizing
transferring
treating
trenching
trimming and shaping
tunnel excavating
underpinning

Think About It!

1. Identify one place in your neighborhood where earthmoving is now going on. Why is it being done?
2. Why do earthmoving contractors like to have their equipment in use as many hours a day as possible?
3. The sides of excavations have sometimes caved in and workers have been caught inside. Has this ever happened in your community? What should have been done to prevent this?
4. What must be done to support an existing structure if earth is removed very close to its foundation? Have you ever heard of an incident in which damage was done to an existing building? If so, when and where did it take place?

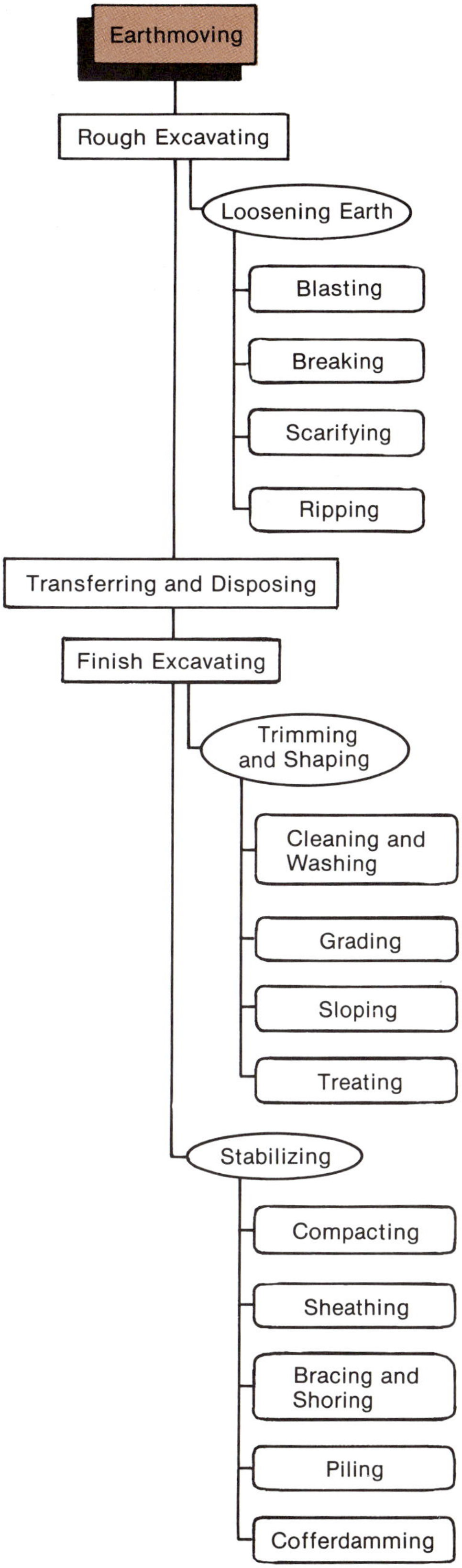

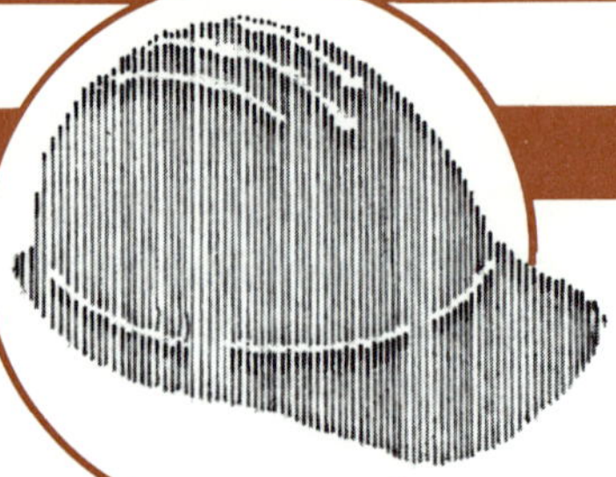

Reading 31

Setting Foundations

The Purpose of a Foundation

All structures rest upon *foundations*. See Fig. 31-1. Usually foundations are below the surface of the ground. For this reason, they are called *substructures*.

A foundation supports a *superstructure*. This is the part of the structure that is above ground. The foundation must support the weight of the structure. If it does not, the structure will sink into the soil. See Fig. 31-2.

A foundation is much like your foot is when you are standing. Study Fig. 31-3. If the soil is hard, your shoe will rest on the surface (A). In snow or mud, your foot will sink into the soft surface. To keep from sinking, you need a support (B). On a very soft or spongy surface, you need stilts. The stilts go down through the spongy surface and touch firm soil (C).

Fig. 31-1
A sturdy foundation is necessary to support the weight of the superstructure. The foundation being constructed will support a large building.
Lone Star Industries, Incorporated

Freelance Photographer's Guild

Fig. 31-2
The Tower of Pisa was not placed on a firm foundation. Because of this, it tilted. Now it is more than four meters from a vertical or plumb position.

Fig. 31-3

Shoes, snowshoes, and stilts are much like the different types of footings of a structure.

A. Shoes make contact with a firm surface (spread footings).

B. Snowshoes spread weight over a large surface (raft footings).

C. Stilts (pile footings) reach a firm surface.

Parts of a Foundation

A foundation (substructure) can be divided into three parts.

- Bearing surface.
- Footing.
- Upright supports.

The *bearing surface* is the earth upon which the foundation rests. It may be soil or bedrock. The bearing surface may be far below the surface of a swamp, river, or ocean.

The *footing* is the flat part of the foundation. It spreads the load of the structure above it.

Upright supports may be walls or piers (masonry columns) that rise above the footing. They form the rest of the substructure. See Fig. 31-4.

Fig. 31-4

The parts of a foundation.

Types of Footings

Three kinds of footings are commonly used. These are

- spread footings,
- raft footings, and
- pile footings.

See Fig. 31-5.

Spread footings are used on bearing surfaces of rock or hard soil. Spread footings are like pads. They may be long and rather narrow. Under piers or posts, they may be square.

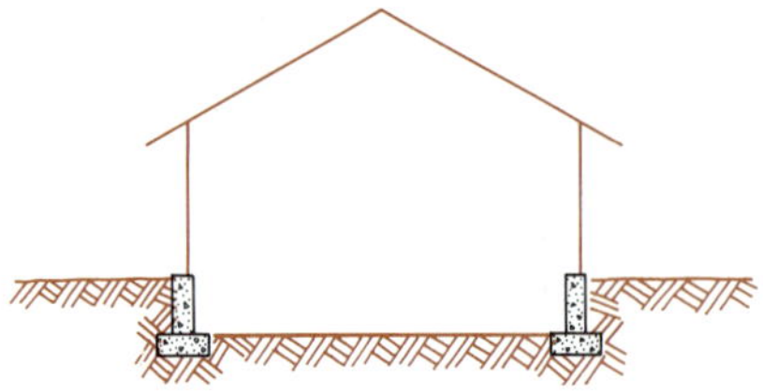

Raft (slab) *footings* are used in soft soil. They are often called floating footings. Resting on these footings, structures appear to float on the ground. Raft footings are used in earthquake zones. They can support structures that vibrate a lot. (Spread footings would not do this well.) Raft footings spread the weight of a structure over a large area. Most often they are used to support small, lightweight structures. These footings cover a larger area than the structure they support.

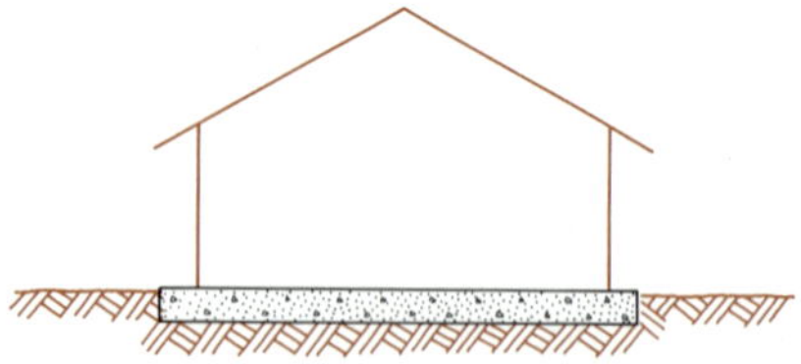

Pile footings are used in very soft soil. For example, they are used in sand or marshy land. They also support structures built over water.

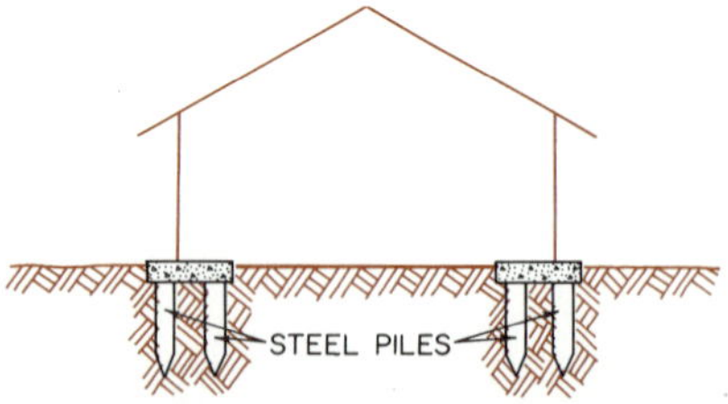

The piles must be driven down to find a good bearing surface. Then a footing is placed on top. This footing resembles a spread footing. It is called a *pile cap*.

Foundation Materials

Concrete is the most common foundation material. It is strong. It can support heavy loads. Steel rods or wire may be buried in concrete. This *reinforcement* makes it even stronger. See Fig. 31-6.

Crushed rock or gravel may be used for footings. These materials are used in highways and airport runways.

Natural materials, such as wood or stone, may be used for piers, piles, or foundation walls. These foundation members may also be made of steel.

How Foundations Are Set

On most sites, some earthmoving is done. Earth may be moved to control water. Moving earth allows the water to drain away from the excavation. See Fig. 31-7. (If earth cannot be moved, pumps may be used.)

Often topsoil is of high quality. It may be moved off to the side. After the structure is built, the stored topsoil is put back. It makes a good layer for landscaping.

The bearing surface must support the footing. If it is hard and solid, no additional work is done. If it is not, the earth may be *compacted* (pressed together). This makes loose soil hard and firm.

Fig. 31-6
This chimney foundation under construction will be made of concrete. Steel reinforcement must be put into place before the concrete is poured.

Fig. 31-7
Water often collects in excavations. Before a foundation can be built, the water must be drained or pumped from the excavation. Here, it is being pumped.

Next, footings are placed on the bearing surface. Most footings are made of concrete. Concrete footings are made in six steps.

1. Forms are set in place. If forms are wooden, carpenters build them.

2. Reinforcing ironworkers (sometimes referred to as rodsetters) place steel reinforcement in the forms where needed.

3. Concrete is mixed by a mixer operator. This may be done on the site. Or, concrete may be moved by truck to the site.

4. Concrete laborers and cement finishers place the concrete in the formwork.

5. Cement finishers compact the concrete inside the forms. Then they finish (smooth) the surface of the concrete.

6. The concrete is allowed to harden. Then laborers remove the formwork.

Sometimes, the superstructure rests directly on the footing. A highway is an example of this. Concrete is placed over compacted fill. At other times, upright parts are placed on the footing. Concrete upright parts are built by following the six steps just described.

Upright parts may also be made of stone, block, or brick. If so, *masons* lay foundation walls or piers on the footings. Ironworkers may fasten metal columns to the footings.

The foundation must be kept as dry as possible. Plans must be made to provide permanent drainage. For example, drain tile are usually placed near the base of the foundation. They will help keep the basement dry. See Fig. 31-9.

Fig. 31-9
Drain tile provides permanent drainage. Notice the placement of the drain tile near the base of this spread foundation.

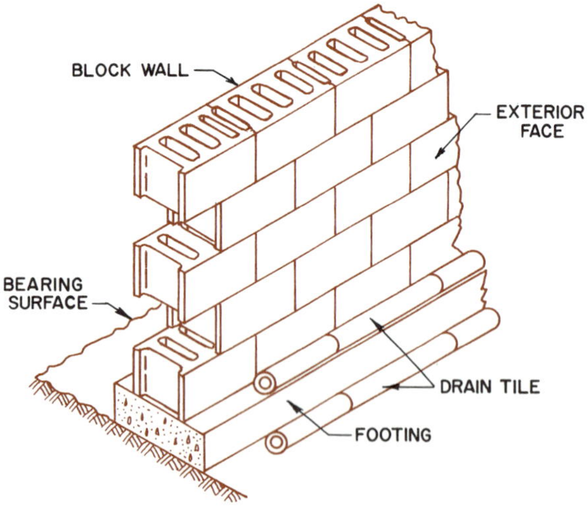

When the concrete has cured completely, *backfilling* is done. Earth is put back against the completed foundation. See Fig. 31-10. Fill is scooped in a little at a time. It is compacted carefully to avoid damaging the substructure.

After backfilling, the substructure is complete. The footing and foundation walls are in place. Now work on the superstructure can begin.

Fig. 31-10
Backfilling is the last task in completing work done on the foundation of a building.

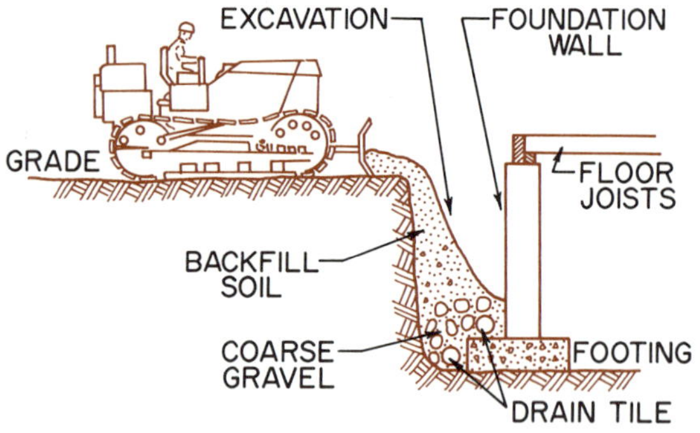

Summary

Foundations are called substructures. They support the load of superstructures. A foundation has three major parts. These are the bearing surface, the footing, and the upright supports.

The bearing surface must be solid. It may be bedrock. Or, it may be firm, hard soil. If the soil is not firm, it must be compacted.

Footings rest on the bearing surface. They may be made of natural or manufactured materials. Most often, footings are made of concrete, crushed rock, or gravel.

Most foundations have walls or piers. These parts rise above the footings.

Foundations are most often made of concrete. Steel reinforcement may be used to strengthen concrete. Many people in many different jobs work together to set foundations.

After the foundation is set, backfilling is done. This completes the substructure.

Terms to Know

backfilling	masons
bearing surface	raft footings
compacted	reinforcement
footing	spread footings
foundations	substructures
pile cap	superstructure
pile footings	upright supports

Think About It!

1. What type of footing do you think is under your home? your school? a nearby bridge? Do you think they are all the same type?
2. Why is the Tower of Pisa leaning? (Refer back to Fig. 31-2.)
3. Do most homes in your community have basements? If many do not have basements, is it because there is an excavation problem? What do you think the problem is?
4. Why is it important to strip off topsoil and save it during the clearing operation?

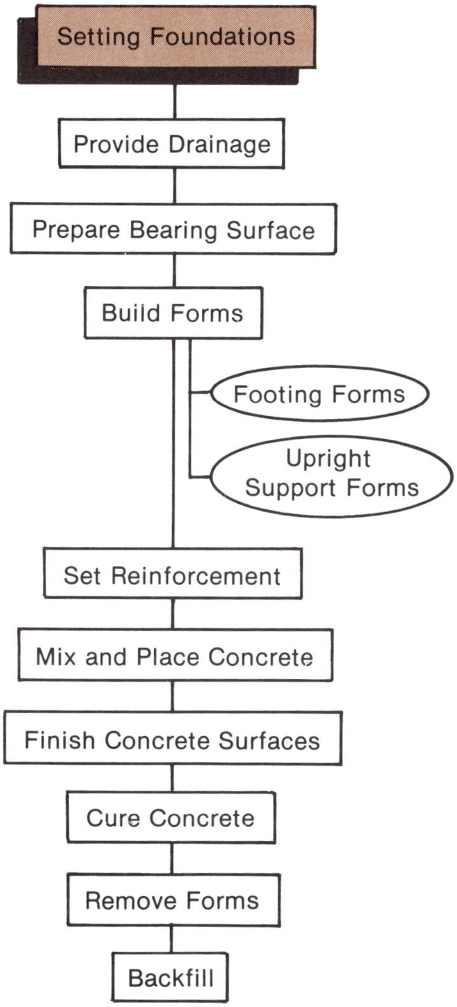

Reading 32

Building Forms and Setting Reinforcement

The Purpose of Forms and Reinforcement

Concrete is a very strong material. It is used often in construction. *Forms* shape freshly placed concrete. They shape concrete foundations or substructures. Forms may also be used in building concrete superstructures.

Before concrete is placed, steel *reinforcement* may be set in the forms. Reinforcement strengthens the concrete.

This reading describes forms and form building. It tells why and how steel is set in forms. This reading mainly concerns building forms for foundation (substructure) work. See Fig. 32-1. In Reading 36 you will read about framing concrete superstructures.

Fig. 32-1
The first steps in constructing a concrete foundation are building forms and setting metal reinforcement. When these tasks are completed, the concrete can be placed in the forms.

Betty Medsger

Forming Concrete

Concrete is made of stone, sand, and water, bound together by cement. Wet concrete is like mud. But shortly after it is mixed, it begins to set (become firm and dry). Soon it becomes as hard as stone.

Concrete is placed in forms while it is still wet and formable. The forms give shape to the mixture as it sets and cures (changes chemically during setting).

The forms are removed after the proper length of time. This time varies. Different types of concrete require different setting times. Weather conditions also affect setting times.

Forms must be built to be removed easily. The concrete must not be damaged while forms are being removed.

Formwork must be strong. Often, workers stand or climb around on it. And the weight of the concrete must not push forms out of shape.

Reinforcing Concrete

When concrete has hardened and cured, it has high *compressive strength*. This means that it is not easily compressed (packed down). Concrete may chip, but it will not dent. It resists heavy loads and heavy blows.

Concrete has low *tensile strength*. Tensile strength is the ability to resist being pulled apart. Setting steel reinforcement in concrete increases its tensile strength. Reinforced concrete supports loads more safely.

Forms and Reinforcement

Concrete foundation forms are usually made of wood. They may be made of steel or other materials. Often they are made of both wood and steel. Reinforcing materials are always made of steel.

Characteristics of Form Materials

Wood forms are lightweight. They are easy to handle. Wood is usually less expensive than other form materials. Also, the size and shape of wood are easily changed. This can be done on the site with hand tools.

Two kinds of lumber may be used to build forms. One kind is dimension lumber. The other kind is plywood or other wooden sheet materials. See Fig. 32-2.

Dimension lumber sizes (in inches)* are 2 × 4, 2 × 6, 2 × 8, or larger. Also 1 × 4, 1 × 6, and 1 × 8 lumber is made. This lumber comes in various lengths between eight and sixteen feet.

Sheet materials, such as plywood, are used to form large, flat surfaces. They come in sheets four feet wide and eight to ten feet long. Thicknesses from 1/4" to 1" are common.

Steel forms can be reused. Panels are made in standard sizes. They are heavy and expensive. Panels of steel (and other materials) are manufactured off the site. See Fig. 32-3.

*U.S. Customary measures are used to express standard sizes and measurements related to standard sizes.

Fig. 32-2
Wood forms may be built from dimension lumber or from sheet materials.

Characteristics of Reinforcement Materials

There are two common types of reinforcement steel.

- Rod
- Mesh

Rods of steel used for reinforcement are called *rebar*. These may be any diameter from 1/4" to well over 1". Each size rod has a number. For example, No. 1 is 1/8" rod, No. 2 is 1/4" (2/8") rod, and No. 3 is 3/8" rod. Each rod is deformed. Deformed rods are not smooth on the outside. Ridges of different patterns have been pressed into the rods at the steel mill. See Fig. 32-4. Smooth rods

Fig. 32-3
Other types of forms may be used.

A. Steel *B. Manufactured molds* *C. Fiber-form material*

Fig. 32-4
Steel reinforcement is available in different shapes and sizes for different purposes.

Wire Reinforcement Institute

Fig. 32-5
A roll of reinforcing mesh is being set in place for a concrete slab.

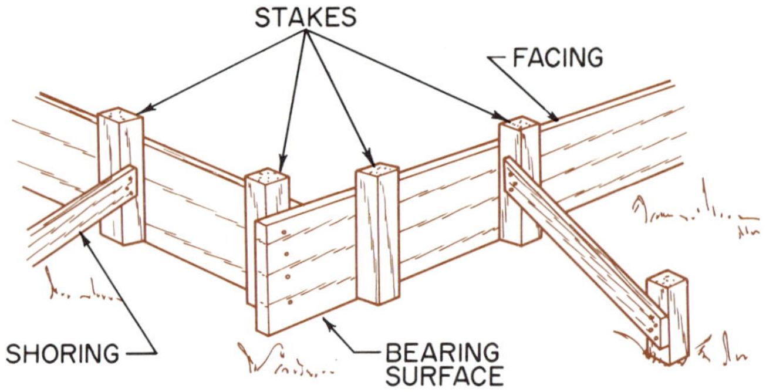

Fig. 32-6
A form for a thin slab foundation.

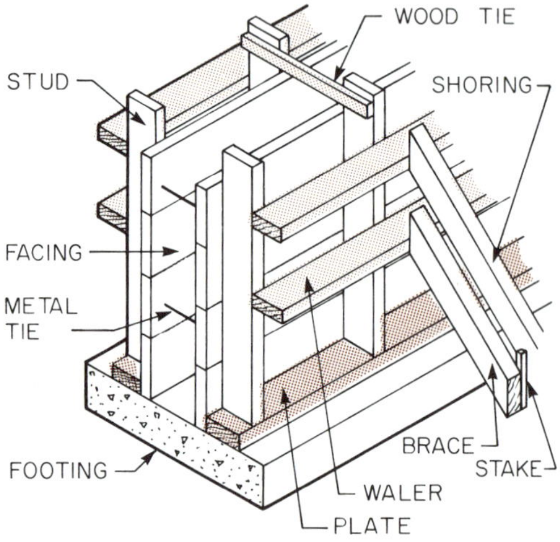

Fig. 32-7
Wood form for a foundation wall.

might pull through the concrete. But concrete hardens around the ridges of the deformed rods. These rods cannot pull loose.

Reinforcing mesh looks something like a wire fence. Generally, it is heavier. Wires cross each other at right angles to form squares. They are welded together at the joints. Mesh is delivered to the site in large rolls. See Fig. 32-5. Rolls of mesh are four to six feet wide. Sometimes they are wider. They may be up to one hundred feet long.

Building Forms

Some special forms are built in factories. These are delivered to the construction site. Other formwork is built partly on the site. The form building described here is usually done on the site.

Carpenters build the wooden forms. They lay out and cut dimension lumber to size. Simple forms, for thin slabs of concrete, are made of three parts — facing, shoring, and stakes. See Fig. 32-6. The forms are assembled on top of the bearing surface.

Forms for upright supports of foundations are more complex. See Fig. 32-7. Forms will be placed on top of the footing. The forms have seven main parts. They are plates, studs, walers, ties, braces and shores, stakes, and facing.

Reusable panels may be built up. Figure 32-8 shows formwork made of these panels. Hairpin ties (steel fasteners) hold the panels in position.

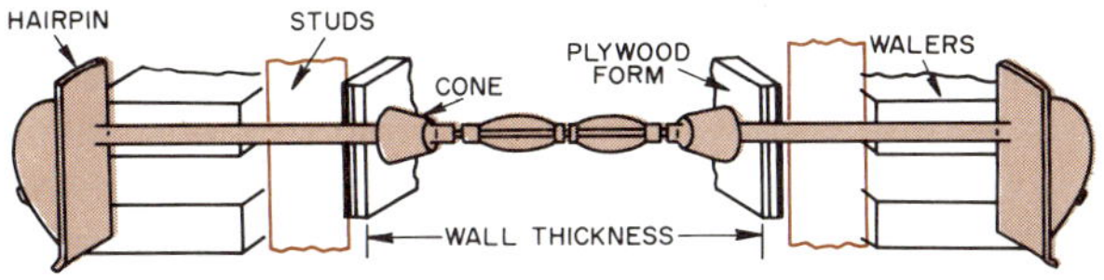

A. Hairpin tie

B. Reusable panels

Fig. 32-8
Formwork may be made of reusable panels. Form ties are used to set the distance between facings. Then the thickness of the wall will not vary.

Fig. 32-9
This is one method of setting steel mesh reinforcement. A worker uses a hooked rod to pull the mesh up into newly placed concrete.

Setting Reinforcement

Steel reinforcement is set differently for different types of forms. In simple forms, the steel may be set last. Study Fig. 32-9. Mesh is simply placed on the earth inside the form. As the concrete is placed, the mesh is pulled up into it. A hooked rod is used to place the mesh just below the center of the concrete.

Rod and mesh reinforcement are placed in concrete slabs. The steel must be supported above the bearing surface. The steel may be held up by pieces of brick. See Fig. 32-10. When the bearing surface is hard, bent pieces of metal called *chairs* support the steel. See Fig. 32-11.

Fig. 32-10
Bricks or small blocks of concrete may be used to keep mats of reinforcing steel off the ground.

Fig. 32-11
Reinforcing steel for a concrete slab is placed before the concrete is poured. The steel may be placed on chairs. (See steel in foreground.)

The Daily Pantagraph

Most forms must have at least two facing surfaces. Consider a wall that has two surfaces. One side of the form is put up first. *Reinforcing ironworkers* set the reinforcing steel (rebar) in position. See Fig. 32-12. Then the other side of the form is built. Sometimes, such as for columns, the steel is built up high. Then the forms are built around the rebar. See Fig. 32-13.

Forms must be *tied* together. Refer again to Fig. 32-8. Study the tie in this illustration. Forms must also be *braced* and *shored*. They must be carefully *plumbed* and *aligned*. This means they must be straight vertically and horizontally.

Summary

Wet concrete is placed in forms. It takes the shape of the forms as it hardens. Most forms are made of wood. Some are steel or other materials. Wood forms may be built in place. This is done piece by piece. Large wooden panels can be made on the site. Large metal panels are usually brought to the site.

Concrete has high compressive strength. Without reinforcement, it has low tensile strength. Steel reinforcement gives concrete greater tensile strength. Most concrete in structures is reinforced with steel.

Steel rod and mesh reinforcement are placed in forms. The forms are carefully tied together. They must be braced and shored to withstand heavy weights. They are plumbed and aligned. Finally, concrete is placed in the forms. After the proper hardening or curing period, forms are removed.

Terms to Know

aligned	rebar
braced	reinforcement
carpenters	reinforcing ironworkers
chairs	reinforcing mesh
compressive strength	sheet materials
dimension lumber	shored
forms	tensile strength
plumbed	tied

Fig. 32-12
When building a wall, reinforcing steel is placed in position before the forms are completed.

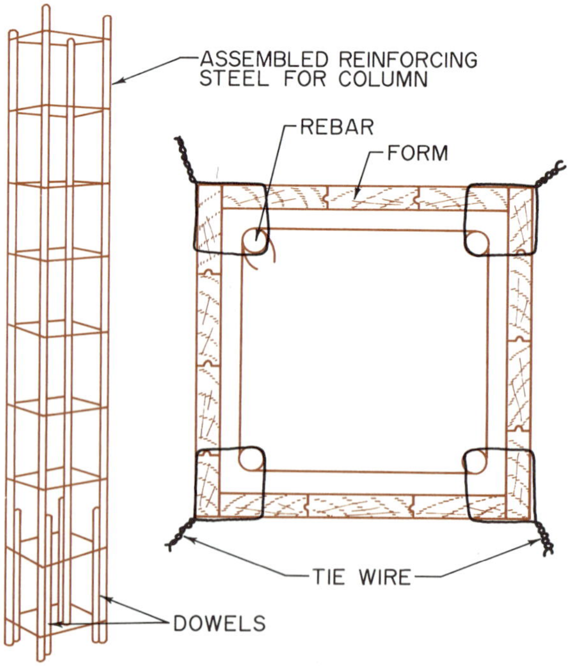

Fig. 32-13
Reinforcing steel in a pier may be preassembled. The form is built around the rebar.

Think About It!

1. Look around you or out the window. Can you identify at least two concrete surfaces that were shaped against forms? Were these forms foundation forms or forms for the superstructures?
2. Why are reinforcing steel rods deformed? Would smooth-surfaced reinforcing steel work as well? Why or why not?
3. Does reinforcing steel in concrete allow the solid concrete mass to "bend"? In a dictionary or encyclopedia, look up "tensile" or "tensile strength." Does the meaning of these terms give you a clue?

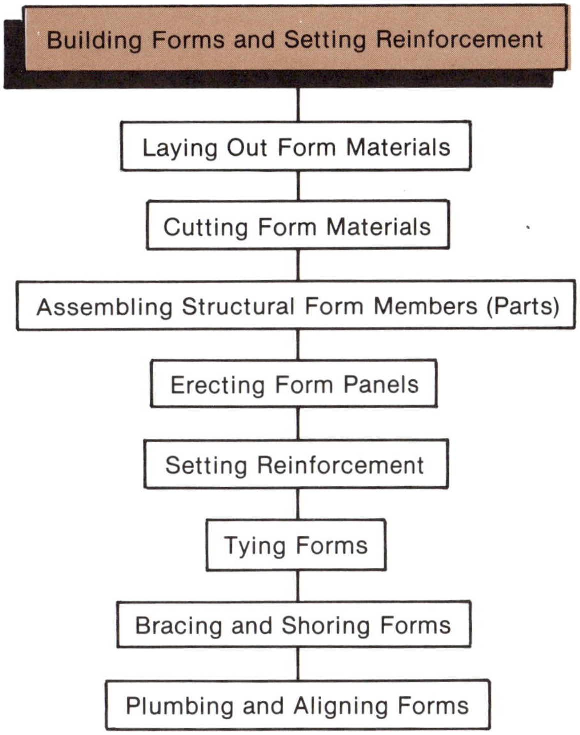

Reading 33

Mixing, Placing, and Finishing Concrete

Concrete

The forms that will shape the concrete have been built. Reinforcing steel has been set in the forms. The forms and reinforcement have been tied together in position. They have been plumbed and aligned. The framework has been carefully braced and shored. Now concrete may be put into the forms.

This reading describes the *ingredients* of concrete. These are the materials that are combined to make concrete. In addition, the practices of making and installing concrete will be discussed. These practices are *mixing, placing,* and *finishing.*

Fig. 33-1
The ingredients for concrete must be mixed together. Sometimes this is done by hand, using a small mixer.

Ingredients of Concrete

Concrete is a mixture of five materials. See Fig. 33-1. These are

- coarse aggregate,
- fine aggregate,
- cement,
- water, and
- admixtures.

Admixtures are not always used.

Coarse aggregate is the first important ingredient. See Fig. 33-2. Gravel or crushed rock is used. Gravel (small stones) is found in the soil. Most gravel is shaped like balls or spheres. It has rounded edges. Crushed rock has sharp edges. It is manufactured by breaking rock or stones into pieces. See Fig. 33-3. Rock used for crushing must not be too soft. If it is, it will crush to powder.

Fig. 33-2
Aggregates of various sizes are used in concrete.

A. Coarse aggregate is gravel or crushed rock.

B. Fine aggregate is sand.

Coarse aggregate is *graded* (sized). Generally the smallest size is about 6 mm. The largest aggregate is usually about 64 mm.

Fine aggregate in concrete is sand. It is graded or screened into different sizes. See again Fig. 33-2.

Both coarse and fine aggregate must be washed. Washing removes particles of rock dust or clay.

Fig. 33-3
A rock crusher reduces the size of the material.

Fig. 33-4
When needed, aggregate is taken from stockpiles. It is then mixed with cement.

After it is prepared, aggregate is stockpiled (stored in piles) until it is needed. See Fig. 33-4.

The *cement* used in concrete is portland cement. This is made from a type of limestone. The limestone is crushed with some clay or shale. The limestone and clay mixture is fired in a special kiln (furnace). See Fig. 33-5. The high temperature of the kiln produces a lightweight aggregate called a *clinker*. The clinker is ground into a powder. The powder is mixed with a small amount of a mineral called gypsum.

When water is mixed with cement, it forms a paste. It sets (becomes firm) in about two hours. A *chemical reaction* occurs during this time. The reaction is called *hydration*. The water must be clean and pure. Oil or dirt in the water will interfere with hydration.

Admixtures are additives. This means they are an extra addition to the concrete. They give concrete some very special characteristics. Some admixtures help hold air bubbles in the concrete. Others thicken the mix. Some are added to slow hardening. Still others speed up hardening. Admixtures help make concrete easier to work with. They extend the usefulness of the mixture.

Fig. 33-5
The inside of a cement kiln under construction.

Lone Star Industries, Incorporated

Fig. 33-6
At a ready-mix plant, concrete is specially mixed for each job.

Lone Star Industries, Incorporated

Fig. 33-7
The container on a concrete truck rotates. This action prevents the concrete from setting until it is placed in the forms.

Mixing Concrete

Concrete is mixed by weight or volume. It is sold by the cubic yard. Persons who order concrete must know the volume they will need. First they figure the number of cubic feet of space inside the forms. Then they find the number of cubic yards this equals. They calculate by dividing the number of cubic feet by twenty-seven.* Someday measuring will be done using the metric system. Then concrete will probably be sold by the cubic meter.

*There are twenty-seven cubic feet in a cubic yard.

Concrete can be *batched* (mixed) in several ways. The materials are mixed together in the correct *proportion* (ratio). A common mix is 1:2:3. This is one part cement, two parts sand, and three parts coarse aggregate. For very small jobs, concrete can be mixed by hand. Look again at Fig. 33-1. The ingredients are placed together in proper proportion. They can be combined in a box called a mortar box. Then they are mixed with a hoe.

Batching may also be done in a small rotary mixing machine. Ingredients are placed in the drum of the machine. The drum rotates (turns round and round). This action mixes the ingredients.

Most concrete, however, is mixed at large "ready-mix" plants. See Fig. 33-6. At the concrete mixing plant, each part of the batch is measured. Different types of concrete are mixed in different proportions. The person ordering the concrete must know what is required. Each job may need different batching.

Mixing-plant workers figure the weight of each ingredient. They even weigh the water and admixtures. They do this for any amount or kind of mix or batch.

Ingredients may be put in a rotating container (mixer) on a truck. The concrete is mixed as it is hauled to the site. See Fig. 33-7. Sometimes concrete is mixed at the plant, then put into the container. The rotating action of the mixer then only *agitates* the mix (keeps it moving).

When concrete arrives at the site, it is ready to use. Sometimes water is added just before the concrete is placed.

Extra-large jobs often have a mixing plant on site. Remote jobs may also need them. The concrete is not transported over long distances.

Placing Concrete

Concrete must be put in position very carefully. The term placing is used to describe the careful way concrete is put in place. The concrete must not be dropped or dumped. If it is, the coarse aggregate may separate from the mixture and sink to the bottom of the form. Coarse aggregate must be well distributed in the mix.

Concrete flows from the truck down a *chute*. It may be placed directly into forms. Often it must be moved to the forms. Hoppers (large containers) may be used to hold or move concrete. Other devices used to move concrete include wheelbarrows and buggies. See Fig. 33-8. Concrete may also be moved in large buckets. The buckets are lifted to high places by cranes. See Fig. 33-9. Conveyors (moving belts) are used to move concrete. Concrete can also be pumped through hoses. Hoses allow concrete to be moved to forms in high places.

Once the concrete is in the forms, it is moved around to fill the space. Thick concrete must be agitated. This *consolidates* (settles) it. A *vibrator* is used. See Fig. 33-10. Settling is especially important near the facing surfaces of forms. Vibration will move the sand and cement next to the form. When forms are removed, the surface of the concrete will be smooth. And any holes or air pockets near coarse aggregate will be filled.

After the concrete has been placed, it must be *leveled*. To level concrete, a *straightedge* is rested on *screeds*. A screed is a guide to help workers make a level surface. Screeds must be on the same level. The sides of forms may act as screeds. Or, screeds may be set parallel to each other, two or three meters apart. This is done in large, open

Fig. 33-9
Buckets of concrete are moved by cranes.

Fig. 33-8
Concrete flows from a hopper into a concrete buggy. The worker then moves the concrete to the forms where it is needed.

Fig. 33-10
A vibrator is used to settle the concrete around the forms and reinforcement.

areas. A long piece of straight wood or metal (straightedge) is moved back and forth across the concrete on the tops of the screeds. Doing this creates a level surface. See Fig. 33-11.

Concrete sets soon after leveling. The screeds are removed when the surface is stiff but not hard. Any grooves left by the screeds are patched.

Before finishing is done, the surface is *floated.* Floating leaves a smooth mixture of sand, cement, and water on the surface. A cement finisher uses a *wooden float.* A float has a long handle. See Fig. 33-12. It is moved back and forth on the surface. Sand and cement may be applied to the surface. This is a *top dressing*. It is floated smooth. The floated surface may be left as the final finish.

Fig. 33-11
Workers place the ends of a straightedge on screeds. Then they move the straightedge back and forth across the concrete. This provides a level surface.

Fig. 33-12
A float may be used to smooth the surface of concrete.

Illinois Laborers' and Contractors' Training School

Finishing Concrete

Often a smooth finish is desired for concrete. This finish can be produced by hand or with a machine. A smooth finish is called a *steel trowel finish*. A cement finisher moves a steel trowel back and forth over the surface. Or, a finishing machine is used to make a troweled surface. See Fig. 33-13. The surface may be troweled several times.

Edges of concrete along forms may be finished with an *edger*. An edger provides a smooth radius on the edge of concrete slabs.

Fig. 33-13
Troweling is a finishing process. It may be done by hand or with a machine.

A. Hand troweling

B. Troweling using a finishing machine

Koehring Company

Some concrete is given a *brush finish*. First, the surface is troweled one or more times. Then, before the concrete has hardened, a large brush is pulled across the surface. The brush produces shallow, parallel lines in the concrete. Such a process provides a relatively skid-free surface.

When the concrete is set, the forms are removed. This is called *stripping the forms*. Laborers do not remove the forms until concrete has properly set. In some cases, the concrete is allowed to fully *cure*. Curing is the process by which the concrete hardens and becomes strong. Concrete that has set but has not cured is called *green concrete*.

Moist curing is often done by spraying water on the concrete surface. Then the surface is covered with earth, straw, sand, burlap, or other materials. This keeps the surface damp. See Fig. 33-14.

Concrete should be kept moist for several days. Concrete cures naturally during a period of fourteen to twenty-eight days. Curing may be speeded up with special admixtures. Special, expensive processes can cure concrete in twenty-four hours or less. These processes are used only when it is worth the extra cost.

Formed surfaces are concrete surfaces that touch the forms as they are curing. They cannot be seen until the forms are stripped. Sometimes concrete is not vibrated enough. Small holes remain in the formed surfaces. See Fig. 33-15. These holes are patched and filled to provide a smooth surface.

Fig. 33-14
Covering fresh concrete with plastic will prevent loss of moisture. This helps the curing process.
Wire Reinforcement Institute

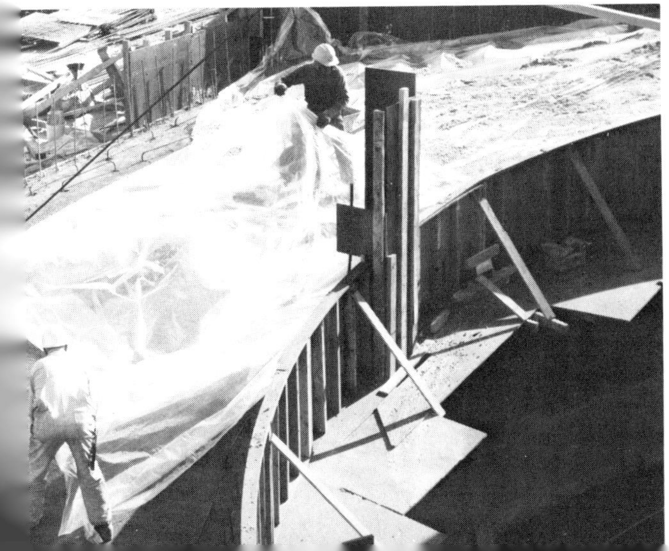

Special finishes are applied to some concrete. See Fig. 33-16. A *stoned finish* is made by rubbing the surface with a special tool. A *sack-rubbed finish* is made by applying cement and sand. *Special coatings* may be applied. They may be sprayed on the surface. *Asphalt* is used to waterproof concrete.

Fig. 33-15
Formed surfaces may have holes that must be patched.

Fig. 33-16
The surface of concrete may be finished in many different ways.

A. Brush finish
Portland Cement Association

B. A special waterproof coating
Dow Corning

Summary

Concrete is made of five ingredients. These are coarse aggregate, fine aggregate, cement, water, and admixtures. Batches of concrete are made by mixing the ingredients in certain proportions.

The concrete is placed in forms. Then it is leveled and finished to produce the desired surface. Forms are removed when the concrete is set. Then concrete must cure to become hard and strong. Special surface treatments are sometimes applied.

Think About It!

1. What is the difference between cement and concrete?
2. Look in your local telephone directory. Find the name and address of two "ready-mix" concrete mixing plants in your community.
3. Look around your home or school and find several different concrete surfaces. Identify at least one steel trowel finish and one float finish.
4. List two reasons why some hardened concrete surfaces must receive a final surface finish.

Terms to Know

admixtures
agitates
asphalt
batched
brush finish
cement
chemical reaction
chute
clinker
coarse aggregate
concrete
consolidate
cure
edger
fine aggregate
finishing
floated
formed surfaces
graded
green concrete
hydration
ingredients
leveled
mixing
moist curing
placing
proportion
sack-rubbed finish
screeds
special coatings
steel trowel finish
stoned finish
straightedge
stripping the forms
top dressing
vibrator
wooden float

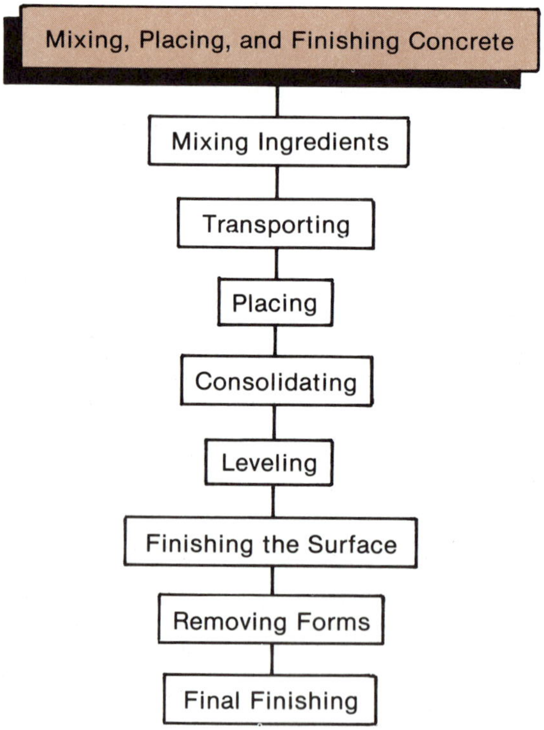

Reading 34

Building Mass and Solid Wall Superstructures

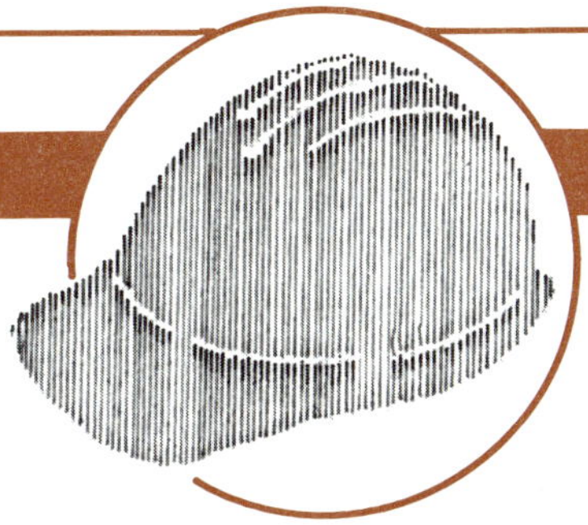

Structures

Structures are designed for different purposes. Buildings are structures. They enclose space. Bridges are structures that span rivers. They are usually designed to carry traffic. Towers are structures, too. They may serve several purposes. Some towers make it possible for people to observe an area from a high place. Other towers support electrical transmission lines.

Each part of a structure must be designed to suit the purpose of the whole structure. The two main parts of a structure are the foundation or *substructure* and the *superstructure*. The foundation is built first. Then the superstructure is built. See Fig. 34-1.

Fig. 34-1
Every structure has a substructure and a superstructure.

Patent Scaffolding Company

Superstructures

A superstructure is the part of a structure that rests on the foundation. Usually, it is the part that rises above the ground. But in some structures, the superstructure is not clearly defined. Dams and highways are examples. And others, such as tunnels, are completely underground. Still, each structure is supported by a foundation or substructure. And the part above the foundation may be considered a superstructure.

Types of Superstructures

There are four types of superstructures. These are

- mass,
- solid wall,
- frame, and
- air-supported.

Mass superstructures are made of large bodies of materials. They may be constructed of earth, brick, stone, or concrete. Examples are dams and dikes. Very little space (if any) is designed for use within a mass superstructure. See Fig. 34-2.

Fig. 34-2
This earthen dam is a mass superstructure. It has an earth embankment and a concrete spillway.

Niagara Mohawk Power Corporation

Fig. 34-3
Solid wall construction has been used for many, many years. For example, the Taj Mahal shown here has a solid wall superstructure. It was completed in the mid 1600s.

Fig. 34-4
Frame superstructures can support much weight. Skyscrapers have steel frame superstructures. Shown here under construction is the Sears Tower in Chicago.

Sears, Roebuck and Company

Solid wall superstructures enclose a space. The walls are made of masonry or reinforced concrete. They are solid and thick. These are *load-bearing walls.* They support the weight of the floors and roof above them. Many old buildings were of this type. An example is the Taj Mahal in India. It was built more than 300 years ago. See Fig. 34-3.

In a *frame superstructure,* the framework or frame is much like a skeleton. It functions in the same way as the bones of your body. It provides shape and support. Frames may be made of steel, reinforced concrete, or wood. A frame superstructure may or may not be covered. See Fig. 34-4. Buildings commonly have frame superstructures. The weight of the inside and outside covering (skin) is supported by the frame. Steel-framed buildings may be built over 305 m high. For example, the Sears Tower has a steel frame. It is over 443 m tall, and has 110 stories. Concrete-framed buildings are usually built to a height of only 30 to 60 m. Wood-framed structures are rarely built more than three stories high.

Air-supported superstructures are made of lightweight materials. They may even be made of fabric or plastic. These superstructures enclose large spaces. See Fig. 34-5. Center supports are not needed. These superstructures are held up by air pressure against the inside of the structure. Air-supported superstructures are not very common.

In this reading you will learn about mass and solid wall superstructures. You will learn more about steel-, concrete-, and wood-framed superstructures in the next three readings.

Fig. 34-5
This unusual structure has an air-supported superstructure. It is one of the terminal buildings at the Los Angeles International Airport.

Mass Superstructures

Many useful structures, such as dams, are built of solid, hard masses of materials. Earth, stone, brick, and concrete are materials used in mass superstructures.

Earth

Earth may be used to build dams. Earth embankments are built along rivers for flood protection. Earth is also used to shape highways and waterways. This material is inexpensive and is available in large quantities.

Using earth to build mass superstructures usually involves five steps.

1. Good soil is selected.
2. The earth is excavated.
3. The material is moved or hauled to the site.
4. The material is mixed and spread on a firm foundation.
5. The earth is compacted to form a solid mass. See Fig. 34-6.

Stone

Another natural material used in superstructures is stone or rock. It may be combined with the soil in the superstructure. A mixture of stones and soil forms good dams.

Large blocks of rock may be cut in a *quarry*. These pieces can be used for very large or massive (thick and large) superstructures. A monument is an example. See Fig. 34-7. Quarried rock also makes good *breakwaters* and *jetties*. These structures extend out into an ocean or large lake. They

U. S. Department of the Interior

Fig. 34-6
An earthen dam, such as this one being constructed, is a mass superstructure. The earth must be compacted. This makes it solid.

Fig. 34-7
The Washington Monument in Washington, D.C., is a mass superstructure made of stone. It is almost 170 meters high.

American Airlines

Fig. 34-8
Stone blocks can withstand much pounding by waves. Many breakwaters are made of stone blocks.

protect the land from waves. This reduces *erosion*. See Fig. 34-8. The stones may simply be placed together in a mass. Or, they may be fastened together with a joining material called *mortar*.

Brick

Brick is made from clay. When fired in a kiln (special oven), the clay becomes hard. *Adobe* brick is made of sun-dried earth and straw. Brick is often used in mass superstructures. Structures built of brick last for many years.

Concrete

Concrete is a hard, durable material. It is used often in mass superstructures. It is used in highways and airport runways.

To construct a highway or runway, workers first build an earthen foundation. The soil is compacted or pressed until it is firm. Crushed rock or gravel is added to the foundation. Forms are set. Or, slip forms are used. Slip forms move along with the machine that places the concrete. Either way, the concrete is placed and then cured. See Fig. 34-9.

Fig. 34-9
Concrete for a highway may be formed as it is placed.

Sometimes *bituminous concrete* is used for highways or runways. It is made by mixing heated (melted) asphalt with sand and gravel. Bituminous concrete may be built up in layers. It may cover a slab of concrete. A *paver* spreads the warm mixture. Steel-wheeled *rollers* pack it down. When bituminous concrete cools and hardens, it forms a smooth, hard mass. See Fig. 34-10.

Solid Wall Superstructures

Heavy, solid walls are often called bearing walls. They carry the load of the floors and roof. They may be made of masonry or reinforced concrete.

Framed walls may be made of some of the same materials as solid walls. But solid walls are different from framed walls. Solid walls do not often rise more than two or three stories.

Masonry Materials

Masonry is used in solid wall superstructures. It is made by joining small units of material with mortar. Brick, block, and stone are common masonry materials.

Bricks can be laid in many different patterns. Load-bearing walls made of brick are thick and heavy. Bricklayers place each brick on a layer of mortar. They must be careful to lay the bricks straight and plumb.

Concrete blocks may be used to make walls. Lightweight blocks such as cinder blocks may also be used. They are laid in much the same way as bricks.

Fig. 34-10
City streets may be made of bituminous concrete. The warm mixture is placed into a paver and spread in an even layer. A roller packs the concrete and makes the surface smooth.

Fig. 34-11
Bricks and blocks are placed and leveled carefully. Bricklayers stand on scaffolding inside the building.

See Fig. 34-11. Mortar is used to fasten the blocks together. Blocks are larger than bricks. Therefore, work goes faster with blocks.

Many years ago it was common for solid walls to be built of stone. They were often a meter or more thick. In buildings today, stone is used most often as a veneer (outside surface) or for other decorative purposes. It is seldom used to make solid walls.

Reinforced Concrete

Solid walls may be made of reinforced concrete. It may be cast (set) in place. Or, wall sections may be made off the site. They are transported to the site and set in place. See Fig. 34-12.

Summary

The superstructure is built on top of the foundation or substructure. Its design is determined by the function of the whole structure.

Mass superstructures may be made of earth, stone, brick, or concrete. Mass superstructures are commonly used for such structures as dams and highways. They may also be used for monuments and other massive structures.

Fig. 34-12
Reinforced concrete solid (or bearing) walls may be cast off the site and transported to the site when needed.

Solid walls support the load of the roof and floors. They are thick and heavy. Solid walls may be made of masonry or reinforced concrete. Most solid walls help enclose the space within buildings.

Air-supported superstructures are made with lightweight materials. They are held up by air pressure.

There are several types of framed superstructures. They will be discussed in three of the next readings.

Terms to Know

adobe
air-supported superstructures
bituminous concrete
breakwaters
brick
concrete
erosion
frame superstructure
jetties
load-bearing walls
mass superstructures
mortar
paver
quarry
rollers
solid wall superstructures
substructure
superstructure

Think About It!

1. Classify the following superstructures as (a) mass, (b) solid wall, (c) frame, or (d) air-supported.
   a. Your home.
   b. Your school.
   c. Television tower.
   d. Highway overpass.
   e. Earthen dam.
2. Why are highways or airport runways considered mass superstructures? Can you give reasons why they might be considered substructures?
3. Sometimes brick and stone are used in building construction but do not make up a solid wall superstructure. How, then, are they used? Give one example of such a structure.

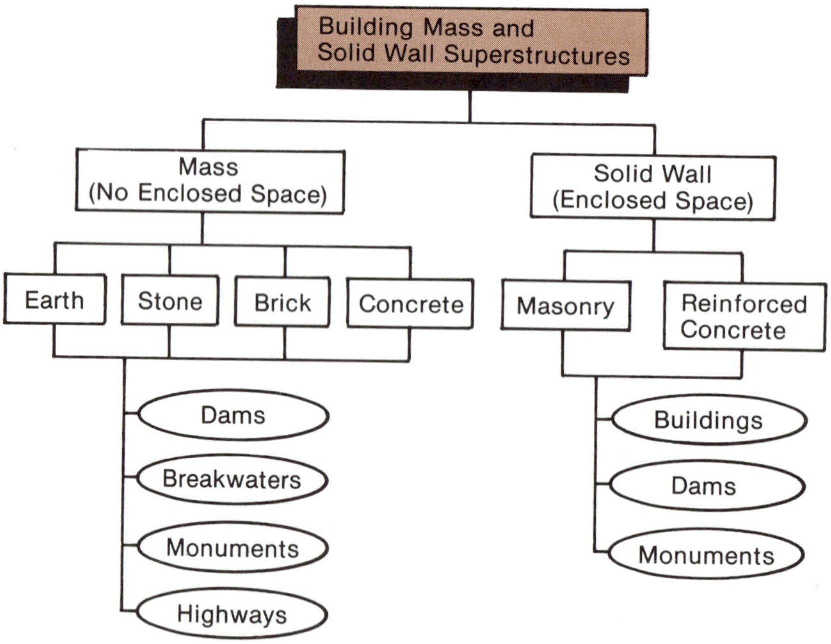

Reading 35

Erecting Steel Frames

Structural Steel

Steel frames are most often erected (built) at the construction site. They are assembled one piece at a time. See Fig. 35-1. Each piece of steel must be the proper size and shape. *Structural steel* is usually made in a *shop*. The pieces are sized according to an engineer's design drawings. The pieces are then made according to detailed working drawings called *shop drawings*.

The steel pieces used in frames are called *shapes*. See Fig. 35-2. Shapes are made in steel mills. They are made in long pieces and are cut to fit. Cutting is done with power saws, shearing machines, or cutting torches. Cross sections of common shapes look like the letters I, H, T, U, and L.

Fig. 35-1
Many constructed works have steel frames.

The assembling of structural steel is often subcontracted to a *structural steel erector*. This subcontractor employs *ironworkers*. Ironworkers *rig*, *handle*, *fasten*, and *plumb and brace* the parts of structural steel frames.

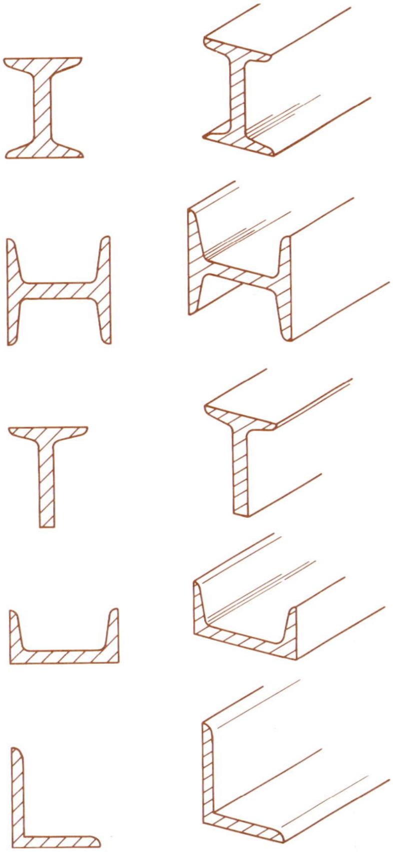

Fig. 35-2
Shapes are assembled to make steel frames. The ends or cross sections of these steel pieces look like letters of the alphabet.

FMC Corporation

Fig. 35-3
Cranes are used to handle the parts of structural steel frames.

Fig. 35-4
Derricks are used to build a structure level by level. They carry steel parts to the level being built. Then they are moved upward to complete the next floor of the structure.

Rigging and Handling Steel Shapes

A *crane* is commonly used to set steel frame members. Cranes move on wheels or crawlers. See Fig. 35-3. Cranes cannot reach high parts of very tall buildings or towers. For these high parts, *derricks* are used to set the steel. See Fig. 35-4.

A derrick has two main parts. One is called a mast. The other is called a boom. The *mast* is a tall pole. It is held vertically upright by legs or by steel wire ropes called *guy wires*. The *boom* is another long pole. It can move up, down, or sideways.

The derrick is set on the framework above ground level. The boom of a derrick reaches over the side of the steel framework. It raises and lowers a cable. The cable has a hook that can be lowered to the ground. Steel shapes are hung from the hook. Then the shapes are lifted into place.

As much of the framework as can be reached by the boom at that level is finished. Then the derrick is taken apart. It is moved up into position on the steel that has just been set. Work continues.

Workers called *riggers* hang the steel shapes on the hoisting (lifting) cable. Rigging must be done in a certain way. It must be done safely. And the pieces must be hoisted into the correct position. Then they can be fastened in place. See Fig. 35-5.

Bethlehem Steel Corporation

Fig. 35-5
Riggers attach steel shapes to hoisting cables and guide the pieces into place using a tag line.

Rigging requires special knowledge. Riggers must know about hoisting frames, cables, and ropes. They must also know special hand signals. These signals are important. Often the crane operator is out of hearing range. Riggers must use hand signals to tell the crane operator what to do.

Setting the Supporting Steel

Each part of a steel frame supports other parts. The bottom pieces of the frame must be set first.

Anchor bolts are set in concrete foundations. They are used to fasten steel frames to the foundations. Steel plates are welded to the undersides of the lowest *columns* (vertical pieces). These steel plates are called *base plates*. Holes drilled in the base plates match the positions of the anchor bolts.

Each piece is lifted by a crane. The base plate is set on the foundation. The anchor bolts must stick up through the holes in the base plate.

Base plates must be set at the correct height. This may be 25 to 50 mm above the foundation concrete. Also, the plates must be level.

Shims are used to level base plates. Shims are small, square pieces of steel plate or sheet. These may be different thicknesses. They are placed between the concrete and the bottom of the base plate. Shims are set under each corner of the base plate. They provide the thickness needed to make the base plate level. Then steel nuts are drawn down tightly against each base plate. The nuts lock the shims in place. They also anchor the column.

Setting the Connecting Steel

The connecting pieces of the frame are placed next. These horizontal pieces are called *beams*. Connecting pieces can also provide support.

Each piece of steel is shown by a number on the shop drawing. A shop drawing is used to guide the work. The pieces are numbered from the drawing. Ironworkers select the right piece by its number. A crane is used to lift each piece. Riggers use a *sling* to attach the piece to the crane. The sling is made of wire rope.

A length of fiber rope is tied to one end of the piece. This is a *tag line*. Look again at Fig. 35-5. An ironworker holds the tag line. It is used to guide the piece of steel into position.

Fastening Steel Shapes

Each ironworker carries a *spud wrench* on a leather belt. The spud wrench has jaws on one end. The jaws fit around a nut or the head of a bolt. On the other end is a handle about 305 to 455 mm long. It tapers down to a rounded point.

The holes in the end of each beam must match those in a connecting piece fastened to the column. An ironworker aligns and fastens the pieces. One hole in the beam is lined up with the correct hole in the connecting piece. The ironworker pushes the tapered wrench handle through the two holes. This also lines up the other holes in the two connecting pieces.

Each ironworker has bolts and nuts in a pouch hung from his or her tool belt. Two ironworkers are at opposite ends of the beam. Each ironworker puts a bolt through each hole of the connection. Nuts are put on the threaded ends of the bolts. At the same time, both workers use the jaws of their spud wrenches to turn the nuts tight. See Fig. 35-6. They make all connections in the same way.

The pieces or sections are fastened temporarily with bolts. After they are set straight, they must be bolted, riveted, or welded together permanently.

A *torque wrench* is used to bolt pieces permanently. A torque wrench operates somewhat like a

Fig. 35-6
Ironworkers must carry tools and supplies with them. Here a worker is using a spud wrench to tighten the nuts and bolts that connect the steel shapes.

very slow electric drill. It turns the nut down hard on the bolt. The two pieces of steel are pressed tightly together. This torque (pressure) is applied according to specifications. These are the written instructions that tell the builder how to do the job.

Occasionally, pieces are riveted together. Rivets look like bolts without threads. The smaller ends are hammered to form heads. Riveting is not used much today.

Many frames are welded together. Welding is done around the edges of the connecting pieces.

Plumbing and Bracing

Before pieces are fastened permanently, they must be set straight. This work is called *plumbing and bracing*.

Workers plumb a column by hanging a *plumb bob* on a string beside it. The face of the column is lined up with the string. If it lines up, the column is plumb. If it does not line up, the shims under the base plate may have to be changed. Shims are added or taken away until the column is plumb. See Fig. 35-7.

The beams must also be leveled. Columns may need to be raised or lowered to level the attached beams. A tool called a *level* is used to align the horizontal pieces.

The plumbed, leveled frame is braced (held in position) with guy wires. These may be fastened from the bottom of one column to the top of another column. Guy wires may also be fastened from a column to a stake. A long steel stake is driven into the ground. It is put some distance away from the column.

At this stage the first part of the framework is plumbed, leveled, and temporarily braced. Next, *grout* is forced in around the shims. Grout is a mixture of sand and cement. It is put between the base plate and the concrete foundation. It helps hold the base plate in place. Very little water is added to grout. Grout sets in much the same way as concrete. Once the grout has set, workers can finish putting up the steel frame.

Steel is set section by section until the top is set in place. When all steel is in place, the structure is said to be "topped out." The occasion may be celebrated. In the United States, an American flag is often fastened to the highest part of the frame.

Fig. 35-7
Columns and beams must be plumbed and braced.

Summary

Steel frames are assembled by ironworkers. A crane is used to set each piece in place. Riggers attach the pieces to the lifting cable. Steel shapes may be connected by bolting, riveting, or welding. First, the pieces are fastened temporarily. Then the steel frame is plumbed and braced with guy wires. Then it is fastened permanently.

Steel for the upper levels of tall structures is set with a derrick. The derrick is placed at the level of the last steel erected. It raises steel for the next level. Then the derrick is moved up, level by level. This continues until the steel frame is "topped out."

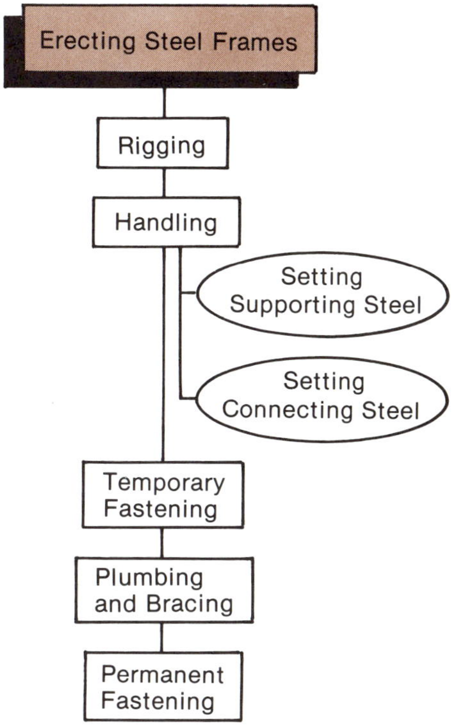

Terms to Know

- base plates
- beams
- boom
- columns
- crane
- derricks
- fasten
- grout
- guy wires
- handle
- ironworkers
- level
- mast
- plumb and brace
- plumb bob
- plumbing and bracing
- rig
- riggers
- shapes
- shims
- shop
- shop drawings
- sling
- spud wrench
- structural steel
- structural steel erector
- tag line
- torque wrench

Think About It!

1. Is there a steel-framed building being erected near your home or school? If so, briefly describe the work being done.
2. What are the basic kinds of work done by ironworkers? Try to name at least four.

Building Concrete Frames

Forms for Concrete Frames

Many buildings have *concrete frames.* Some parts of concrete frames may be manufactured at a plant. These are hauled to the site. At the site, they are lifted into place by a crane.

Most concrete frames are built at the site. See Fig. 36-1. First, the forms are built. See Fig. 36-2. Forms for frames are much the same as forms for foundations. Foundation forms were discussed in Reading 32. The same kinds of materials are used. Steel or wood forms are used most often. Concrete is placed into the forms. The inside surface of the formwork shapes the frame. After the concrete hardens, the forms are removed.

Materials for Concrete Frames

Concrete for superstructure frames is made in the same way as concrete for foundations. But there is one difference. Stronger concrete is needed.

The strength of concrete can be measured. It is measured by the weight of the heaviest *load* it can support after twenty-eight days. By that time, the concrete has cured. Some concrete frames must support 2000 pounds on each square inch. Others may need to support more weight. If so, even stronger concrete can be made.

To strengthen concrete, more cement is added to the mixture. That is, the proportion of cement to aggregate is increased.

Fig. 36-1
Forms are built to shape cast-in-place concrete frames.
Patent Scaffolding Company

Fig. 36-2
The form for a concrete column must be strong. Note how the walers fasten the wood panels tightly together. Walers and bracing provide support when the heavy concrete is placed inside the form.

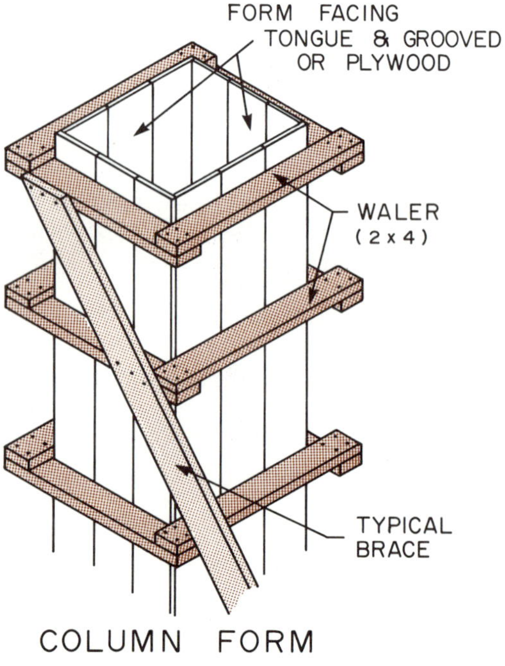

For foundations, large masses of concrete are used. For concrete frames, *columns, thin slabs,* and *beams* are used. These are much smaller in mass than the concrete used for foundations. As in foundations, reinforcing steel is used to keep the concrete from cracking. However, more steel is used in frames than is used in foundations.

Concrete Columns and Walls

Vertical supports for concrete frames are columns or walls. As you know, steel columns are fastened on concrete foundations with anchor bolts. *Concrete columns,* too, are anchored (fastened) to the foundation. But they are anchored in a different way. Pieces of reinforcing steel, called *dowels,* are set in the foundation concrete. See Fig. 36-3. They are placed where the column will be built. The dowels extend above the surface of the foundation. The surface around them is left rough. This will help the concrete in the column to bond (stick) to the concrete in the foundation. Long steel reinforcement bars for the column are tied to the dowels. The bars are held in position by *hoops* of smaller steel bars. Look again at Fig. 36-3.

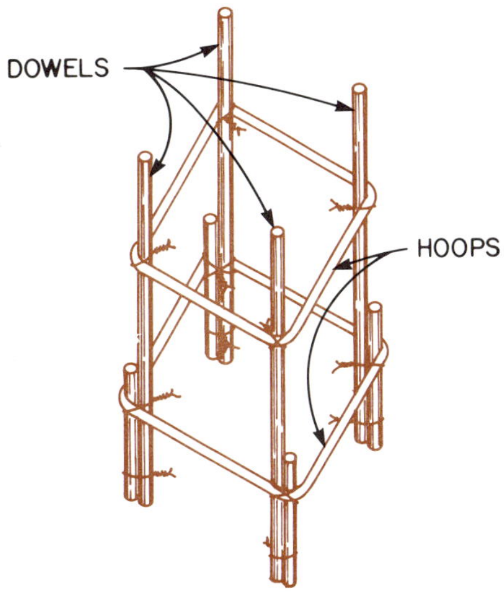

Fig. 36-3
Vertical steel bars are used to reinforce concrete columns. These bars are tied to dowels set in the foundation.

Building Forms for Columns and Walls

Columns may be square, rectangular, or round. The sides of forms for square or rectangular columns are made of panels. Small pieces of *waler* fasten the panels together. Look again at Fig. 36-2. The panels overlap at the corners. They are nailed together. Metal bars called *column clamps* may be used to fasten column form panels. Sometimes, *form ties* are used. Then the forms are plumbed and braced. Round columns are set in curved metal forms like those used for piers. (See Reading 32.)

Concrete walls are built in the same way as foundation walls. Forms with two parallel facings are built. This means these facings are the same distance apart at all points. When the wall is poured, its thickness will not vary.

Column and wall forms are built to the level of the beams. Then beam forms are built.

Concrete Beams

The connecting pieces for a concrete frame are called beams. These may be any size or shape. They may connect two columns or two pieces of wall. Or, they may connect one column and one piece of wall.

Providing Support for Poured Beams

Concrete foundations are supported on bearing surfaces or footings. But concrete beams are above the ground. Supports must be provided for the beams until the concrete hardens. Then the concrete can support itself.

You learned earlier that *shores* are diagonal supports. Shores are also used to support overhead concrete forms. But these shores are vertical posts. Usually, they are made from *dimension lumber.* Some shores are made from pieces of metal pipe. Each of these has a long threaded bolt at one end. This makes it possible to adjust the length of the shore.

Shores are spaced to support the load of the form and concrete. The bottom of the shore may rest on the ground. Foundation concrete or another beam may support it.

A piece of lumber is nailed across the top of each shore. It is called a *T-head.* It is fastened at a right angle to the beam. The result looks like a "T." The top piece is 635 to 762 mm longer than the width of the bottom of the beam. It is braced to the shore with diagonal braces. Carefully study Fig. 36-4.

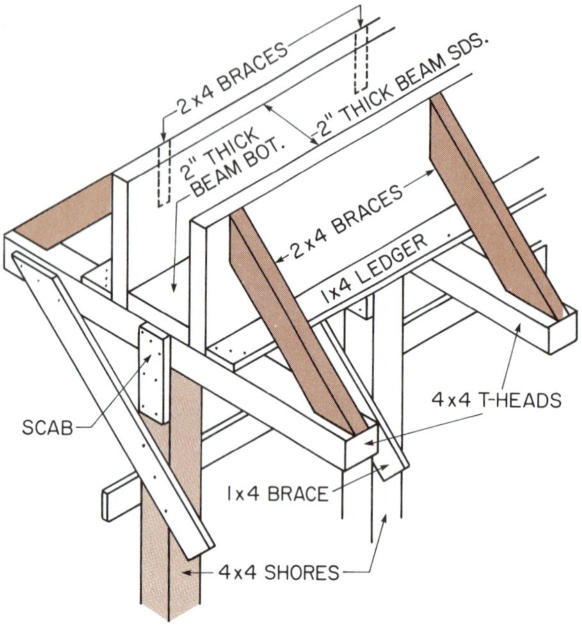

Fig. 36-4
Forms for horizontal concrete beams require bracing and shoring. The forms support the heavy concrete until it cures. Notice how the form is constructed.

Building the Beam Form

The form for a concrete beam is made with three pieces of dimension lumber. One piece is used for the beam bottom. Two other pieces are used for the beam sides.

After the beam form is made, a bar of reinforcing steel is set in it. This bar is called a *trussed bar*. See Fig. 36-5. It is bent at a 45° angle near the ends. The central part of the bar is placed near the bottom of the beam. The bent ends are placed near the top of the beam. A large hook is on the end of each bar. It hooks into the reinforcing steel in the column. Trussed bars may be fastened together with smaller bars, like the hoops in a column.

Suspended Slabs and Slab Forms

A concrete frame for a structure may be built just of columns and beams. Or, a concrete floor slab can be made as part of the beams. If this is done, a supported form must be built for the slab. This form is built between the beam forms. See Fig. 36-6. It is called a *slab form*.

A slab form is built in several steps. Look again at Fig. 36-6. First, shores are placed in rows about 1200 mm apart. Then a wooden beam called a *stringer* is laid on top of them. Other pieces are placed on edge and nailed to the stringers. These pieces are called *joists*. Sheets of plywood are nailed over the joists. These form the facing for the underside of the concrete slab. Where the slab meets the beam, the side of the beam form is cut down. This allows room for the plywood to be nailed to the top.

Next, straight bars of reinforcing steel must be set for the slab. They are set at intervals (spaced apart) on the plywood. The intervals are about 100, 150,

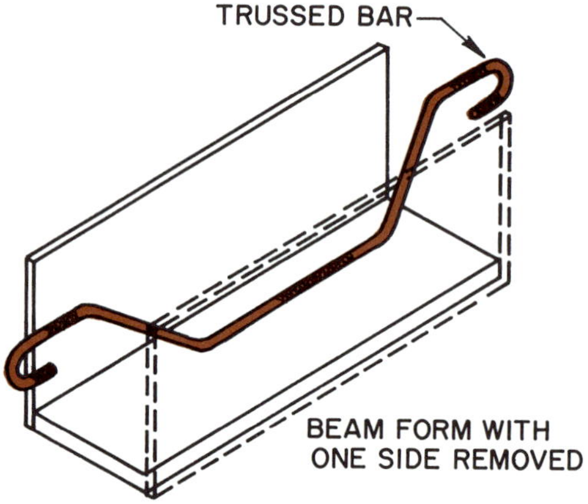

Fig. 36-5
A trussed bar is set in the beam form. The hook on the end of the bar will hook into the reinforcing steel in the columns.

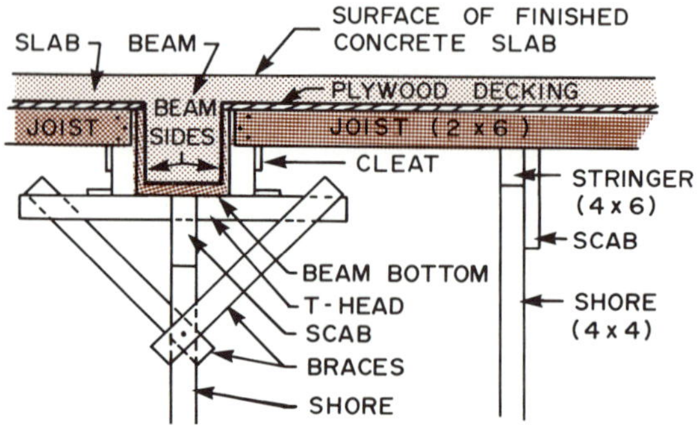

Fig. 36-6
Slabs and beams are typically cast in place together.

or 200 mm in both directions. Where the bars cross, they are fastened together. They are then set on small wire supports called *chairs*. The chairs rest on the surface of the plywood. Placing the steel bars on chairs keeps them above the bottom of the concrete slab.

Placing the Concrete

Concrete for the frame of a structure is placed in the forms. This is done in much the same way as placing concrete for a foundation.

Beams, slabs, and columns or walls are formed together. First, the concrete for the frame is placed in column forms. Concrete shrinks (gets smaller) as it sets. Therefore, concrete for the frame is placed in forms shortly before concrete for beams and slabs is placed in forms. This gives it time to shrink before the other concrete is placed.

If the structure is not very high, a crane may be used for placing concrete. The crane must have a long boom. Concrete is moved in a *concrete bucket*. The bucket is hung from a hook on the crane's lifting cable. See Fig. 36-7.

For taller structures, a *hoist* may have to be used. A hoist is like an elevator in a tall building. It is a steel framework set against the side of the structure. An enclosed platform is attached to the steel framework. Cables raise or lower the platform. Hoists can stop at any level of the structure. A *concrete buggy* is filled with concrete. It is pushed into the hoist cage at ground level. Then the cage is raised to the level at which the concrete is needed. The buggy is removed from the cage. It is pushed over *runways* to the location where the concrete will be placed.

On many jobs, concrete is moved through hoses. See Fig. 36-8. Concrete can be moved thirty meters or more above the pumping unit.

Fig. 36-7
A crane lifts a large bucket of concrete into position above a form.

Finishing Concrete Frames

Some concrete surfaces in structural concrete frames are left exposed (without a covering). They are finished in the same way as foundation concrete. A steel trowel is used to give slabs a smooth finish. Then tile or other flooring can be put over them. Column surfaces will usually be covered by other materials. The undersides of concrete slabs and many beams are usually covered, too. Therefore, they do not need to be finished by rubbing or troweling.

Fig. 36-8
Concrete can be pumped through hoses. Hoses are especially useful for placing concrete in high or other hard-to-reach places.

Removing the Forms

Foundation forms may be removed from the foundation concrete within a day or two. This is not possible for beam and slab concrete. Concrete above ground must be strong enough to support its own weight. This takes one or two weeks. Then forms can be removed.

Concrete used in foundations or frames must be cured. It is kept moist by sprinkling or coating with curing compounds. Coating reduces evaporation. Concrete takes a month to cure fully.

Precast Concrete Shapes

As you learned earlier, some parts of concrete frames are made at manufacturing plants. See Fig. 36-9.These pieces are called *precast concrete shapes.*

Some precast concrete shapes are called *prestressed concrete.* Most concrete is reinforced with reinforcing rods. Prestressed concrete is not. It is reinforced with strong wires. The wires are stretched very tightly in the forms. Then the concrete is placed in the forms, around the wires.

Prestressed concrete is more expensive than concrete reinforced with rods. But it is also stronger. The stretched wires inside the concrete try to draw back to their original shape. This creates a pulling force (stress) inside. This internal (inside) force helps the concrete resist pressures from outside.

Precast or prestressed concrete beams are often used for highway bridges. The beams are placed side by side on the foundation. Then road paving is placed over them.

Precast or prestressed planks can be used instead of concrete slabs. Concrete planks are 610 or 914 mm wide. They may be supported by concrete beams. Or, they may be supported by beams in structural steel frames. Sometimes planks are supported by masonry walls.

LaBarge Incorporated

Fig. 36-9
Precast and prestressed concrete shapes are set in metal forms at manufacturing plants. After the concrete has cured, the shapes are transported to construction sites.

Summary

Concrete frames for superstructures are usually built on the site. Some parts may be manufactured at a plant. Materials used for concrete frames are like those used for foundations. The difference is that the concrete for frames is stronger.

Supports for concrete frames are usually columns. They are connected by beams. The beams rest on top of the columns.

Columns are made in the same way as foundation piers. Forms for concrete beams and slabs are supported by shores. Shores support the forms and concrete until the concrete is hard enough to support its own weight. Forms for structural frames must remain in place for one to two weeks.

Concrete buckets lifted by cranes are used to place concrete in forms. Concrete buggies lifted by hoists also may be used. And concrete can be pumped through long hoses.

Precast concrete shapes have many uses. They may be used as beams for highway bridges. Or, they may be slabs for steel or concrete frames. Precast concrete shapes are also used as T-beams for roofs or structures. Some precast concrete is prestressed. Prestressed concrete is reinforced with strong wires under tension.

Terms to Know

beams
chairs
column clamps
columns
concrete bucket
concrete buggy
concrete columns
concrete frames
dimension lumber
dowels
form ties
hoist
hoops
joists
load
precast concrete shapes
prestressed concrete
runways
shores
slab form
stringer
T-heads
thin slabs
trussed bar
waler

Think About It!

1. Why do you think the concrete used in concrete-formed superstructures must be stronger than the concrete used in foundations?
2. What are some advantages of using precast concrete shapes? In the future, do you think more or fewer precast structural members will be used?

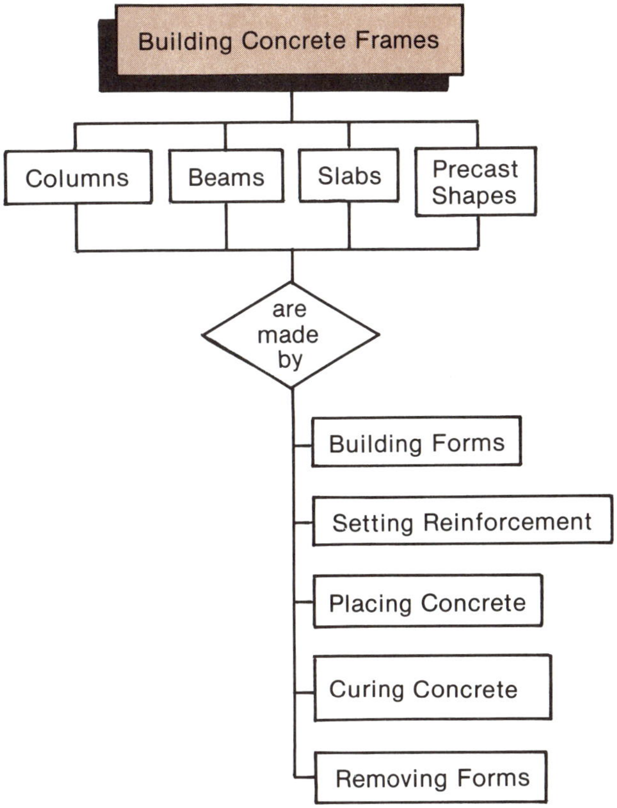

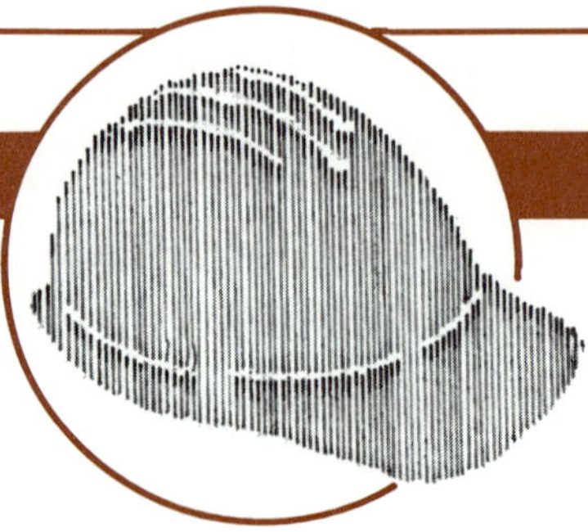

Reading 37

Building Wood Frames

Framing Members

Wood frames are commonly used for small buildings such as houses. A wood frame (skeleton) has three major parts. It has floor framing, wall framing, and roof framing. Most floor framing consists of horizontal members (pieces) called *joists*. Most wall framing is constructed of vertical members called *studs*. And most roof framing is made of sloped members called *rafters*.

Years ago, labor and lumber were plentiful. Framing was made of heavy *timbers* (4 × 4s, 6 × 6s, 8 × 8s). Timbers are large pieces of wood. Because nails were scarce (hard to find), timbers were joined with wood pins.

Today, labor and lumber are more expensive. Modern framing members are smaller pieces of lumber (usually 2 × 4s, 2 × 6s, and 2 × 8s). And nails are plentiful. They are used to join the pieces of lumber.

Fig. 37-1
An anchor bolt set in a foundation wall.

Building the Frames

There are five major steps in erecting wood frames at the site.

1. Laying out and marking locations and positions of framing members. This is done according to the working drawings.
2. Marking off the lengths of lumber.
3. Sawing the lumber.
4. Assembling the parts.
5. Leveling and plumbing the frame.

Setting Sills

The wood framing must be secured to the foundation. This is done with *anchor bolts*. They are set in concrete along the center of the top of the footing, slab, or wall. One bolt is set about every 1525 or 1830 mm. The bolt's threaded end is pointed up. See Fig. 37-1.

Once anchor bolts have been set, construction of the frame can begin. A *sill plate* is made to be placed on top of the foundation. See Fig. 37-2. The sill is a horizontal piece of lumber. It is the bottom frame

Portland Cement Association

Fig. 37-2
A sill plate is fastened to the foundation wall. It is placed over the anchor bolts and secured by nuts.

member. Holes for the anchor bolts are marked and drilled in the sill. A thin layer of *grout* is placed on top the foundation. Grout is a mixture of sand, cement, and water. It is used to level the sill. The holes of the sill are lined up over the anchor bolts. The sill is placed down over the bolts on top of the grout. Then nuts are fastened to the threaded ends of the anchor bolts. This holds the sill in place.

Termite Shields

In some areas, termites (wood-eating insects) are a problem. They can eat their way through wood framing. The structure becomes weak. Repairing framing members weakened by termites can be very expensive.

Termites may get to the wood through cracks in the concrete. Or, they may build a mudwalled tunnel around the concrete.

In areas where there are termite problems, a *termite shield* is used. It is made most often from a strip of sheet copper. The metal is placed on the foundation before the sill is attached. See Fig. 37-3. The shield is placed the full length of the sill. It extends over the sides of the foundation. The termites cannot eat through metal. The shield prevents them from burrowing into the wood frame.

Assembling Joists

Joists are planks set on edge. They provide a base upon which flooring materials can rest. Joists rest on sills. They extend between opposite foundation walls. Joists look like floor beams. But they are laid much closer together. Generally, joists are set sixteen inches *on center* (16″OC). See Fig. 37-4. Joists are usually 2 × 6s, 2 × 8s, 2 × 10s, or 2 × 12s.

Headers are nailed to the ends of joists. They hold the joists together. Headers are pieces of lumber of the same dimensions as joists. Headers may be set flush (even) with the outside edge of the sill.

Bridging holds the center of the joists firm. See Fig. 37-5. It keeps them from bending or twisting. Bridging connects the joists. And it helps spread a heavy load. A load placed on one joist will be shared by the joists connected to it.

Diagonal bridging (at angles) is usually made from 1 × 4 boards. Bridging may also be made of short lengths of studs or joists. These may be used for diagonal bridging, too. Or, they may be used for solid bridging. Look again at Fig. 37-5. This is an example of solid bridging. Specially made pieces of metal may also be used to make bridging.

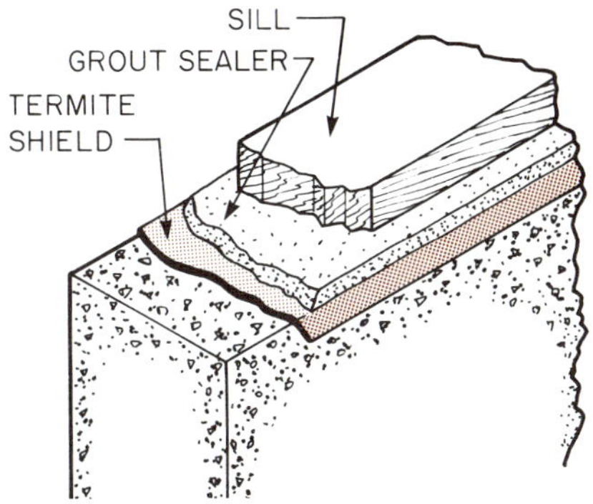

Fig. 37-3
Sheet metal may be placed under the sill as a termite shield. Termites cannot get through the shield to weaken or destroy the wood frame.

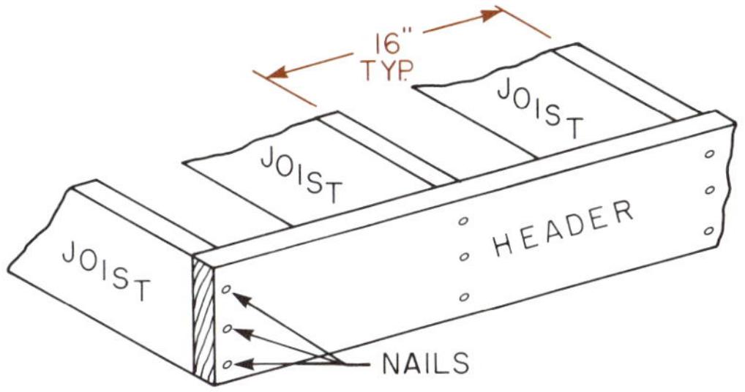

Fig. 37-4
Joists provide support for the floor.

Fig. 37-5
Bridging holds the joists in place.

Assembling Girders

Sometimes there is a great distance between foundation walls. The joists need more support. A *girder* is used to support them. A girder is stronger than the joists. It can safely span a greater distance. In a wood frame, a girder may be a steel I-beam. Or, a girder can be made by nailing or bolting wood planks together.

Girders may rest on sills, *foundation posts*, or steel pipe columns called Lally® columns. See Fig. 37-6. Sometimes joists rest on top of the girders. Or, girders are placed at the same level as joists. Then the joists are butted against the girder. This means the ends are placed flat against the girder.

If wooden girders are used, *ledgers* (wood strips) are nailed to the sides of the girders. Then joists are set on the ledgers. Joists can also rest in special metal hangers. The joists are *toenailed* to the girder. See Fig. 37-7.

When steel girders are used, planking is bolted to the top of the girder. Joists may rest on planking. See Fig. 37-8. They are toenailed to the planking.

Joists also may be butted to steel girders. A ledger is bolted to the face of the girder. The joists are notched to fit the girder and ledger. Then the joists are toenailed to the wood ledger.

Fig. 37-6
Girders support the floor joists. Girders are supported by foundation walls or posts or by Lally® columns.

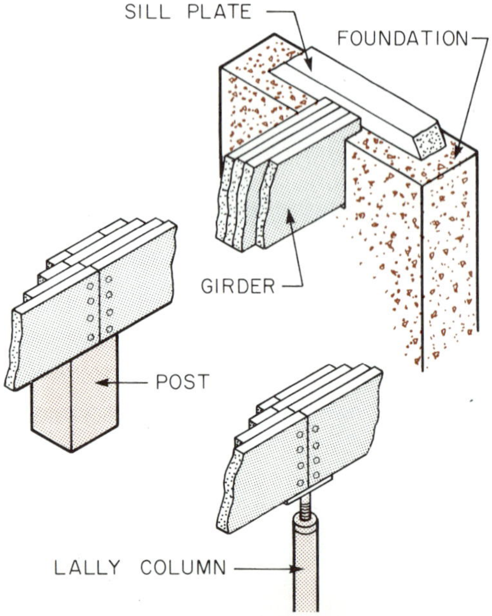

Fig. 37-7
Assembling wooden girders.

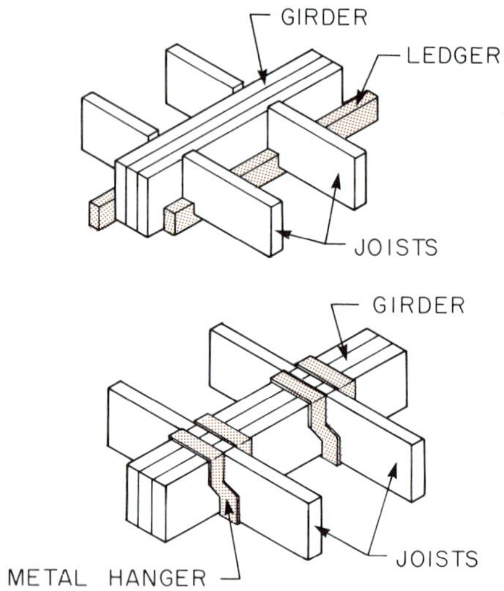

A. *Joists can be set on ledgers or in metal hangers.*

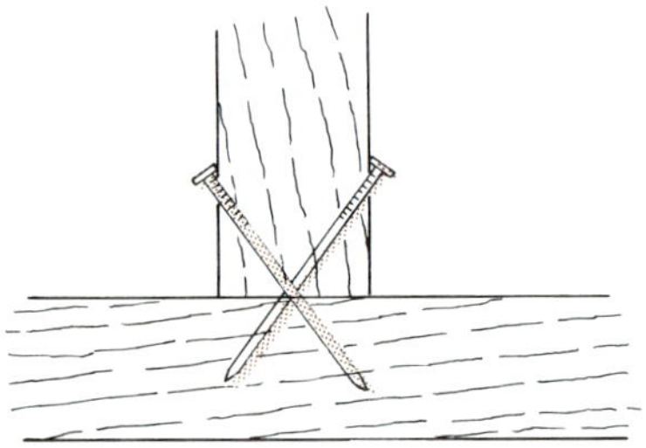

B. *Joists are fastened to the girder by toenailing. Nails are driven at an angle into both pieces.*

Fig. 37-8
Joists are toenailed to planking when steel girders are used.

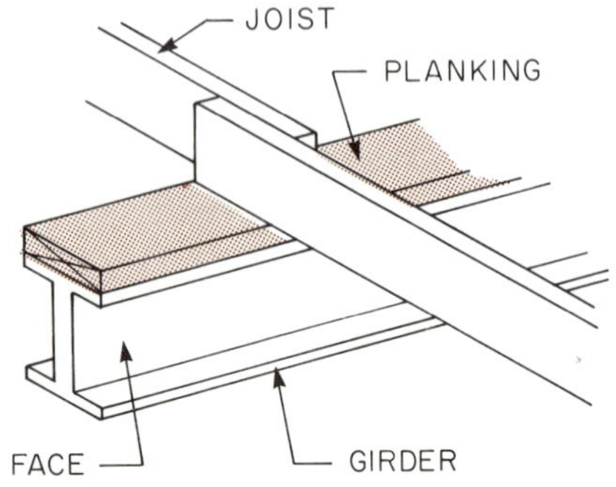

Laying Subflooring

Next, a *subfloor* is nailed to the top of the joists. It covers the joists and makes them rigid. It is also called rough flooring. A second floor is always placed on top of the subfloor. Boards or wooden sheet materials are used for subflooring. If boards are used, they are laid across the joists on a slant. Sheet materials are most often used for subflooring. They are laid perpendicular (at right angles) to the joists. See Fig. 37-9.

Assembling Wall Frames

After the subflooring has been placed, walls for the first floor are laid out. The exterior walls will be put up directly on top of the outside foundation walls. Interior walls are usually put up over inside foundation walls or girders. First-floor walls are usually framed with 2 × 4 lumber. This lumber does not measure a full 2″ × 4″. When dressed (finished), it measures 1½″ × 3½″. But pieces this size are called 2 × 4s.

A piece of 2 × 4 lumber is nailed flat to the subflooring. It is nailed down exactly where the wall is to be. This piece of lumber is called the *soleplate*.

To complete the wall framing, studs are cut to the needed length. Studs are pieces of 2 × 4 lumber. They are stood on end along the soleplate. Like joists, they are spaced 16″ OC. Then they are toenailed in place. A 2 × 4 is nailed to the top of the studs. This is called a *top plate*. A second top plate is nailed over the first. Care is taken that the joints in the second plate are not directly above the joints in the first plate. Staggering the joints makes the double plate like one continuous piece. See Fig. 37-10.

There is another way to build a wall. The wall frame is assembled on the subfloor. Then it is tilted up and nailed in place. The double plate is nailed on. The double plate overlaps the adjoining wall. This ties the walls together and stiffens them.

Bracing

Bracing the studs of the wall frame keeps them from bending or twisting. There are several common ways to brace studs. One way is with *fireblocks*. These are short lengths of 2 × 4 lumber. Fireblocks are placed horizontally between the studs. They are also known as *draft stops*. They keep out drafts. This slows the spread of fire in a burning building. Fireblocks may be nailed in line (in a straight row). Or, they may be staggered above and below a line. This makes nailing easier.

Another method of bracing studs is by using *diagonal bracing*. Braces run from the top corners of the frame to points on the soleplates. The braces are at an angle of about 45°. This makes a

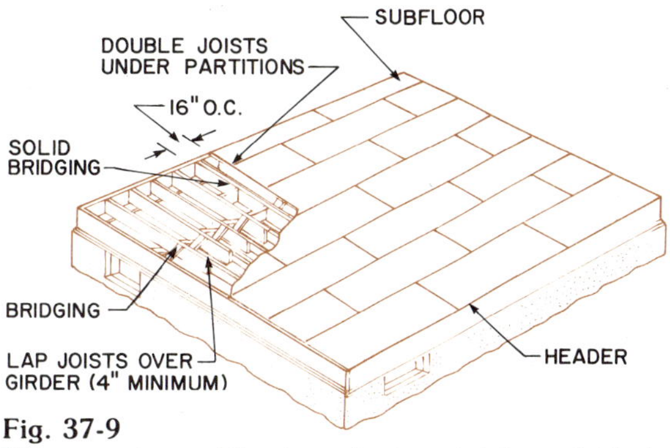

Fig. 37-9
When used as subflooring, sheet materials are laid at right angles to the joists.

Fig. 37-10
Construction techniques can make structures stronger. Staggering the joints of the top plates is an example.

Fig. 37-11
The plywood bracing on the corners of this home adds strength to the structure. Note that this home is completely covered with insulation sheathing.

triangular frame within the wall section. It makes the frame rigid and solid. Notches are cut in the studs. These notches are the thickness of the diagonal brace. The brace fits inside them. Then it does not interfere with the installation of sheathing material.

Today, plywood is commonly used for bracing. A sheet of plywood is placed on each corner of a building. See Fig. 37-11. Sometimes the entire exterior of the frame is covered with plywood. This makes it very strong.

Sheathing

Material used to cover exterior wall frames is called *sheathing*. Boards or manufactured sheet materials, such as plywood, can be used for sheathing. Rigid insulation is often used. Sheathing is nailed to the wall studs. It makes the frame more rigid. Finished siding material can be attached to the surface of sheathing.

Building Additional Stories

A second story may be built in the same way as the first story. Joists and headers are placed on the top plate of the first story. They are covered with subflooring. Walls are built on top of this. Additional stories may be added in this same way.

Assembling Roof Frames

Most wood-framed buildings have sloped roofs. The top of a roof's slope is called the *ridge*. The bottom of the slope is called the *eave*.

The roof is framed using rafters. These are pieces of lumber set on edge. Rafters commonly are 2 × 4s, 2 × 6s, or 2 × 8s. They rest on the double plates and meet at the ridge. A notch is cut where the rafter rests on the double plate. This is done so that there will be a flat edge resting on the double plate. Usually the rafter goes a foot or more past the exterior of the wall. This holds the roofing that goes beyond the outside wall.

Rafters are set in pairs. They slope down each side of the roof. Each pair meets at a *ridgeboard* along the top. A ridgeboard provides a straight line that helps brace the rafters. If the rafters are long, they are braced by *collar beams*. See Fig. 37-12. A collar beam is a piece of lumber nailed between a pair of rafters. It is placed about halfway down the rafters.

Instead of conventional rafters, trussed roof members may be used. A roof *truss* is a triangular piece. It is usually made of a combination of rafters and ceiling joists. A truss may be made on the site. Or, it may be manufactured off the site. It is installed as a unit. See Fig. 37-13.

Roof frames are covered with *roof decking*. See Fig. 37-14. Roof decking serves the same purposes as sheathing and subflooring. It covers the roof frame and makes it rigid. Roof decking may be boards or sheet materials. These are nailed over the rafters. They completely cover the rafters except when wood shingles are used. Then space is left between sheets of decking. This saves material. And it allows better air circulation. Air circulation is needed to dry the shingles after wet weather.

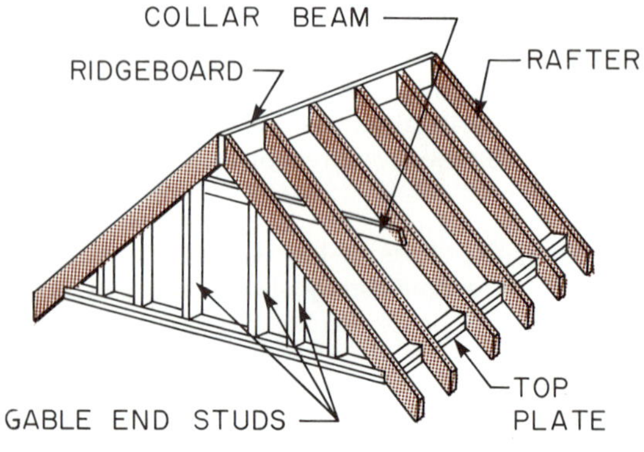

Fig. 37-12
A sloped roof and its framing members.

Fig. 37-13
A roof truss is a combination of roof members. It is very strong and can be installed quickly.

Fig. 37-14
This worker is installing roof decking over the rafters.

Summary

Wood framing is used for small buildings such as houses. Precut dimensional lumber or manufactured sheets are brought to the site. They are used as framing materials. The main framing members are joists, studs, and rafters. Refer to Fig. 37-15.

Anchor bolts are set in the foundation wall. Then a sill is laid over them. This begins the framing process. Floor joists are laid on sills. These joists cross the space between the walls. They are braced with bridging. Girders provide support. They make it possible to use shorter joists. Subflooring is placed over the joists.

To frame a wall, the soleplate is nailed to the subfloor first. Then studs are nailed to the soleplate. Next, a top plate is nailed on top of the studs. Studs may be braced with fireblocks or diagonal bracing. After the wall frame is put up, sheathing is applied. Sheathing covers the exterior walls and makes them rigid.

Roofs are framed with rafters or trusses. Roof decking is nailed over the roof framing.

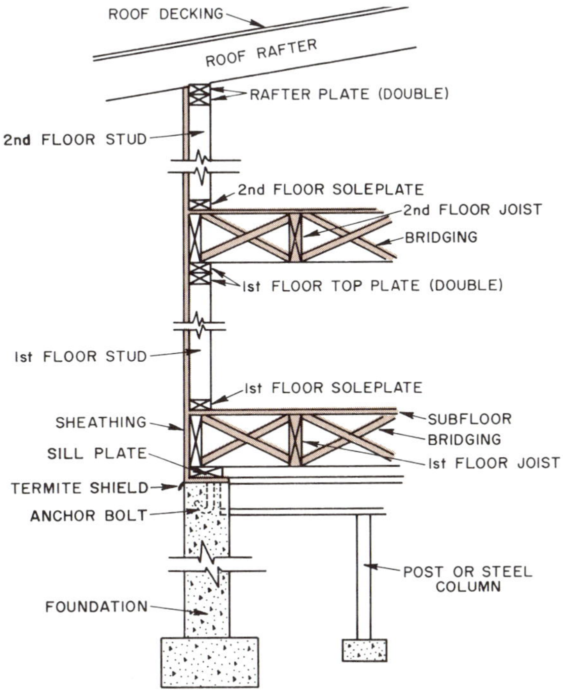

Fig. 37-15
A cross section of a house and its framing members.

Terms to Know

anchor bolts
bridging
collar beams
diagonal bracing
draft stops
eave
fireblocks
foundation posts
girder
grout
headers
joists
ledgers
on center (OC)
rafters
ridge
ridgeboard
roof decking
sheathing
sill plate
soleplate
studs
subfloor
termite shield
timbers
toenailed
top plate
truss

Think About It!

1. Why is wood framing used for structures such as houses? Are wood-framed structures common in your neighborhood? If not, why not?
2. In the future, wood-framed structures may be built completely in a factory and transported to the site. What advantages, if any, can you see in such a plan? What disadvantages?

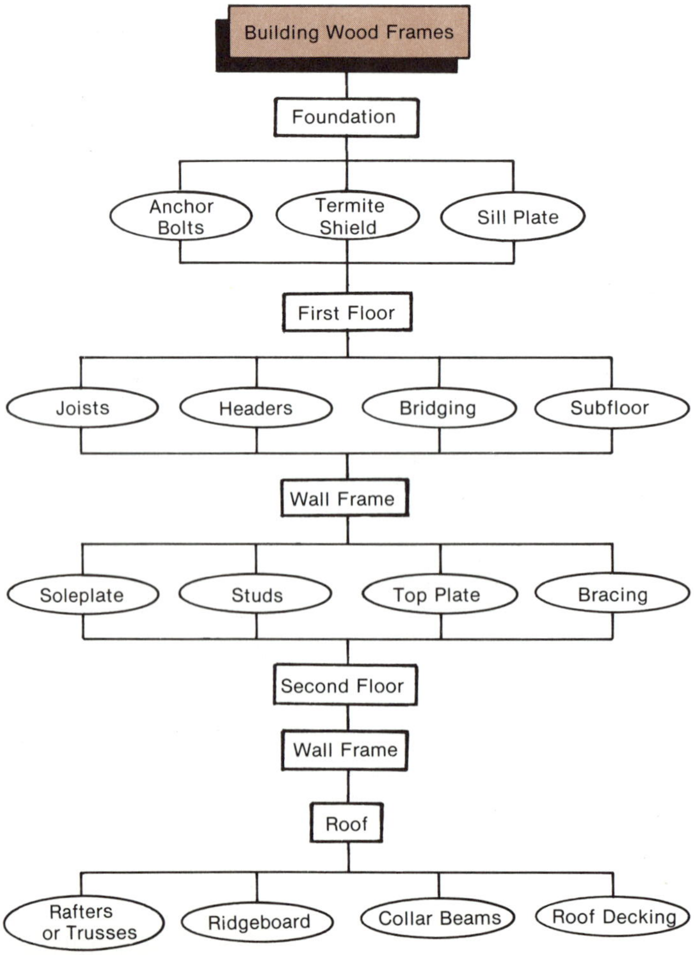

Reading 38

Installing Utilities – Outside

Utility Systems

Utilities are systems for providing water, sewage disposal, gas, electricity, and telephone service. Utility systems are constructed. For example, many construction projects are required in building a city water system. Pumping stations must be built. They bring water from its source. Treatment plants are needed to purify water. Water towers are required for water pressure and storage. And networks of pipes must be built to carry water throughout the city. Utilities are also called *services*.

Almost all constructed works require utilities. For example, tunnels must be lighted and ventilated. Highways need traffic controls and drainage. A bridge often needs lighting. And controls may be needed to raise or open the bridge.

Utilities are installed in many kinds of structures. However, most utility installation work is done in buildings. Utility installation is often the major cost of building construction. In other readings you will learn about installing utilities inside buildings. This reading is about installing *outside utility systems*. These include cross-country and city networks. These networks make utilities available to customers. See Fig. 38-1.

Engineers design outside utility systems. A variety of contractors and skilled workers install the systems. Often, engineers and work crews are employed by the utility company.

Fig. 38-1
Utility systems provide people with needed services. Utility lines are often run underground.

Water Systems

Water is usually drawn (taken) from rivers, reservoirs, or wells. It is pumped through a *pipeline*. When the water reaches a *treatment plant*, it is purified. Then it is piped through *water mains*. Water mains are large pipes. See Fig. 38-2. They

Fig. 38-2
Checking the size of precast mains.

Lone Star Industries, Incorporated

Fig. 38-3
People should know where the entrance lines to their homes are. Then, in an emergency, they can shut off the water quickly.

branch into all parts of a *service area.* They supply water to people in homes or businesses. Water mains also supply fire hydrants.

Water is pumped through the mains. Or, in some locations, it flows by gravity. The pipes are usually run under the streets. Pipelines and mains are installed by *pipefitters.*

Plumbers tap the main to supply users with water. This means that they bore a hole in the main. A smaller pipe leads from the tap to the user. This pipe is called the entrance line.

Usually, meters and shutoff valves are placed in entrance lines. See Fig. 38-3. Meters record the amount of service provided.

Sewage Systems

A *sewerage* is a system of drainage lines or pipes. These pipes or lines carry off sewage (waste water) or excess rain. Sewage usually flows by gravity. Sometimes, however, it must be pumped. A branch of the system is called a *sewer.* The two types of sewers are sanitary sewers and storm sewers.

Sanitary Sewers

Sanitary sewers carry sewage from buildings. Entrance lines are tapped into these mains. See Fig. 38-4. Mains usually lie under the streets. Or, they are placed underground beside the streets.

Sanitary mains lead to a sewage treatment plant. There the waste water is purified and discharged. The *effluent* (discharge) usually flows into a stream.

Sewers are not usually metered. Most often, users pay a fixed rate for sewage removal.

Storm Sewers

Storm sewers carry away storm sewage (water from the streets). Drains are installed in low areas along streets.

Unlike sanitary sewage, storm sewage is not usually treated. It flows directly into natural drainage systems, such as streams.

Electrical Systems

Electricity is a form of energy. It involves the movement of electrons. Electrons are the smallest particles of electricity. Ordinary electric current is a controlled flow of electrons. They move through a conductor, such as a wire.

Power plants generate (produce) electricity. They change other kinds of energy to electric energy. For example, falling water can be used to create electric energy. Fossil fuels, such as coal, petroleum, and natural gas, are burned to create energy. Atomic fission (the splitting of atoms) is another source of energy. See Fig. 38-5.

Voltage is the measure of electric force. This force pushes electric current through wires. Over a hundred thousand volts of electric current are transmitted (sent) from a power plant. See Fig. 38-6. Cross-country lines (wires) are held up by high towers. Near a city, the lines enter a substation.

Fig. 38-4
Lines from buildings are tapped into sewer mains.

Public Service Electric and Gas Company

Fig. 38-5
A nuclear power plant produces electric energy.

Here *transformers* reduce the electric force in the line. The amount reduced varies. The power is distributed through wires.

Wires may be strung on poles or buried underground. Underground wires are usually placed in a duct or *conduit*. Conduit is piping. It protects the wires. Conduit may be made of metal, plastic, concrete, or tile.

Near users, transformers in another substation reduce the voltage again. Most homes, stores, farms, and industrial plants require only 240 volts. Large industrial plants may require higher voltage.

At the building, the utility company installs the service. The "drop" (from wires above ground) or lateral service (from wires below ground) is made up of three wires. These lead from a small transformer to the meter. See Fig. 38-7. From the

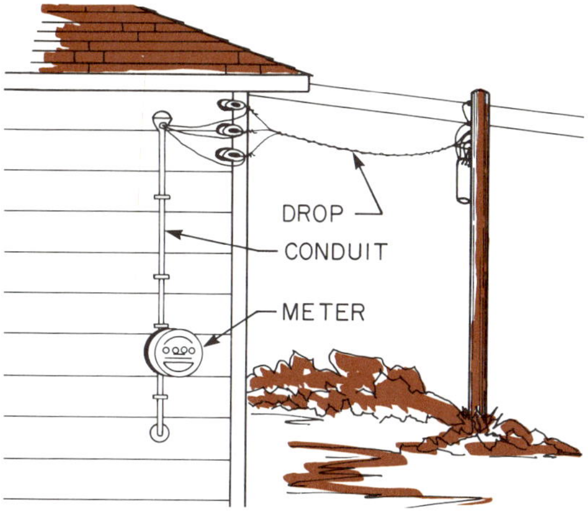

Fig. 38-7
Before electricity reaches a building, it enters a transformer. There, voltage is reduced. The drop is three wires that lead from the transformer to the building.

Fig. 38-6
Electricity is produced near a source of power. Then it is transmitted from the power plant through high-voltage transmission lines. The high voltage is reduced at transformer stations. It must be reduced again for use in homes and industry.

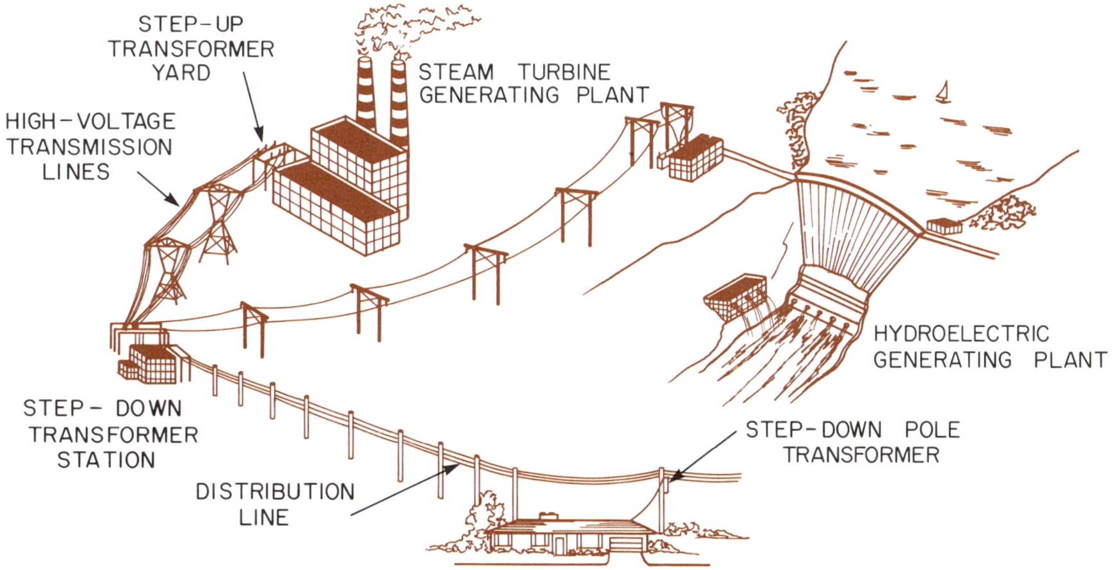

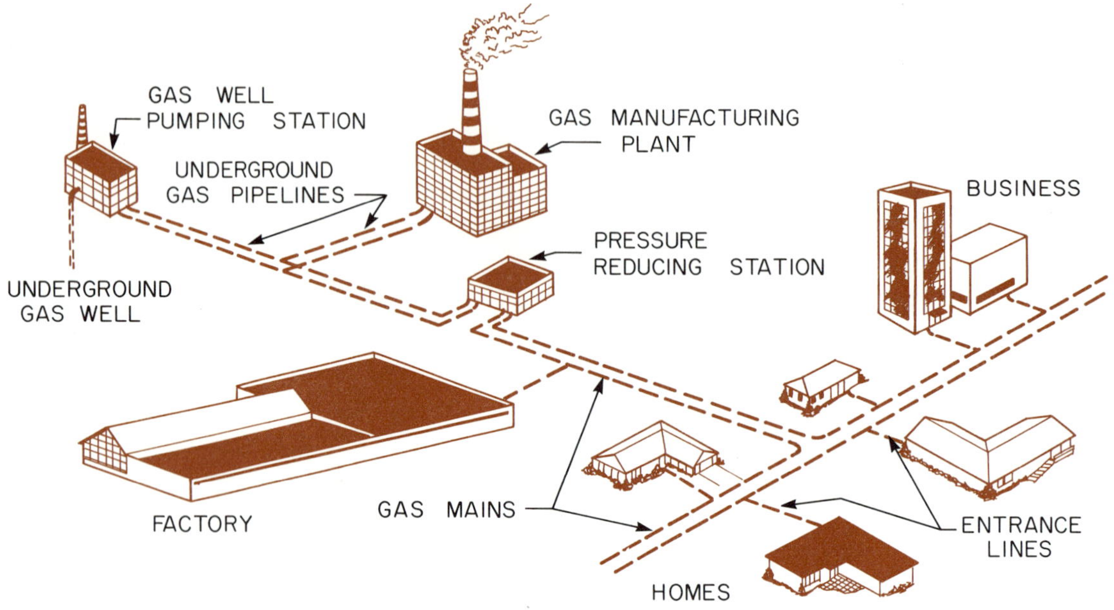

Fig. 38-8
Gas for fuel is pumped from wells or manufacturing plants into pipelines. Pressure in the pipeline is reduced in stations near users. Also, a pressure reduction valve and a shutoff valve are installed close to each meter.

meter, the wires are connected to a fuse or circuit breaker box. Circuit breakers are safety devices. Suppose, for example, that several electric appliances are run at the same time on one circuit. The electricity needed by the appliances may be too great for the circuit to carry safely. If so, it will overload and begin to heat up. The overload will cause the circuit breaker to open. This stops the flow of electricity. A circuit breaker prevents fires and damage to the circuit.

Cross-country and city line work is done by employees of the electrical company. These linepersons are responsible for work leading to the meter. The user hires other electricians (wirers) to install the inside circuits.

Gas Systems

Most fuel gas exists in a natural form underground. Some is manufactured from substances such as coal and oil. High-pressure pumps force the gas from the source through pipelines.

Near cities, pressure is reduced. Gas is then distributed through gas mains. See Fig. 38-8. Gas mains are placed under the streets. To serve users, a gas company employee taps a main. An entrance line connects the tap to each meter. The pressure is reduced again near the meters. Shutoff valves are placed in the lines. Users should know where these valves are.

Telephone Communications Systems

A set of wires connects each telephone to a system. See Fig. 38-9. These wires enter a building through a *drop*. The drop is installed from a connection at a nearby telephone cable. Cables may be installed above the ground on poles. Or, they may run underground through conduits.

Many cables lead to a telephone exchange (office). At the exchange, cables are connected to relays (switches) and dialing equipment. Calls are processed (handled) through operators or through automatic equipment. People at different central telephone exchanges may need to communicate to complete calls.

Telephones are usually not metered. Users pay a fixed rate for the service. Extra charges are made for long-distance calls. Also, users are responsible for the telephone lines on their property.

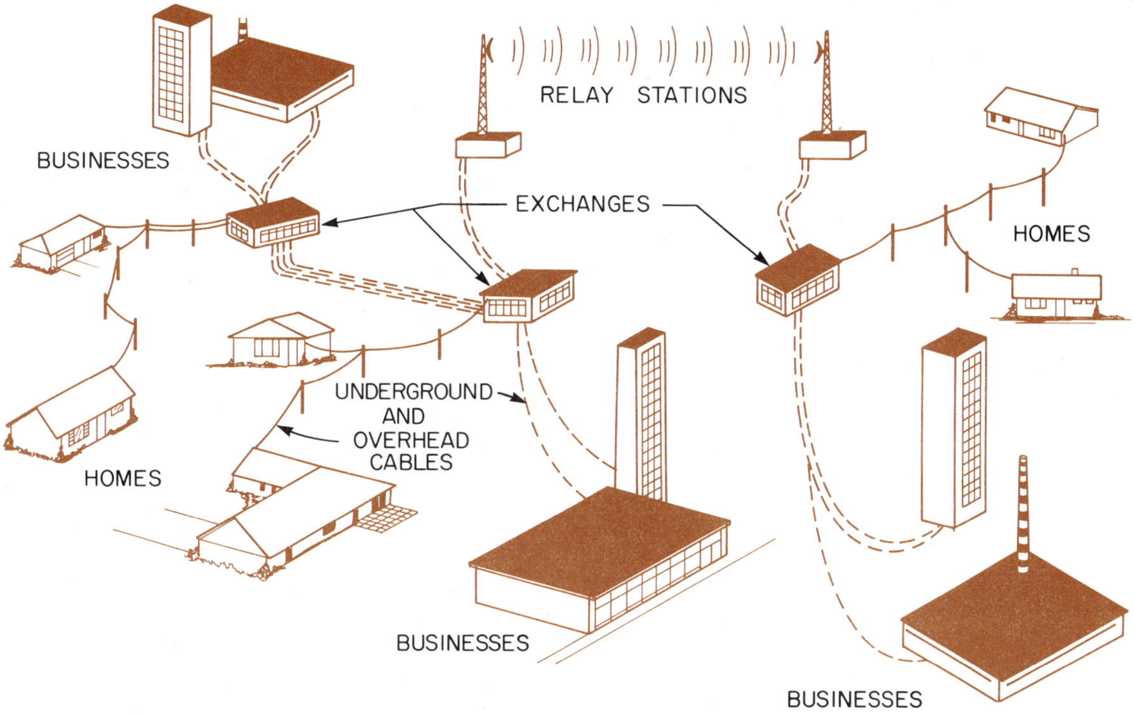

Fig. 38-9
Your telephone is part of a very complex system. Much outside construction is required to build the total system.

Transmitting Calls

A telephone message usually is transmitted in the same way as an electric current. The speaker's voice *modulates* (shapes) the current. A telephone receiver changes the modulated current into sound waves.

In some newer installations, glass fibers replace some of the wires. Electric current is changed into light waves. The glass fibers transmit light waves. Messages are carried by these light waves. The light waves are changed back into a modulated electric current to complete a call.

A message can also travel through the air as radio waves. Electric current carried by wire can be changed to short radio waves called *microwaves*. Microwaves do not travel very far. *Relay stations* are needed. Relay stations are located on towers or high buildings. They use disc-shaped antennas. See Fig. 38-10. Each relay station receives the radio signal and relays (sends) it to the next station. Or, the signal may be sent to a telephone exchange. Here, the microwave message is changed back to a modulated electric current. Then this current finally is changed into sound waves.

Fig. 38-10
During a long-distance telephone call, the sound of your voice may travel as microwaves from relay station to relay station.

Another method of sending telephone messages is by satellite. Satellites are positioned several hundred miles out in space. They function as microwave relay stations for long-distance calls.

Telephone and Other Communications Systems

Telephone lines are often used as part of other communications systems. For example, radio and television systems often lease (rent) telephone lines. The radio and television signals are sent through the air. But part of their journey may be by telephone.

Maintaining Utility Systems

Utility systems require maintenance. Usually a utility company services (takes care of) the outside system. Each company keeps records of the locations of all lines, shutoff valves, and connections. Users, too, should know the locations of shutoff valves for their utilities. This is important in case of emergencies or for general repairs. Users buy and maintain much of each system that is within their property.

Even the briefest break in a utility service can cause many problems. Utility companies know this. They maintain high standards on new construction. They also make sure repairs are made. Repair crews work in all kinds of weather. They must be ready to work any hour, day or night.

Summary

Under the average city street, there is a network of utility lines. Water, sewer, and gas mains are hidden there. Electric and telephone cables may also lie under the streets.

Utility systems branch out across the countryside. They may be above ground or buried. Some parts may even be out in space.

Utility companies install and service transmission lines. Their responsibility for this usually ends at the meter. From the meter to the inside of the structure, the user must have the system installed and serviced.

All of us depend upon utility networks for essential (necessary) services. That makes them important. They must be installed and maintained carefully.

Terms to Know

conduit
drop
effluent
microwaves
modulates
outside utility systems
pipefitters
pipeline
plumbers
relay stations
sanitary sewers
service area
services
sewer
sewerage
storm sewers
transformers
treatment plant
utilities
voltage
water mains

Think About It!

1. What utilities service your home? What other utilities are used in your community?
2. Assume that all the utilities you use are suddenly shut off. What are some of the problems you will have?
3. Where are the main shutoff valves for your utilities? (List their locations.)

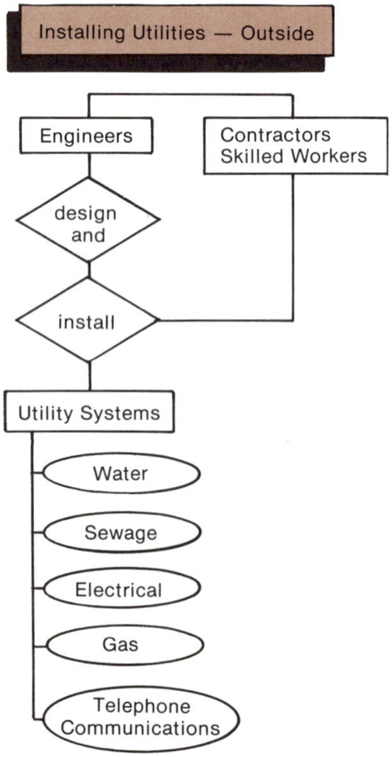

Reading 39

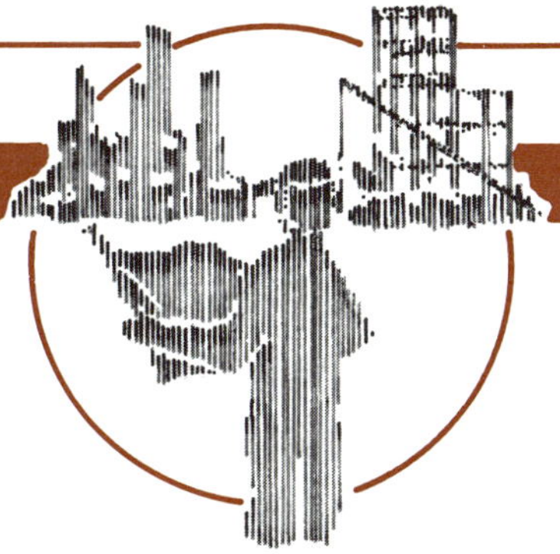

Installing Utilities — Inside

Utilities

The principal public *utilities* are water, natural gas, electricity, and sewerage. *Utility systems* are installed in structures. They provide services to the people who use the structures. Utility lines bring the service to the structure. Some public utility lines are buried underground. See Fig. 39-1. Others (such as electricity) are hung between poles above ground. The right to locate utility lines across private property is called an *easement*. Utility lines are generally located just outside the front and rear property lines. They are then brought across the private property. Finally, they are installed inside the structure to provide the services needed.

Niagara Mohawk Power Corporation

Fig. 39-1
A city cannot operate without the services of utilities. Many utility lines (pipes, wires) are hidden underground.

Roughing In Utilities

Most utility lines cannot be seen inside a building. They are enclosed inside floors, walls, and ceilings. *Roughing in* means placing parts of utility systems inside the building skeleton before it is enclosed.

Two major types of parts are roughed in. One type is the *ducts* or *pipes*. Solids, liquids, or gases flow through them. Ducts and pipes are important parts of the building's *mechanical systems*. The mechanical systems move heat or water into, around, and out of the structure.

The other part that is roughed in is the *wiring*. Wires are important parts of the building's *electrical system*. Wires carry electricity. They also connect other parts of the electrical system. For example, they connect lights with switches.

Roughing in is done in a special order. Sheet metal ductwork is put in first. See Fig. 39-2. Ducting is normally larger and less flexible than piping or wiring.

Large pipes are placed in the structure at about the same time as ductwork. The rest of the piping is usually done after ductwork is installed. See Fig. 39-3. There are many different pipes in a structure. For

Fig. 39-2
The system for heating and cooling is roughed in first.

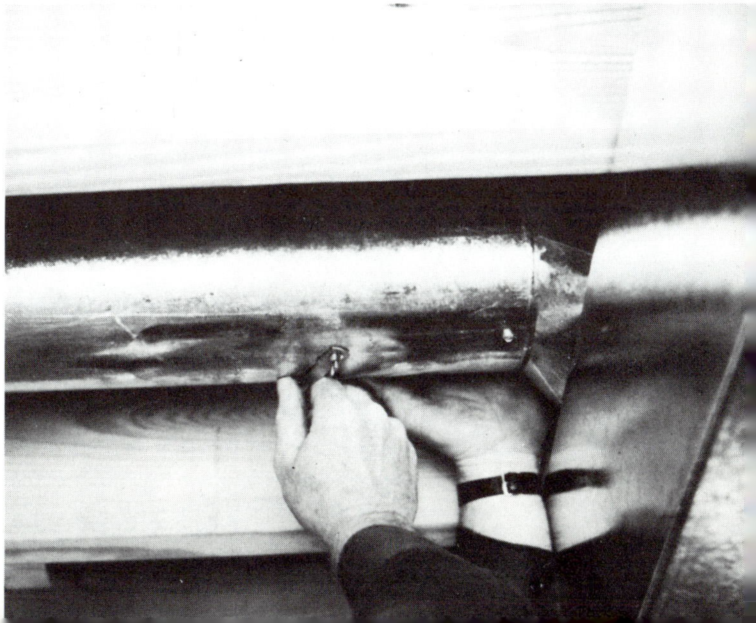

example, some may carry waste. Others may carry water or natural gas.

Since wiring is so flexible, it is roughed in last. Wires are installed in walls, under floors, and in open spaces above the ceilings. See Fig. 39-4.

In custom-made buildings, utilities are roughed in on the site. In modular construction, some utilities are placed in the units in a factory. The units are assembled on the site. At that time, the utilities of separate modules are connected.

Controlling the Inside Climate

People want to control the *inside climate* of structures. This is done by heating and cooling the air. It may also involve controlling the air's moisture content. And the air may be cleaned.

Contractors must know how much space in the building needs to be heated or cooled. They also need to know how *energy efficient* the structure will be. To do this, they figure the expected *heat loss*. Heat loss is calculated from four main factors.

- Sizes of rooms.
- Sizes and kinds of windows.
- Sizes and kinds of doors.
- Sizes and kinds of walls, ceilings, and floors to be exposed to the outside air.

Prevailing winds and average temperatures of the area must also be considered. All these figures are compiled (gathered). From them, the heat loss is calculated. Then the contractor or engineer can determine what size heating and cooling unit is needed. See Fig. 39-5.

Heating

There are two kinds of heating. They are called *dry heat* and *wet heat*. Dry heat may be *hot-air heat*. Air acts as the agent (carries the heat). Dry heat may also be *radiant heat*. Radiant heat is produced by electrical units. Wet heat is produced by the steam or hot water in pipes and radiators. The steam or water is the agent.

Fuels such as natural gas, oil, and coal are used to produce heat. However, these fuels are becoming more expensive. It takes fuel to produce electricity. Electricity, too, is becoming an expensive source of heat.

The heat of the sun may be used to help heat structures. *Solar energy* may be used in *active systems* or *passive systems*. In active systems, the sun's heat energy is collected and moved through the structure. The energy is used to heat water or air. In passive systems, the sun's energy is absorbed by the building. Heat is radiated as the interior of the structure cools.

In active systems, heat from the sun may be collected by *solar panels*. See Fig. 39-6. Water or air is warmed by the heated panels. Warm water moves to the interior of the structure through pipes. Warm air is forced inside through ducts.

Fig. 39-3
This plumber is preparing a connection by applying adhesive to a pipe.

Fig. 39-4
This electrician is running wiring through holes that have been marked and drilled.

Fig. 39-5
Heating and cooling specialists install units such as this furnace with an air-conditioning unit on top.

Berry and Company

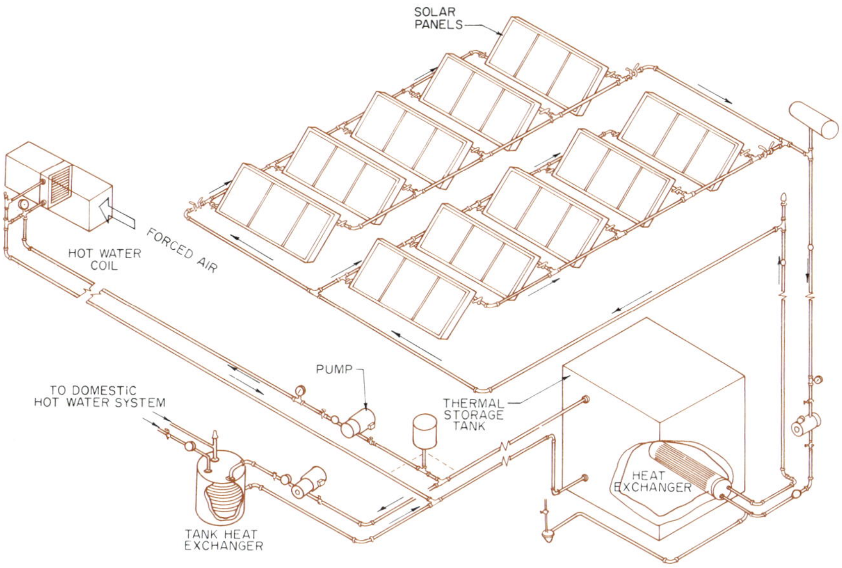

Solar Energy Products, Incorporated

Fig. 39-6
Solar panels and piping.

Solar heat can be stored. Heated water from solar collectors can be stored in special tanks. This warm water is pumped through the heating system as needed to heat the structure. Rocks or pebbles may be used to store heat. These are placed in special storage units. Warm air from solar heat collectors is blown through the unit. The air warms the rocks. When heat is needed in the structure, fresh air is forced through the unit. This air is heated by the rocks. It is then moved through the structure to provide heat.

In a passive system, heat energy is collected by the structure itself. The structure must be placed so that the maximum use of the sun is made. Windows let the sun's energy inside. The windows facing south let in the greatest amount of sunshine. Energy from the sun warms the walls and floors. Then the warm walls and floors heat the interior space.

Cooling

Most cooling is done by moving cooled air through sheet metal ducts. Special air-conditioning units are installed to cool the air. See Fig. 39-7.

Like heating, cooling requires energy. But commonly used fuels are becoming more expensive and scarce (hard to find). Other energy sources must be discovered. People must search for inexpensive ways of heating and cooling.

Fig. 39-7
Installing an air-conditioning unit.

Singer Company

Other Types of Climate Control

Dust and moisture in the air must be controlled. *Filters* are put in heating and cooling units to remove dust particles from the air. Special devices are placed in heating and cooling systems to control moisture in the air. Some air has too much water vapor (humidity) for comfort. Water vapor may be removed from air by a *dehumidifier*. Moisture needs to be added to very dry air. This is done by a *humidifier*.

The heating and cooling system also *ventilates* the structure. This means that some fresh air is brought in from the outside. Also, stale air is moved out of the building.

Many different workers install the utilities in structures. A ductwork system is installed by *sheet metal workers*. *Plumbers* install water systems and other piping. *Pipefitters* install hot-water and steam heating systems. *Electricians* put in the electrical wires that service the building.

Plumbing and Piping

At least one water outlet must be available at the site very early in construction. Water is used by workers who construct footings and foundations. It is also used in mixing mortar and for some cleaning purposes.

A plumber must study building plans to see where *plumbing fixtures* (sinks and bathtubs, for example) are to be installed. Next the plumber lays out and marks where the pipes will go in the skeleton.

First, the plumber installs the *waste pipe*. A waste pipe is a large drain pipe used to remove waste. It is installed before concrete is placed in basement floors. The plumber starts installing the waste pipe at the *sanitary sewer* connection. This is just inside the foundation wall. All *drain lines* must slope toward the sanitary sewer. See Fig. 39-8. Drain lines run to the fixtures and continue through the roof. In the roof, they act as *vents*. Vents allow fixtures to drain properly. *Traps* are installed in drains. Look again at

Fig. 39-8
Sanitary and storm sewer piping systems.

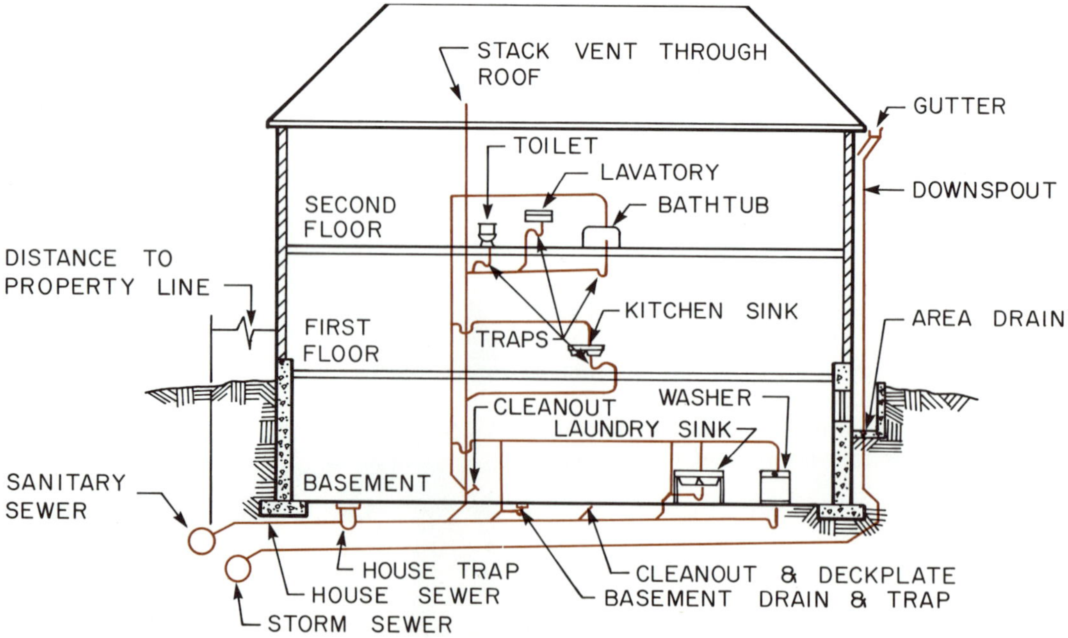

Fig. 39-8. Find the traps in the diagram. Traps keep sewer gas and odors from reentering the building. Without vents, the water seal in a trap would be siphoned (removed).

Waterlines start at the *water meter*. Meters may be located just inside the foundation wall. In warm climates, they may be outside.

Cold-water lines are always installed on the right side of a fixture. Hot-water pipes are always placed on the left side. The outlets of waterlines are capped off. See Fig. 39-9. Then the water pressure is tested. The caps are removed from the waterline outlets as fixtures are installed.

Permanent fixtures such as bathtubs and shower bases must be built into the walls. These are roughed in very early. Then the inside walls and floors are completed.

Electricity

Electric service is hooked up to the site early in construction. This provides electric energy for other persons working on the structure. The first item installed is the *electric meter*. The meter measures the amount of electric current the customer uses. The meter is part of the *service entrance*. Electric energy enters the structure through the service entrance. See Fig. 39-10. For large buildings this entrance is usually underground. For small structures, such as homes, it is often above ground. In well-planned communities, many wires are buried underground. Buried systems are out of sight. They are safer and more trouble-free.

After the ductwork and plumbing are roughed in, electrical work begins. All wires are run from the electric meter to a *central distribution panel*. See Fig. 39-11. This panel contains *circuit breakers*. They protect the electrical equipment and the building. Circuit breakers automatically break (open) a circuit if an electric overload occurs. This prevents fire or damage to appliances and other property.

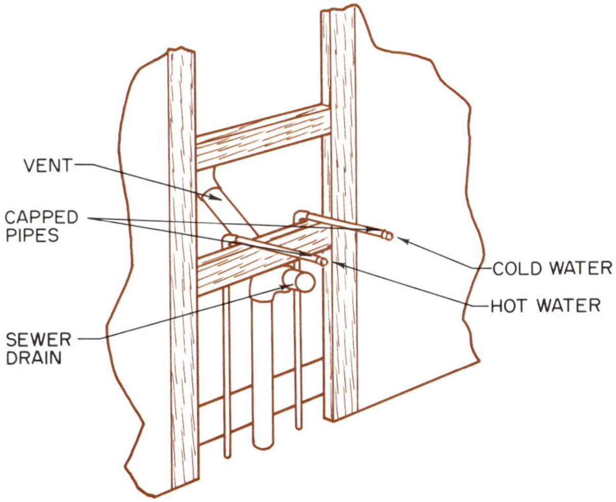

Fig. 39-9
Water supply lines are roughed in for a bathroom.

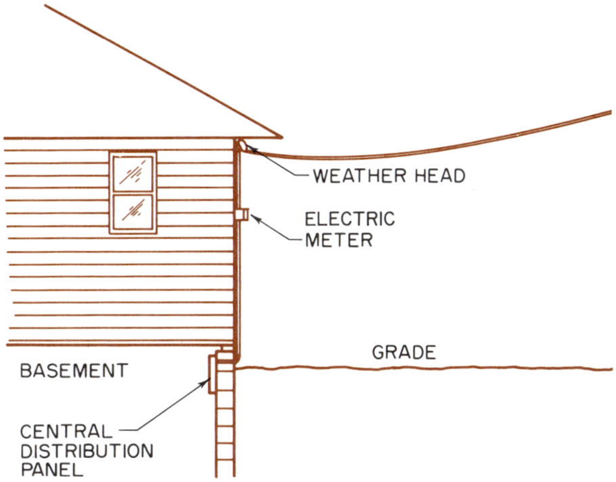

Fig. 39-10
This is a service entrance. Outside poles and overhead wires are not very attractive. Wires may be run underground instead.

Fig. 39-11
All wires are run from the meter to a central distribution panel.

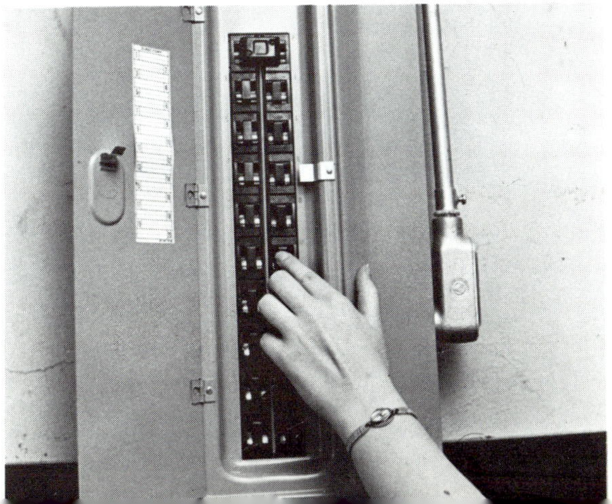

An electrician studies the building plans carefully. The location of electrical boxes is shown in these plans. Drawings in the plans tell where to attach electrical boxes to the walls and ceilings. Insulated electrical wire of the proper size is then run through the skeleton of the structure. Wires may be in special metal tubing called *conduit*. The *feed ends* of the wires are attached to the panel box. They are not hooked up to the terminals. The *load ends* of the wires are coiled up in the outlet boxes. See Fig. 39-12. The circuits are "dead" at this point. No electricity will flow until all electrical fixtures, switches, and outlets have been installed.

Finishing Utilities

Next, interior floor, wall, and ceiling surfaces are applied. They cover the utility lines in the building skeleton. Heating and cooling registers and *returns* are installed. Plumbing fixtures are put in. Finally, electrical switches, light fixtures, and other electrical devices are fastened in position. See Fig. 39-13.

Summary

Public utility lines are placed just outside property lines. They must be brought into a building to provide service. The utilities serve the people that occupy the structure. The most important utilities are water, natural gas, electricity, and sewerage.

Utilities are roughed into the structure. This means they are placed in the skeleton of the building before the framework is enclosed. The work is done in a special order. Ductwork is done first. Ducts carry fresh, warm, or cool air through the building. Next, pipes are installed. Pipes may carry water, waste, natural gas, or other materials. Electrical wires are roughed in last. They provide electrical energy for the building.

Once the utilities are roughed in, the interior surfaces are enclosed. Then fixtures, switches, and outlets are installed.

Fig. 39-12
Wires coiled in an outlet box.

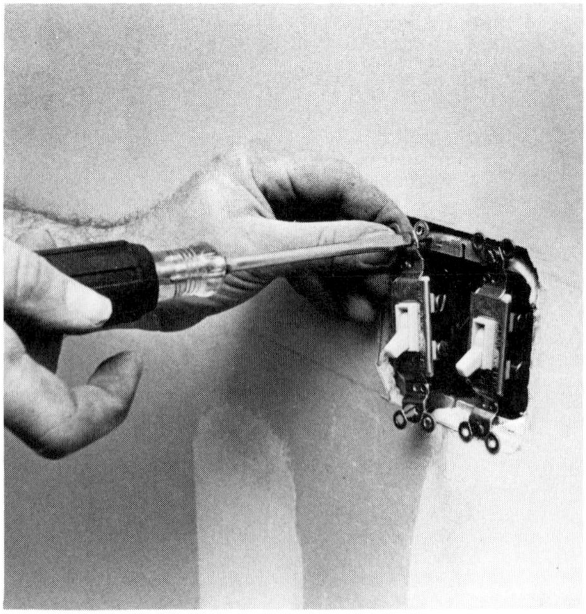

Fig. 39-13
Installing an electrical switch.

Terms to Know

active systems
central distribution panel
circuit breakers
conduit
dehumidifier
drain lines
dry heat
ducts
easement
electrical system
electricians
electric meter
energy efficient
feed ends
filters
heat loss
hot-air heat
humidifer
inside climate
load ends
mechanical systems
passive systems
pipes
pipefitters
plumbers
plumbing fixtures
radiant heat
returns
roughing in
sanitary sewer
service entrance
sheet metal workers
solar energy
solar panels
traps
utilities
utility systems
ventilates
vents
waste pipe
water meter
wet heat
wiring

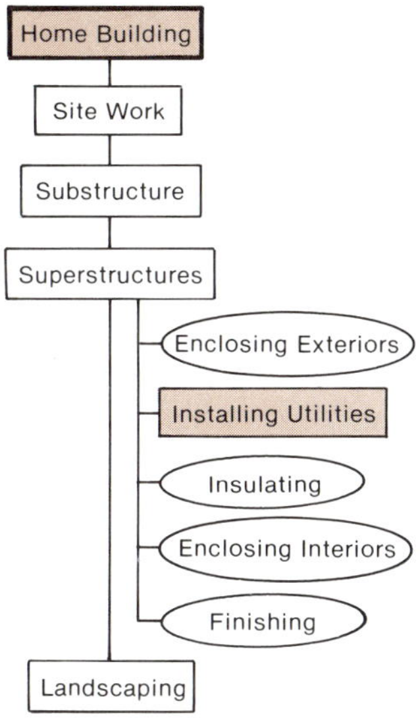

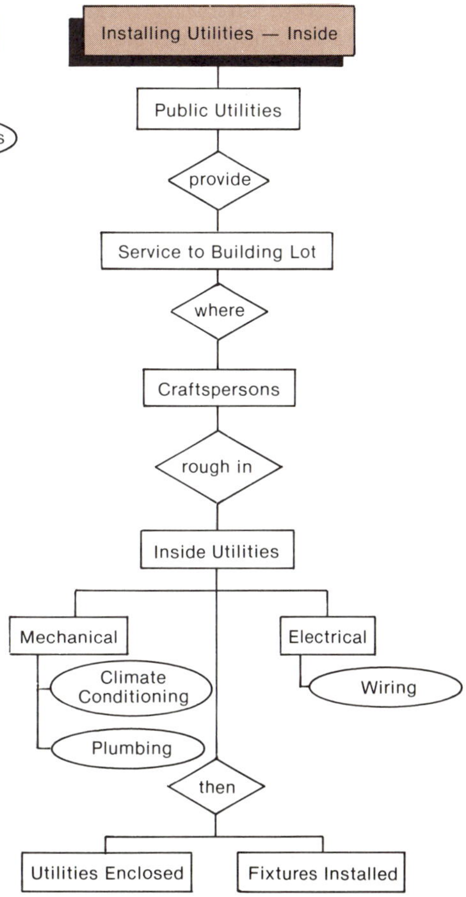

Think About It!

1. Why are heating and cooling utilities installed before plumbing or electrical utilities?
2. In the next thirty years, what changes do you think there will be in heating and cooling utilities? in plumbing? in electricity?

Reading 40

Installing Heating, Cooling, and Ventilating Systems

Treating and Conditioning Air

Buildings enclose space. Air occupies much of this space. People inside buildings need good-quality air for comfort. To be of good quality, air often needs to be treated or conditioned. Air may be warmed, cooled, or purified (made pure). Moisture may be added or removed.

People need treated air. Other things need treated air, too. Plants and flowers in greenhouses need certain temperatures. And some machines work best at certain temperatures. Computers are an example.

Treated or conditioned air must be distributed (moved) throughout the structure. Treating and moving the air is called *air handling*. Systems are installed to treat and move air. See Fig. 40-1. This reading describes the *installation* of these systems.

Air is treated in four different ways.

- Heating.
- Cooling.
- Humidifying (adding moisture) or dehumidifying (removing moisture).
- Filtering (purifying).

Fig. 40-1
Installing sheet metal ducts. Treated air is moved through ducts to various points within a structure.

Some of these treatments may take place at the same time. For example, heated air may also be purified and humidified.

Heating

Air may be heated by a centrally located furnace. Or, it may be heated by separate heating units. Common heat sources are gas, oil, coal, and electricity.

Heat energy from the sun is being used more and more. Heat is absorbed (taken in) by liquids, solids, and air in a structure. *Solar-assisted* heating means that the sun is not the only heat source. See Fig. 40-2. The structure must have *solar collectors*. They may be part of the roof or an outside wall. The collectors take in the heat. The heat is "stored" in liquids or solids and used when needed. This usually involves installing expensive equipment. But the heat source is free!

There are two main kinds of central heating systems. They involve either indirect or direct heating of air.

Fig. 40-2
A warm-water solar-assisted heating system. Water is pumped to the collector and warmed by heat from the sun. The warm water is stored in a tank and sent through the heating system as needed.

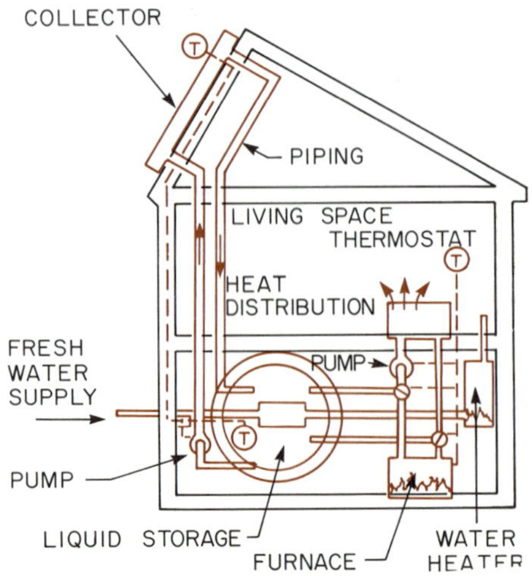

Fig. 40-3
The steam generated by this huge boiler can supply heat for more than fifty multistory buildings.

Indirect Heating

In indirect heating, water absorbs heat from burning fuel or from the sun. The water travels through the building as either hot water or steam. See Fig. 40-3. It moves through pipes. The pipes lead to a *radiator* in each room. The radiator heats the air.

Another method of indirect heating involves the use of heat wires or tubes. The wires or tubes may be installed in the floor or ceiling. They also serve as radiators.

Direct Heating

Most structures built in recent years use direct heating. Fuel or solar energy heats the air itself. The air travels through the building in a system of *ducts*. Ducts are thin-walled pipes made of sheet metal. They may be circular or rectangular in cross section.

A fan or a blower is built into a duct near the heating source. It creates a slight pressure. The pressure forces the warm air slowly through the duct. The warm air enters each room through a *register*. Slow air movement does not create breezes or *drafts*. Drafts make a room uncomfortable.

Often, large buildings have high-pressure duct systems. Smaller ducts are used. This saves space. In these systems, the air moves more quickly through the ducts. But fast-moving air creates drafts. Therefore, the air's speed must be reduced before the air enters a room.

Ductwork that delivers warm air can also carry cool air. However, an air-cooling unit must be built into the system.

Cooling

Air may be cooled by centrally located air-conditioning units. Or, electric window units may be used. Both types of units remove heat from structures.

Central air-conditioning units may run on gas. But more often they run on electricity. Some of the electric units also provide heating. These are called *heat pumps* or reverse-cycle units.

Humidifying and Dehumidifying

Ventilating is letting fresh air in and stale air out. The fresh air is drawn (pulled) in from outdoors. Some outdoor air has too much water vapor for comfort. Water vapor in the air is called *humidity*. On a hot, damp day, humidity makes people feel uncomfortable. A *dehumidifier* takes humidity out of the air. It uses heat and fans or chemicals.

Some air has too little water vapor. A warm, dry room can be uncomfortable. A *humidifier* is often used to add humidity to the air. A humidifier sprays fine drops of water into the air. Or, it may evaporate water into the air.

Filtering

Another way to treat air is by purifying it. The air is passed through *filters*. A filter is a screen of fiber, metal, or plastic. It works in the same way as a window screen. Most insects cannot pass through a window screen. And most dust and dirt cannot pass through a filter. The filter removes dust and dirt from the air in the system.

Some filters are permanent. In these the air may be cleaned by electronic or sound devices. Permanent filters may need to be cleaned occasionally. Other filters are disposable. They are thrown away when they get dirty. Then new filters are installed.

The Air Return System

Used or polluted air must be removed from rooms. This is as important to people's well-being as treating air. Used air is exhausted (let out) through *return air registers*. These open into a second set of ducts. A slight *vacuum* (suction) is created by fans. It pulls used air slowly through the return air ducts. This air may be exhausted to the outdoors. Or, it may be refiltered, reheated, or recooled. Then it is recirculated into the distribution system.

Pneumatic Conveying and Recovery Systems

Sheet metal ductwork has uses other than climate control. It is also used as part of pneumatic

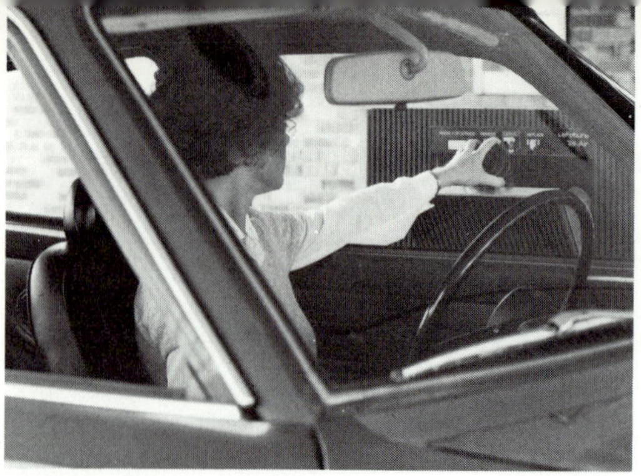

Prairie State Bank

Fig. 40-4
Pneumatic devices may be installed in drive-in bank facilities.

Conrad Sheet Metal Company

Fig. 40-5
Forming ductwork.

conveying and recovery systems. These systems are used by industry to move materials and air.

In a *pneumatic conveying system*, a stream of air carries materials through the ductwork. Many types of materials may be moved in this way. Grain, flour, or nuts and bolts are just a few. They are carried from receiving or storage areas to the locations at which they are needed.

A *pneumatic recovery system* is like a home vacuum sweeper. Waste materials, such as wood shavings, are drawn inside the ductwork. A stream of air carries them to a collection point. There they are settled or filtered out of the air system.

Material moving systems and climate control systems have much in common. They are similar in design and engineering. They are fabricated (made) and installed in much the same way. Also, both systems have blowers, ducts, filters, humidity regulators, and controls. But there is one major difference. Pneumatic systems rarely require a source of heat or cooling. See Fig. 40-4.

People Who Install Ducts

A duct system is carefully planned by a mechanical engineer. The engineer determines the size and location of each duct. The size of the duct depends on the amount of air that is to move through it each minute.

Ductwork is usually done by a subcontractor. This may be either a sheet metal subcontractor or a mechanical subcontractor. Mechanical subcontractors often do both sheet metal work and plumbing.

The subcontractor prepares detailed shop drawings. The drawings show how each part is to be made. The engineer may have to approve the shop drawings. Next, sections of ductwork are made in the subcontractor's shop. See Fig. 40-5. The sections of ductwork are shipped to the site. Then the ductwork is installed.

Sheet metal workers fabricate and install ductwork. They are highly skilled. They also perform other tasks involving sheet metal. For example, they may install gutters and downspouts.

Installing Ducts

Near the furnace or air-conditioning unit, the ducts are large. They generally become smaller as they approach each room. The reduction is gradual. This helps keep the pressure about the same throughout the duct. See Fig. 40-6.

Fig. 40-6
Treated air is sent from the air conditioner throughout the structure. Return air is taken back to the air conditioner, treated again, then recirculated.

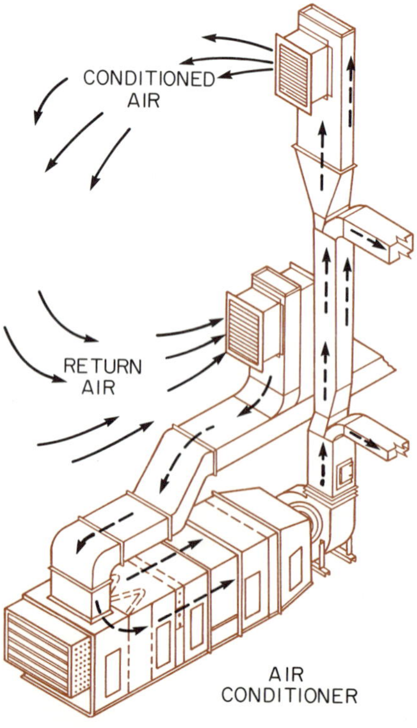

Fig. 40-7
Hangers are fastened to the floor above to hold ductwork in place.

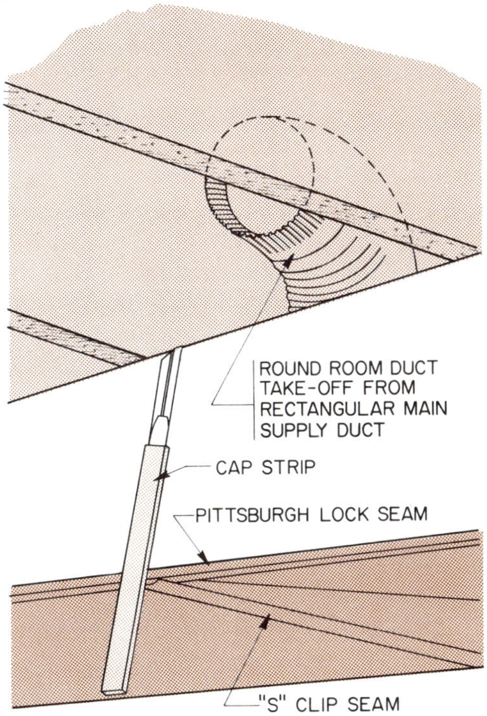

Fig. 40-8
Locking together two sections of duct.

Fig. 40-9
Controls are set to operate heating and cooling units automatically.

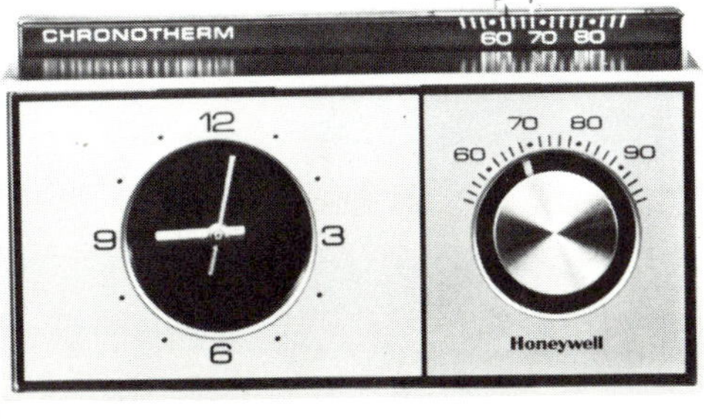

Ducts are hung in place by straps or *rods* and *angles.* See Fig. 40-7. Each section of duct is joined to another. Usually a mechanical connection is used to make the joint. See Fig. 40-8. Some ducts (usually round ones) are made smaller on one end. Two ducts can be joined by slipping the end of one duct inside another duct.

Ducts are placed carefully. They must not interfere with other utility lines. Other lines include water pipes, sewer pipes, and electrical conduits. All the utilities are placed into *chases.* Chases are openings between different parts of a structure. Often main ducts are run between a ceiling and the floor above it. This is usually done in hallways.

Ducts are often hidden in the structure. They are run to each room. Ducts of various sizes are used. Changes and repairs are hard to make after ductwork is completed and enclosed. Therefore, ducts must be installed skillfully.

The air carried in sheet metal ducts is kept at a controlled temperature. It may be warmer or cooler than air outside the duct. Insulation may be needed to reduce heat transfer. That is, it prevents heat from passing through the walls of the ducts. Fiberglass mats or plastic foams are used to insulate ducts. Insulation mats may be fastened to the exterior of the ducts with *adhesives* or *metal fasteners.* Insulation also may be built inside the duct as a lining.

Controls

To make rooms comfortable, heat, cold, and humidity must be controlled. If this is not done, a building may become too hot or cold. Or, it may get too damp or too dry.

The air speed in the ducts must also be controlled. Air must move quickly enough to provide enough ventilation. But it must move slowly enough to avoid drafts.

The air temperature in a building is controlled by an electric device. See Fig. 40-9. This device is called a *thermostat.* It is a combination thermometer and control switch. The switch may be set at a desired temperature. When the temperature varies from the set figure, the thermostat turns on the furnace or air conditioner.

A *humidity regulator* keeps the desired amount of water vapor in the air. It is also known as a *humidistat.* It controls the amount of moisture a humidifier adds to the air. And it controls the amount of moisture a dehumidifier takes out of the air. By doing this, the humidistat keeps the humidity at the desired level.

A *damper* controls the flow of air into a room. It is located at the register or in the duct. The damper is made of one or more metal pieces. These can be moved to block the air flow. In one way, a damper is similar to a sink faucet. The faucet stops the water in a water pipe. The damper stops the air in ducts. A damper can be set for different amounts of air flow.

Summary

Heating, cooling, and ventilating systems make buildings comfortable. They provide clean, warm, or cool air for people. The air is the correct temperature and humidity. There are very few drafts.

Air may be treated directly or indirectly. Centrally located units usually treat the air directly. The treated air flows through ductwork to the area in which it is needed.

Air is exhausted from rooms. It is either discharged to the outside or re-treated.

Ductwork must be planned and installed carefully. Also, proper controls are necessary to provide a pleasant indoor climate. This includes controls for heating, cooling, humidity, and air flow.

Terms to Know

adhesives
air handling
angles
chases
damper
dehumidifier
drafts
ducts
filters
heat pumps
humidifier
humidistat
humidity
humidity regulator
installation
metal fasteners
pneumatic conveying system
pneumatic recovery system
radiator
register
return air registers
rods
solar-assisted
solar collectors
thermostat
vacuum

Think About It!

1. State four ways in which air can be treated.
2. What devices control temperature, humidity, or the flow of air in a room?
3. Why is it easier to install ductwork before installing plumbing or wiring?

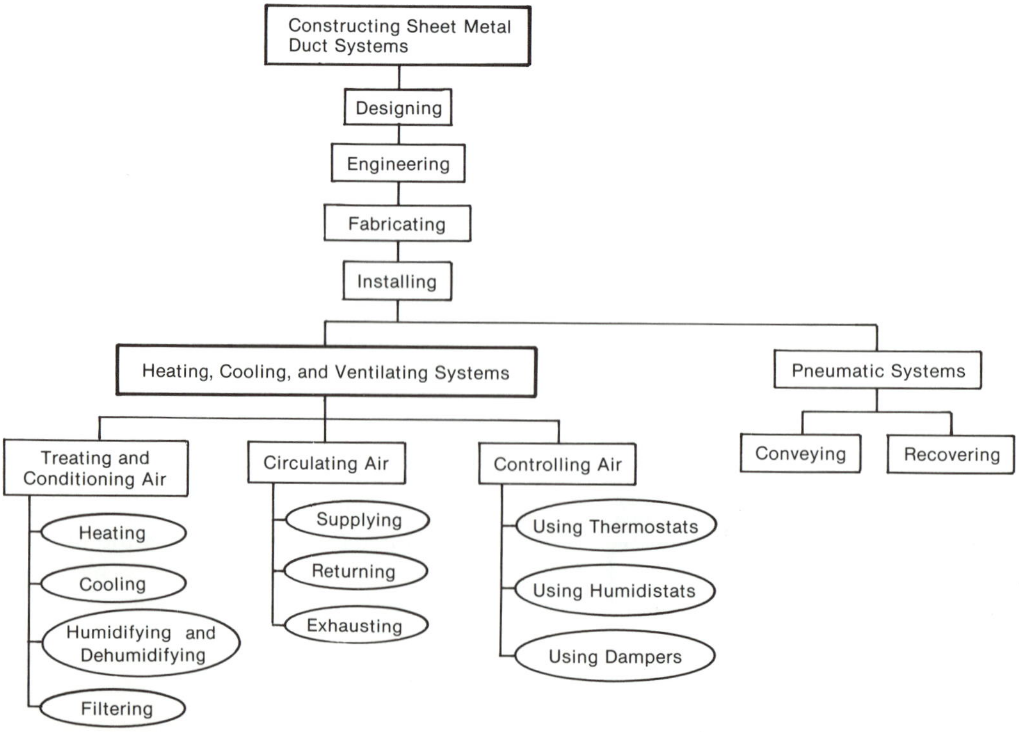

Reading 41

Installing Plumbing and Pipelines

Piping Systems

Liquids and gases are distributed from their sources through piping systems. Inside buildings, piping systems are called *plumbing*. Outside buildings, they are called *pipelines*.

Plumbing systems may use rigid pipe or flexible tubing. Or, they may use a combination of the two. See Fig. 41-1.

Pipelines are generally larger in diameter than plumbing. See Fig. 41-2. One pipeline may serve several buildings. It may branch out to serve a whole city or even several states. Pipelines are built mostly of rigid pipe.

Fig. 41-1
Plumbing systems may be very complicated. They use many kinds of piping, tubing, valves, and fittings.

Atomics International

Major Kinds of Plumbing and Pipeline Systems

Plumbing systems move

- fresh water.
- hot water or steam.
- water for fire protection.
- gases.
- fluids for industrial production.
- sewage.

Fig. 41-2
Pipelines may require related construction. Pipelines placed on these supports can be checked even when snow covers the ground.

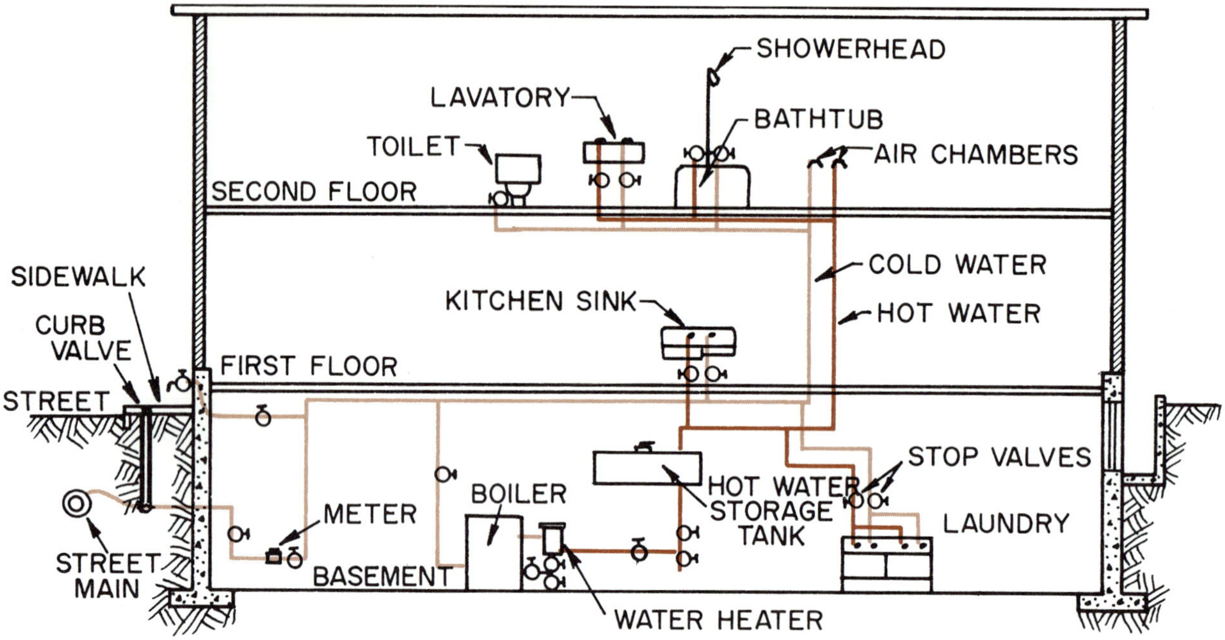

Fig. 41-3
This is a fresh-water plumbing system inside a building. Can you trace the fresh-water flow from the street main to the showerhead?

Pipelines carry many of the same substances as plumbing systems. They carry natural gas, petroleum, fresh water, and steam. They move them from a source to users. Pipelines also carry off sewage from buildings. And they carry excess water from the streets and from buildings. They transport these materials to streams or treatment plants.

Materials used for piping vary. Pipes may be made of copper, steel, cast iron, clay, or plastic. The purpose of the system and standard practices determine the material used. Also, designers, owners, or contractors may prefer one type of pipe over another. All piping selected must meet rules set by building codes.

Building codes set up rules for construction. Contractors must use good building practices. Plumbing and pipeline practices are described in this reading.

People Who Design and Install Piping

When architects design a structure, they choose the fixtures, appliances, and equipment to be used for plumbing. And they specify the type, quantity, and location of each.

Engineers design the piping needed to service the fixtures, appliances, and other equipment. See Figs. 41-3 and 41-4. Engineers also make sure that workers follow building codes and good building practices.

Plumbing work is usually subcontracted. The crew of a *plumbing subcontractor* or *mechanical subcontractor* may do the work. The subcontractor prepares detailed shop drawings of the plumbing. The drawings are approved by the project engineer. They also are approved by government agencies such as the Board of Health. Then the plumbing is installed.

Pipelines are carefully designed by civil engineers. Designing pipelines involves several steps.

1. Planning the routes.
2. Specifying (naming exactly) pipes and controls.
3. Choosing locations for pumping, processing, or treatment plants.
4. Selecting equipment for these plants.

Pipelines are usually installed by mechanical contractors. Some of these contractors specialize in pipeline construction. Often owners have their own employees install pipelines.

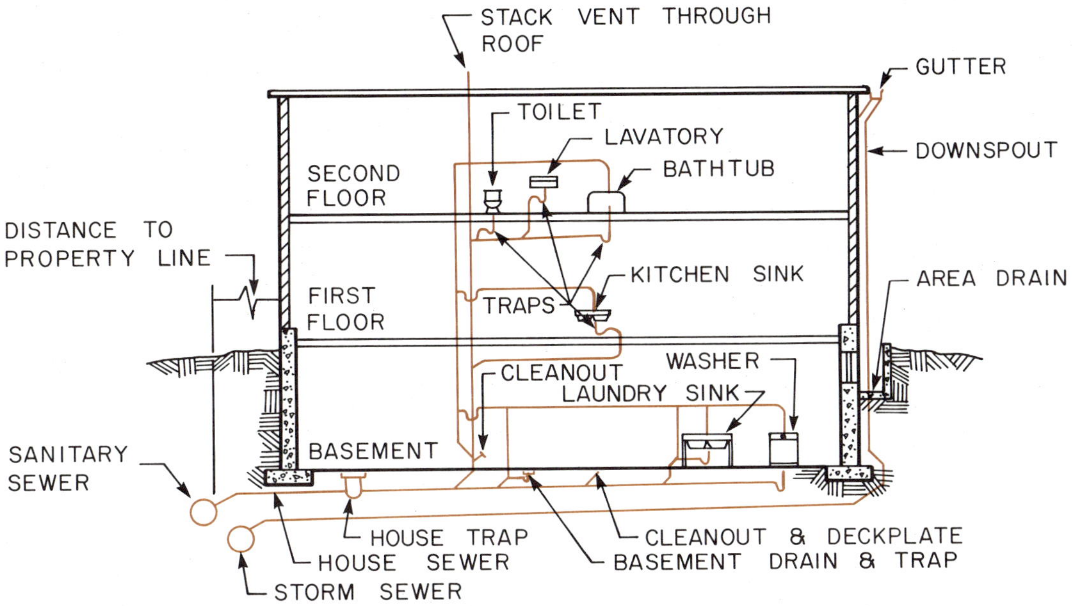

Fig. 41-4
In this drawing, you can see the building's sanitary and storm sewer plumbing systems. Can you trace the drainage from the bathtub to the sanitary sewer pipeline?

Operating engineers operate the construction machinery. This includes backhoes, trenchers, bulldozers, and cranes. *Pipe layers* lay and join the piping. These people may be plumbers, pipefitters, or steam fitters.

Installing Plumbing

At the site, pipe or tubing is cut to the needed lengths. The pieces are joined with *fittings*. See Fig. 41-5. Pipe and tubing are made of different materials. Therefore, they are joined in different ways.

Fig. 41-5
Typical pipe sections and fittings.

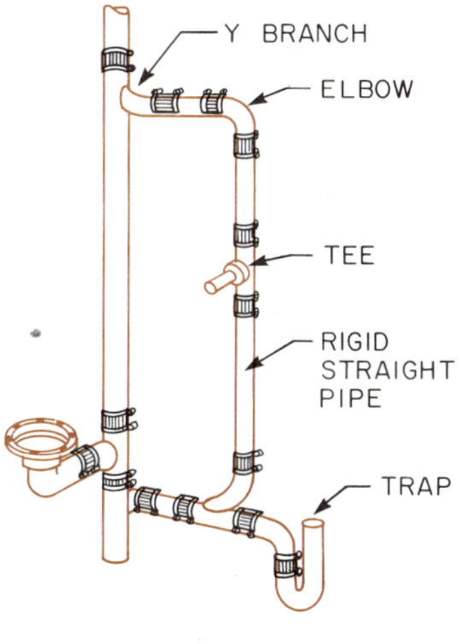

Copper Tubing

On copper tubing, the fittings slide over the ends of the tubing. The joints are sealed with a thin film of *solder*. A joint sealed in this way is called a *sweated joint*. See Fig. 41-6. The soldering metal is called lead solder. But solder is really an alloy (mixture) of lead and tin.

Fig. 41-6
Copper pipe fittings are fastened to pipes by soldering. The solder is melted when applied, but hardens to seal the joint.

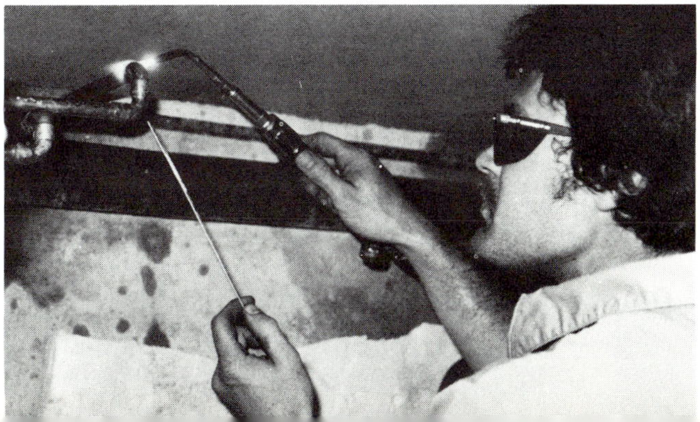

Equitable Gas Company

Fig. 41-7
Large pipes used in plumbing and in pipelines are often joined by welding.

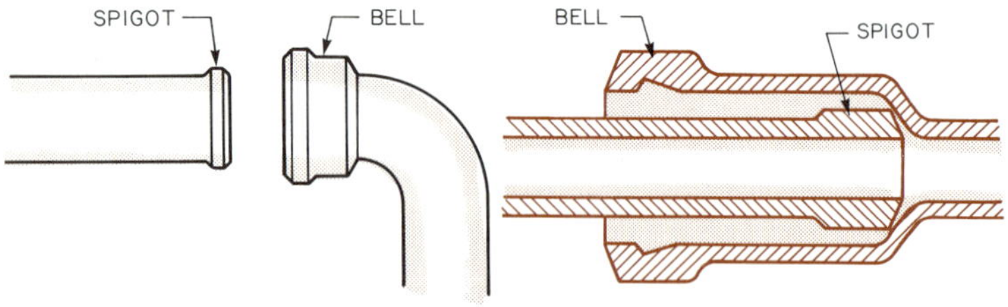

Fig. 41-8
When assembling cast iron pipe, the spigot end of one length of pipe is placed inside the bell end of another. Then it is sealed. The plumber selects the method of sealing according to local regulations.

Fig. 41-9
This plumber is tightening a clamped pipe joint.

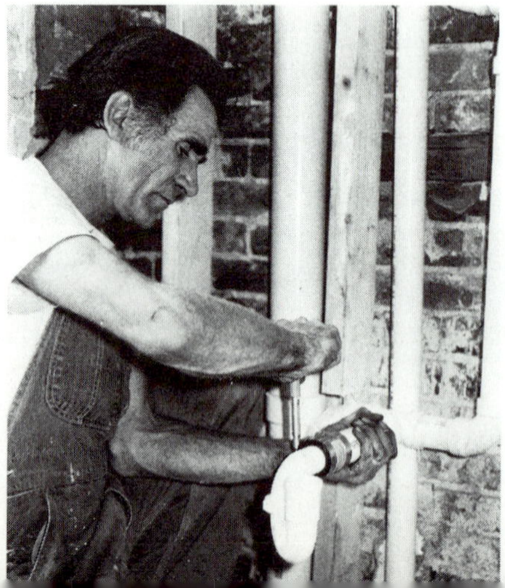

Steel Pipe

Small-diameter steel pipe is assembled by means of threaded joints. Steel pipe fittings are used to join the sections of pipe. Fittings are threaded when they are manufactured. A threading machine is used to cut matching threads into the ends of each section of pipe. Then the pipes and fittings are assembled.

The joints between pipes and fittings must be sealed. A compound (mixture), often called *pipe dope,* is brushed on the pipe threads. Then the pipes and fittings are screwed together. The pipe dope seals the joints and prevents leaks. A special plastic tape may be used in place of pipe dope.

Plumbing must be tailored to (specially made for) each particular building. For this reason, most pipe less than four inches in diameter is prepared at the site. Pipe four inches or larger in size is hard to handle. It is usually prepared in a pipe shop. This pipe is cut and threaded at the shop. Then it is taken to the site.

Pipes larger than four inches in diameter are often welded together. See Fig. 41-7. Also, *flanges* (flat rims) may be used on the ends of pipe pieces. They have holes that allow flanged pieces to be bolted together. Often a part called a gasket seals these joints. They can be taken apart.

Cast Iron Pipe

Cast iron *soil pipe* is used mostly for sanitary drainage. One end of a cast iron soil pipe is flared. This is the bell end. The other end is called the spigot end. See Fig. 41-8. Soil pipe systems are assembled in several ways.

Clay Pipe

Clay pipe is often used for storm drainage. Like cast iron pipe, it has a bell end and a spigot end. The joints of clay pipe may be filled with *cement grout.* This is a mixture of cement, sand, and water. A joint may also be sealed with a gasket or with a rubber ring called an O ring.

Plastic Pipe

Plastic pipe has many uses. And it has an advantage over other kinds of pipe. It is not usually affected by liquids or gases that are corrosive (eat away materials).

Special techniques are required to assemble and install plastic pipe. Mastics (pasty cements) or other cements are often used to join plastic pipe. Also, the pieces may be clamped together. See Fig. 41-9.

Securing or Supporting the Pipe

Piping systems must be fastened to the structure. This prevents strains in the piping. Strains can cause pipes to leak.

Some parts of the pipe run horizontally (parallel to the floor). These runs (sections) are hung from the undersides of floors or ceilings. Clamps or supports called *pipe hangers* hold the piping in place.

To support vertical (up and down) runs of pipe, *backing boards* may have to be installed. Special hangers are used to fasten the piping to the backing boards.

Sometimes a structure is built on a slab. If it is, some piping is installed before the concrete is placed. Parts of the piping extend above the slab. These visible pieces are called *risers.*

Installing Pipelines

Most pipelines are laid in *trenches* and covered with earth. Pipelines are less likely to be damaged if they are run underground. And they are not exposed to the severe temperature changes that may take place above ground.

Only a few pipelines are placed on or above ground. This is a less expensive practice than digging trenches. But pipelines must be located where damage will not occur. And the temperatures must not be extreme.

Trenches are sometimes very deep. If the sides of the trench fall in, workers may be hurt. The trench must be braced to keep the sides from collapsing.

Trenches for pipelines may be dug with machines or hand tools. But they are most often dug with machines. The two types of equipment commonly used are the *backhoe* and the *trencher.*

Using a Backhoe

The backhoe has a long, jointed "arm" on the front of it. A bucket is on the end of the arm. See Fig. 41-10. The end of the bucket is lined with teeth. The arm is stretched out during digging. Then the bucket is dropped into the ground. It rests on its teeth. Then the arm is pulled in. As the arm moves, the teeth bite into the earth, filling the bucket. The bucket is swung to the side of the trench. Then the earth is dumped. Look again at Fig. 41-10.

Using a Trencher

A trencher uses a large wheel to dig a trench. Small buckets are attached to the wheel. As the wheel revolves, the buckets dig into the earth. The buckets dump the earth into a trough. The earth travels down the trough. It is dumped on a pile of earth or into a truck. See Fig. 41-11.

Fig. 41-10
Pipelines for public utilities are often laid in trenches dug beside city streets. Here a backhoe is being used to dig a trench.

Fig. 41-11
This trencher is performing a "cradling" operation. Cradling prepares the foundation for a pipeline.

United States Department of the Interior

Fig. 41-12 Pennsylvania Power and Light Company

These concrete pipe sections were carefully laid in place by a crane. Each section weighs about twenty-four tons.

Laying the Pipe

Pipelines are usually laid with cranes. The piping must be accurately placed in the trench. See Fig. 41-12. This is very important. It is particularly important in sewer lines. Sewer pipes are laid at a slight downward slope. Then the waste water can flow down to the treatment plant. See Fig. 41-13. Sometimes sewage must be pumped. But this is avoided whenever possible.

The Need for Careful Installation

All piping and plumbing must be installed carefully. If it is not, people and property may be harmed by pipeline or plumbing failures. For example, oil

Fig. 41-13

This map shows the water supply and sewage system for a small community. Notice the effluent discharge line. It is at a lower elevation than the community. How will water be raised from the river to the water treatment plant?

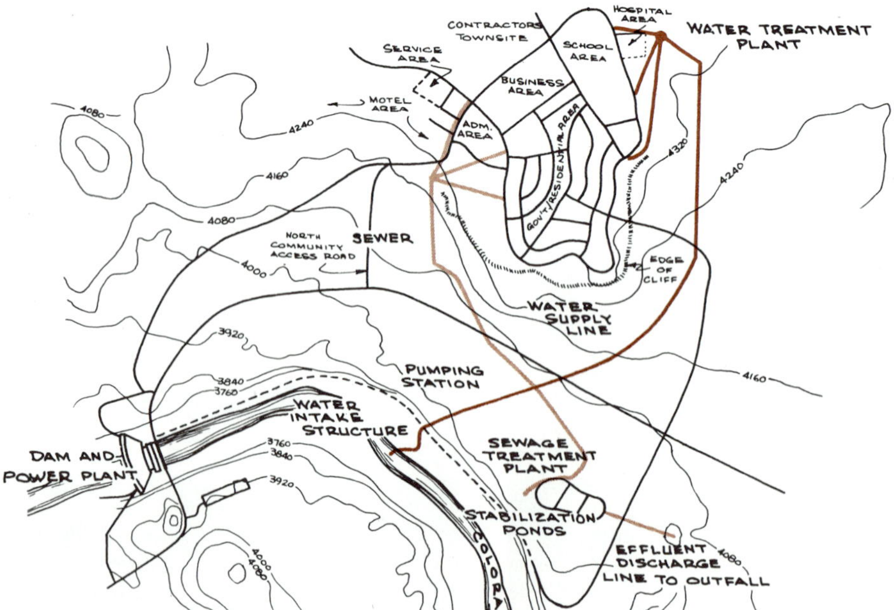

U. S. Department of the Interior

Fig. 41-14
Tamping around a pipe prevents the earth from settling later.

pipeline leaks may damage the land in an area. Plumbing leaks may result in expensive repairs to a home. For example, water-damaged ceilings may have to be replaced. Or, leaking gas from piping may cause a fire or explosion. People may even die from being overcome by sewer gas or other toxic (poisonous) fumes. Making sure that pipes do not leak prevents these dangers.

Backfilling and Tamping

After the pipe is laid, the trench is *backfilled*. This can be done with a bulldozer. Or, it may be done with a crane that has a clamshell bucket. This bucket picks up a load of earth. Then it drops the dirt into the trench.

The earth around the pipe is *tamped* (pounded down). The equipment used is a tamper. Tampers may be hand operated. Many are power driven. See Fig. 41-14. Filling and tamping continue until the surface is level. The earth is also packed above the pipe, at grade level. A roller or the treads or wheels of a heavy tractor do this. Tamping and rolling prevent the earth around the pipe from settling later.

Records

Accurate records must be kept of the locations of underground pipelines. The locations of valves, accesses, pumping stations, and other features must also be recorded. These records are usually in the form of drawings. They are kept in the office of the company or in a government office.

Pipeline locations are often *monumented*. This means permanent marks are placed on the ground. These are made in much the same way as surveying monuments. (See Reading 5 to learn about surveying monuments.)

Pipeline records are used by service workers making repairs. They also are used by persons planning new construction. Persons doing earthwork refer to pipeline records to avoid breaking existing pipelines.

Summary

Plumbing systems distribute liquids and gases throughout a building. They are made of pipes and fittings of various sizes and materials.

Architects determine the number, type, and location of plumbing fixtures, appliances, and other equipment. Engineers design the piping to be used to service these. The actual plumbing work is usually done by skilled workers employed by the subcontractor. The subcontrator directs the work. Plumbing and pipeline workers must work carefully. They must make sure that lines do not leak. The materials also must flow freely through the pipes.

Pipelines are outside of structures. They transport liquids and gases from a source of supply to buildings. Other pipelines move waste from structures and excess water away from the streets and from structures. Materials are taken to treatment plants. Excess water is usually drained into a stream.

Most pipelines are laid underground in trenches. A variety of materials may be used to make the piping and connections. Each kind of piping is installed according to a special technology.

All parts of a piping system must be able to be located for servicing. And builders want to avoid breaking existing lines. Therefore, careful systems records (drawings) are kept. They show the locations of pipelines.

Terms to Know

- backfilled
- backhoe
- backing boards
- cement grout
- fittings
- flanges
- mechanical subcontractor
- monumented
- operating engineers
- pipe dope
- pipe hangers
- pipe layers
- pipelines
- plumbing
- plumbing subcontractor
- risers
- soil pipe
- solder
- sweated joint
- tamped
- trencher
- trenches

Think About It!

1. What are the differences between plumbing systems and pipeline systems?
2. Can you identify a pipeline system in your community that is above ground? below ground? What is each pipeline system used for?
3. Look at some plumbing in your home. What materials is it made of? What does it carry? How many different types of fittings are used?

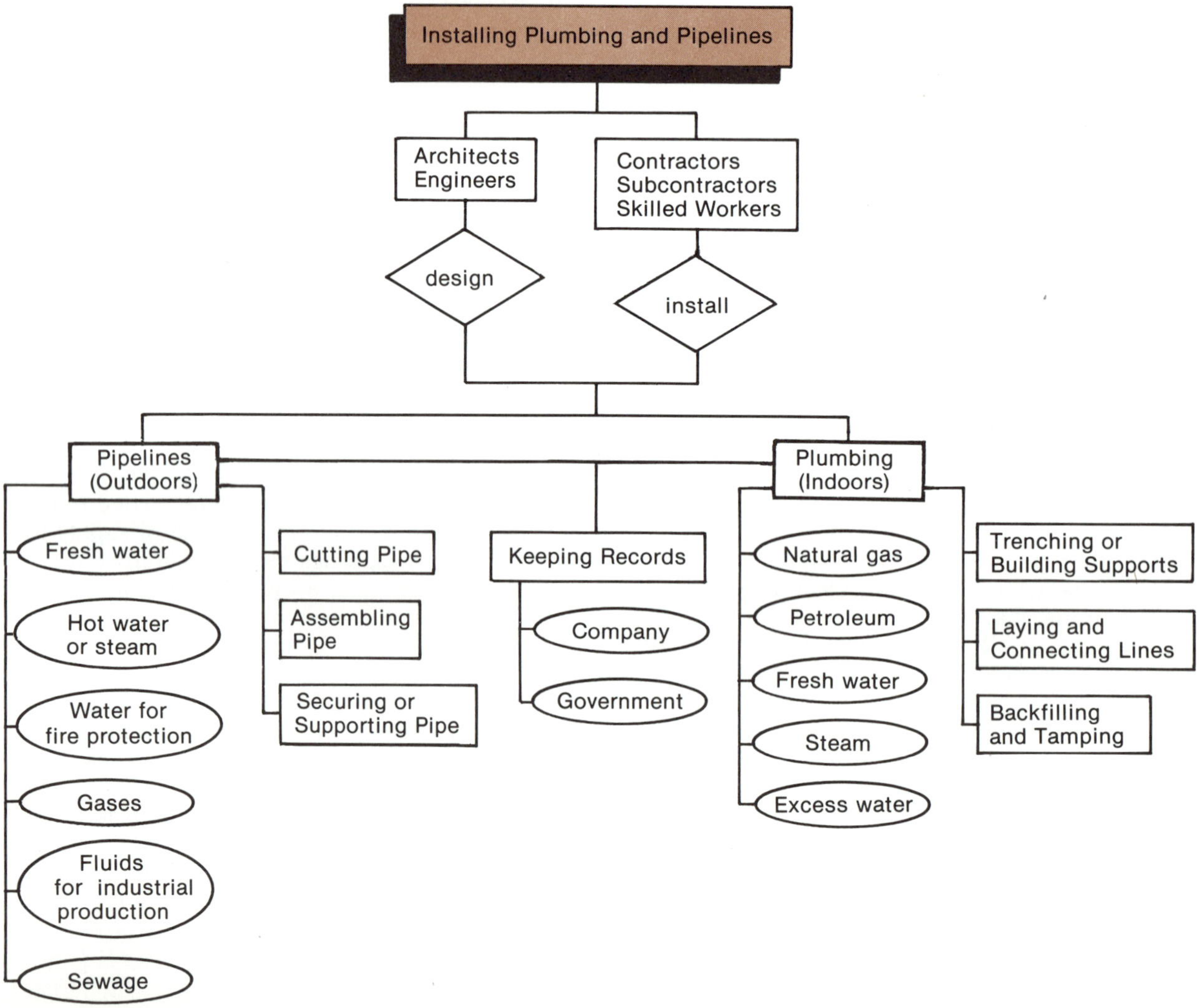

Installing Electrical Systems

Electrical Systems

Electric power is generated (produced) in a power plant. It is supplied to users through two systems. One system involves *outside electrical construction*. It transmits electricity from a generating (power) plant to structures. An earlier reading, Reading 38, told you about outside electrical construction.

The other system involves *inside electrical construction*. This system *distributes* power inside a structure. In another reading, a brief description is given of inside electrical construction. This reading will focus on how electrical systems are designed and installed inside structures.

There are two kinds of inside electrical systems.

- *Power* systems are one kind.
- *Communications* systems are the other.

Electrical utility companies own the power plants. They also own all outdoor transmission equipment. The power is brought to a *service entrance*. This entrance is located on the outside of each structure needing service.

A *meter* is most often installed just outside a structure. It measures how much electricity a customer uses. Generally the utility company owns the meter.

Some communications equipment may be owned by a communications company. But the electrical equipment within the structure usually belongs to the owner of the structure.

People Who Design and Install Power Systems

Electricity can be very dangerous. For this reason, only qualified people design and install electrical systems.

People Who Design

In all large outside and inside jobs, *electrical engineers* design electrical systems. They do not design each piece of equipment. But they select the proper equipment. And they plan the way the equipment should go together. See Fig. 42-1. Civil and mechanical engineers may also be involved.

Each engineer must be *registered* or work with one who is. To be registered, an engineer must pass a state examination. Education and experience help engineers qualify.

People Who Install

A general contractor usually subcontracts electrical installation work. The subcontractor must be *licensed* to do this work. Licenses are issued by the state, county, or city.

Fig. 42-1
Electricity is very useful, but it can also be dangerous. Electrical engineers use their knowledge and experience to design safe electrical systems.

Niagara Mohawk Power Corporation

Fig. 42-2
Skilled linepersons know how to handle "hot" lines safely. They use "hotsticks" when working on energized transmission lines.

Electrical subcontractors must employ qualified workers. These skilled workers are called *electricians*. See Fig. 42-2. Electricians must have technical knowledge and practical ability. They, too, are licensed by state, county, or city authorities. Most *building codes* require that work be done by licensed electricians.

Installing Inside Power Systems

Electric power enters a structure at a service entrance. Electricians connect wires to the meter. The meter is connected with wire to a *distribution panel*. See Fig. 42-3. This panel contains circuit breakers or fuses. Circuit breakers automatically *break* the circuit if an *overload* occurs. This means they interrupt the path of electric current. This protects the electrical equipment and the building from damage. The panel is installed by electricians.

From the panel, independent *branch circuits* are installed. The circuits service the power needs of the structure. Wiring may be done by one of several methods.

Fig. 42-3
Electricity entering a structure passes first through a distribution panel. Wiring is run inside walls, from the panel to all parts of the structure needing power.

Fig. 42-4
Kinds of wire commonly used in homes.

United States Electric Company

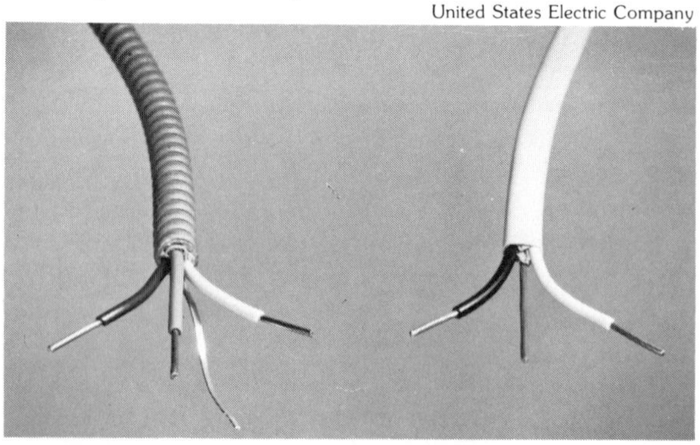

A. Armored cable *B. Nonmetallic sheathed cable*

Cables

In a home, two kinds of wires are commonly used. These are *nonmetallic sheathed cable* and *armored cable*. Both kinds of wiring are flexible (can be bent). See Fig. 42-4. The wires in nonmetallic sheathed cable are covered by plastic material. Armored cable is enclosed by flexible steel called *armor*. Each wire in the cable is *insulated*.

Cable is installed inside walls, ceilings, and floors. It must be fastened in place according to building codes. Cables are connected to outlets such as junction boxes and switch boxes.

Conduit

In non-residential structures, electrical wires are protected by *conduit*. Conduit is thin-walled tubing. It may be rigid (stiff) or flexible. Conduit is made of galvanized steel or aluminum. It carries branch circuits from the distribution panel.

Conduit may be installed and then buried in concrete. It may also be placed in walls, ceilings, and floors. The location of conduit must be planned carefully. It should not interfere with other construction. For example, it should not get in the way of ductwork, plumbing, and other utilities.

Once the conduit is installed, wire is run through it. See Fig. 42-5. A *fish tape* is used to pull wire through the conduit. Then wires are connected to switch boxes and other outlets.

Communications Systems

People can communicate easily face to face. They talk or use gestures such as pointing. But when they are apart, they use communications equipment. See Fig. 42-6. This equipment is also useful in controlling transportation. For example, communications devices are used on lakes and oceans. People use these to control traffic in shipping lanes. Other communications devices make it possible to control air traffic. Communications systems must be designed and installed by qualified people.

Fig. 42-5
In non-residential construction, wiring often is run through pipes called conduit.

Fig. 42-6
Advanced communications systems allow such innovations as "picturephone" conferences. This type of communication service is being developed today. But someday, such systems may be installed routinely in construction projects such as office buildings.
American Telephone and Telegraph Company

People Who Design and Install Communications Systems

Communications systems are designed by electrical engineers. These engineers specialize in electronics. An engineer may work for the operator of a communications system such as a telephone company. Often engineers work for the manufacturer of communications equipment. Engineers also supervise the installation of communications systems.

Technicians install, operate, and maintain communications equipment. See Fig. 42-7. They may be employed by the communications company. Or, they may work for manufacturers or contractors.

A communications company may have its own workers build a communications facility. Or, it may contract the work. For example, a television station may hire a general contractor to build a transmitting tower. Other specialists may install antennas and wiring. Equipment manufacturers usually send their experts to help.

Many types of skilled workers are employed. Linepersons work on outside wiring and equipment. Wirepersons (wirers) work on inside electrical construction. Installers hook up special equipment. See Fig. 42-8. They are all electricians.

Transmitting and receiving devices are used in communications systems. Connecting facilities must be built. This involves installing wire, cable, and antennas. Engineers, technicians, contractors, and skilled workers are required.

Installing Communications Systems

Messages are carried between senders and receivers by land-line or wireless equipment.

Fig. 42-7
Telephone switching equipment is maintained by trained technicians.

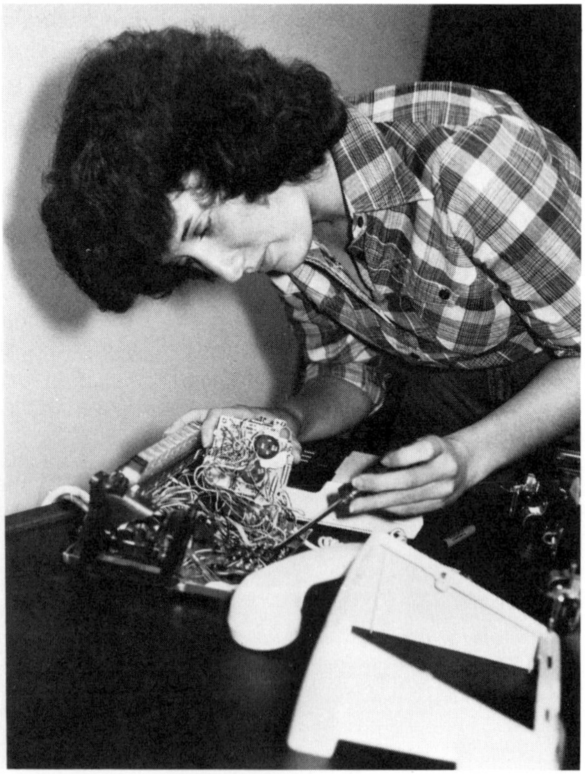

Fig. 42-8
An ordinary telephone is not as simple as it seems. Installers must be knowledgeable about telephone equipment and wiring systems.

Fig. 42-9 Western Electric Company, Incorporated

Microwave equipment is installed in high places. Buildings and power lines cannot interfere with receiving and sending messages.

Land-Line Equipment

Many communications devices transmit and receive messages through *land-line equipment*. This equipment is connected with wires. The telephone is one example. The wires carry a modified electric current. Sound waves are changed to electric waves by a telephone. Another telephone receives the electric waves and changes them back to sound waves.

Special wires are run inside buildings for communications devices. These wires are different from the wires in a power system. They are most often lighter in weight. They usually distribute electricity of lower voltage. The amperage (current) is lower as well.

The common telephone is installed using special wires. They are not connected to the building's power system. The telephone company provides the power service from outside the structure. Telegraph companies also make use of existing telephone lines.

Other communications devices use other special wires or cables. One example is the intercom. Another is a cable television system.

Electricity may need to be transformed (changed) before it can be used in these special wires. A *transformer* changes voltage up or down to usable levels. Power to run these communications systems comes from the power system in the building.

Wireless Equipment

Many messages are sent and received by devices that are not connected by wire. This is communication by a *wireless system*. The communications device changes electric energy into *electromagnetic energy* (waves). These waves are received elsewhere. They are changed back to electric current as they are received. The radiotelephone operates in this way. Radio and television use a combination of land-line and microwave equipment. See Fig. 42-9.

Summary

Electricity is supplied to structures by outside electrical systems. Outside equipment is usually owned by a utility company.

Inside electrical systems distribute electricity inside structures. The owner of a structure owns the electrical equipment within it.

Two kinds of inside systems are installed. These are power systems and communications systems. Homes are wired using cables. Wiring in non-residential structures is protected by conduit. Communications devices are installed in many structures. They may also require separate structures.

Engineers design all electrical systems. Contractors or subcontractors, technicians, and electricians install them.

Terms to Know

armor	insulated
armored cable	land-line equipment
branch circuits	licensed
break	meter
building codes	nonmetallic sheathed cable
communications	outside electrical construction
conduit	overload
distributes	power
distribution panel	registered
electrical engineers	service entrance
electricians	technicians
electromagnetic energy	transformer
fish tape	wireless system
inside electrical construction	

Think About It!

1. Can you trace your school's outside electrical construction?
   a. Power plant.
   b. Transmission lines.
   c. Transformers.
   d. Service entrance.
2. Can you locate your home's three main elements of inside electrical construction?
   a. Entrance.
   b. Distribution panel.
   c. Branch circuits.
3. What communications systems do you have in your home? Do you have both transmitting and receiving devices?

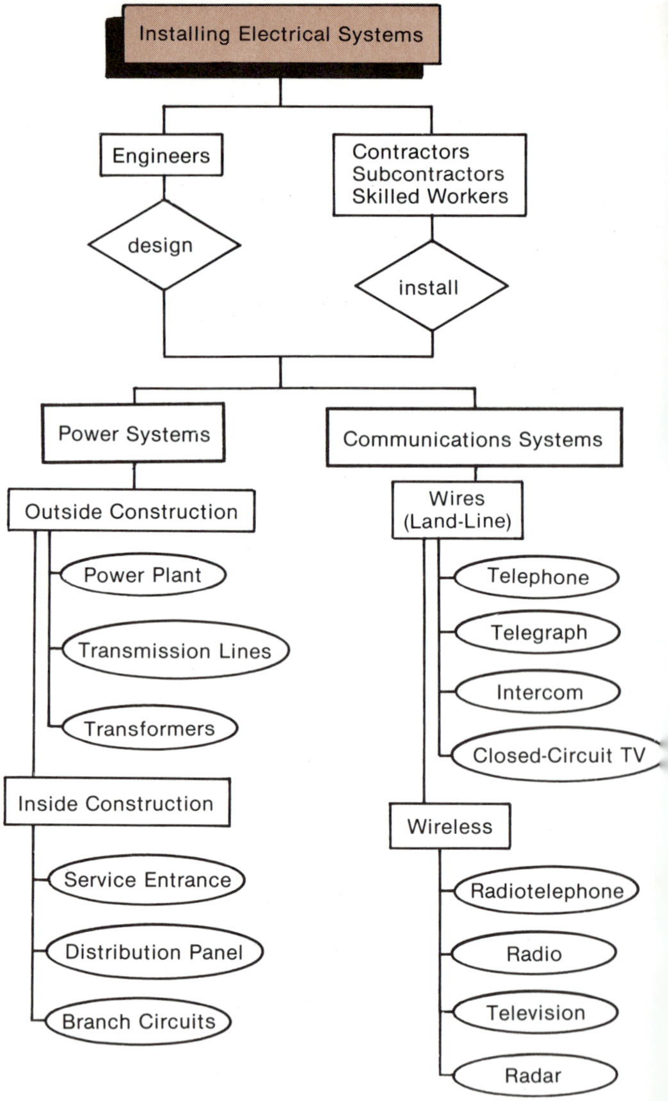

Reading 43

Enclosing Exteriors – Residential

Exterior of the Structure

A *residence* (house or apartment building) needs a firm *foundation*. The foundation is the lowest part of the structure. It is mostly below ground. A foundation provides a *load-bearing* surface. This means it supports the weight of the *superstructure*. *Footings* are the widest part of the foundation. See Fig. 43-1. In structures with a slab foundation or basement wall, the footings support the superstructure. Together, they make up the foundation.

The *skeleton* (frame) of the building rests upon the foundation. The skeleton encloses the space that will form the interior of the building. Usually, both the outside and inside of the skeleton are covered by another material. This reading describes the skeleton and the things done to cover the outside of the skeleton. This outer surface is often called *"skin."*

Building the Skeleton – Walls

The *wall system* is one of the basic elements of a superstructure. They are usually built of wood or masonry. Sometimes they are built of metal. There are four basic types of walls.

- Stud walls
- Panelized walls
- Masonry walls
- Post-and-beam walls

Openings must be left in the walls for windows and doors. Other openings for *vents* and *louvers* must be planned. Vents and louvers allow air to circulate. They are usually located in the attic. If there is no basement, vents and louvers are also located in the foundation.

Fig. 43-1

Footings support the load of the superstructure.

A. *Spread footings are used on surfaces of rock or soil.*

B. *Raft (slab) footings are used on soft soils.*

C. *Pile footings are used to reach firm soil or rock.*

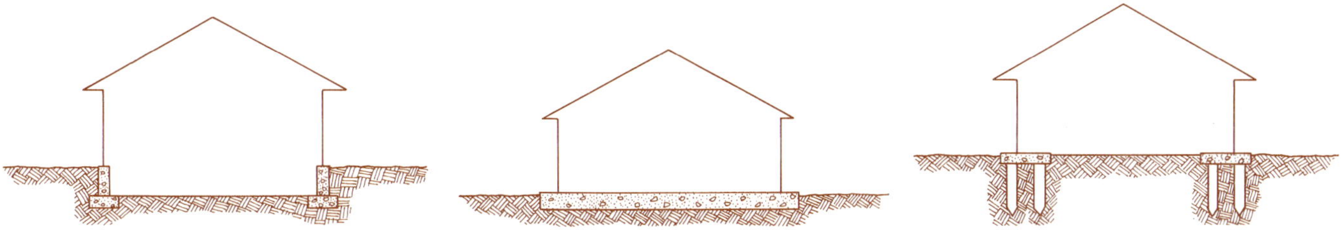

Weyerhaeuser Company

Fig. 43-2
These stud walls have been assembled on the site. The diagonal boards are braces. These make the frame rigid and solid.

Stud Walls

One of the most common types of walls is the *stud wall.* See Fig. 43-2. Studs are vertical supports. They are usually made of lumber. Today, some studs are made of metal. In stud walls, studs are spaced so that standard sizes of siding and panel sheathing can be fastened to them. After the stud wall has been put up, it is covered on both sides by siding or panel sheathing.

A *stud system* carries the load (such as a roof) placed on top of the studs. Space between the studs may be used for insulation. Or, it may be used for heating ducts and electrical and plumbing lines. See Fig. 43-3.

One advantage of a stud system is that it can be put up quickly. See Fig. 43-4. Once a roof has been put on top of the stud system, workers are protected from the weather. Other advantages of a stud system are its structural strength and the low cost of its basic materials. A stud system can be assembled on the site. Or, it can be assembled off the site (in a factory).

Panelized Walls

Panelized walls are *component systems.* They are preframed *wall panels* made of wood or metal studs covered by sheathing. Insulation, electric wiring, and plumbing lines may already be in the wall panels. Windows and doors, prefinished on both sides, may also be in the panels. Panels are made in sizes that are easy to handle. See Fig. 43-5.

Masonry Walls

Masonry walls may be made of concrete block. These are the easiest masonry walls to construct. Brick walls are another common type of masonry wall. A *cavity* (space) may be left between two

Fig. 43-3
Utilities are installed in the stud system.

A. *Water supply lines are installed between studs.*

B. *An electrician uses the studs to run wiring to outlets.*

Fig. 43-4
Stud walls are made on the floor. Then they are tilted up and nailed in place.

Fig. 43-5
This preframed wall panel is lifted mechanically, but only one worker is needed to guide it into place.
United States Steel Corporation

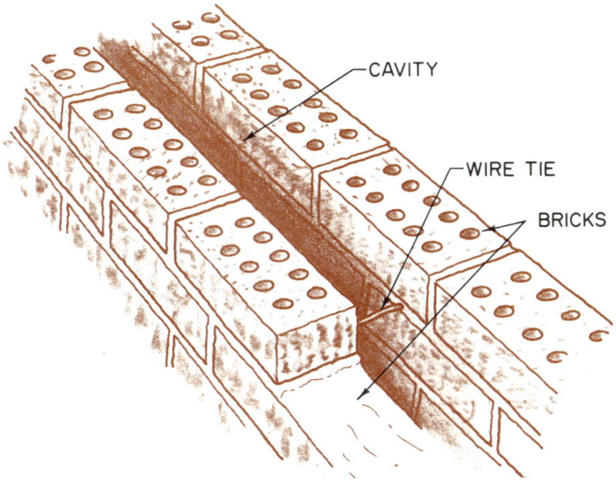

Fig. 43-6
Two masonry walls may be built side by side, tied together with metal ties. The cavity between the walls interrupts the flow of heat, especially when filled with insulation.

separate masonry walls. See Fig. 43-6. Sometimes, both walls are brick. Sometimes, one is brick and the other is made of tile or concrete block. The outside wall may be made of stone. Many kinds of stone may be used.

Masonry transmits heat easily. This can be a problem. In cool weather, heat from inside may be transmitted outside. In warm weather, the sun's heat may be transmitted inside. To reduce this, the cavity can be filled with *insulation*.

Brick or stone *veneer* is a *non-load-bearing* masonry facing. It is put over other structural materials. See Fig. 43-7. Veneer makes the exterior face weatherproof. Concrete block may be left exposed. Or, it may be covered by stucco or other coatings. Masonry walls are strong and have low maintenance costs.

Fig. 43-7
Brick veneer is a durable and attractive covering for a wall.

Post-and-Beam Walls

In a *post-and-beam wall system,* the load of floors, ceilings, and roofs is carried on posts. Wall sections between the posts are non-load-bearing. Beams between posts help support the wall materials. See Fig. 43-8. Posts and beams are attractive. For this reason, they are left exposed. This makes them a part of the structure's design.

Building the Skeleton — Roofs

The *roof system* is another of the basic elements of a superstructure. The roof skeleton is usually made of wood. See Fig. 43-9. In some cases, it is made of concrete or metal. When the roof skeleton is covered, the roof gives protection from weather. A roof can also add beauty to a home.

Styles of Roofs

The roof helps identify a building's *architectural style*. There are many roof types. Roofs may have flat or inclined (sloped) surfaces. Refer to Fig. 43-10 as you read.

Fig. 43-8
Framing for a post-and-beam wall system. Posts and beams support the weight of the structure. Lightweight materials and large areas of glass could be used in non-load-bearing walls.

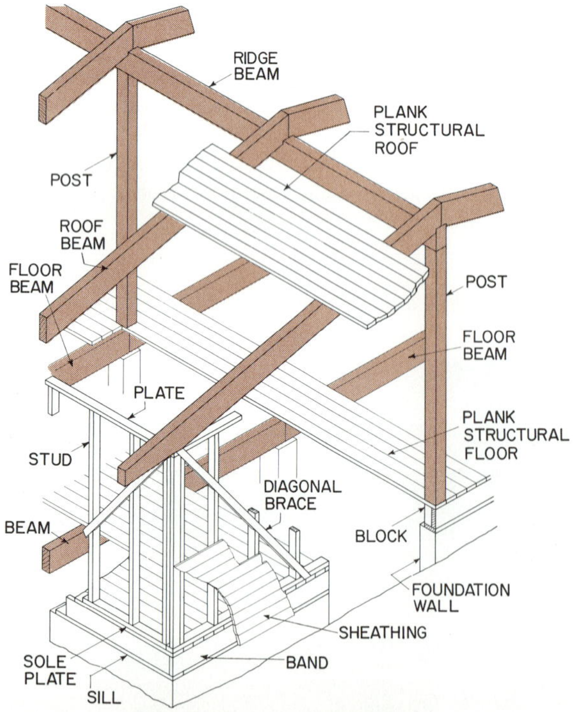

The *flat roof* is the least expensive and easiest to construct. If a flat roof is used in areas of heavy snowfall, it must be made very strong. It should be sealed carefully to prevent leaks.

A *shed roof* is like a flat roof that has been raised on one side. The slope allows moisture to drain off quickly.

The *gable roof* is very common. It looks like a combination of two shed roofs. The two sloped surfaces meet at a center line. The space under a gable roof is often used for an attic.

The *hip roof* has four surfaces that slope up toward the center of a building. This type of roof is often used. In buildings that are square, the four sides of the hip roof form a peak in the center.

The *gambrel roof* has two sides that meet at a center line. Each side has two sloped areas. The upper area has only a slight slope. The lower sloped area is almost vertical. The shallow slope at the top of a gambrel roof makes more of the attic space usable.

The *mansard roof* has four sloped sides. Each side has two sloped areas. The top areas are almost flat. The bottom areas are almost vertical. They form walls. This makes it possible to use the attic space as a living area.

Roof Construction

The principal framing member in roof construction is the *rafter*. Units called *trusses* serve the same purpose. See Fig. 43-11. Trusses are very strong. They may be made of wood or of light metal such as aluminum. Small houses usually have trussed roofs.

Another type of roof can be made with roof beams and planks. Look again at Fig. 43-8.

Fig. 43-9
Typical roof construction.

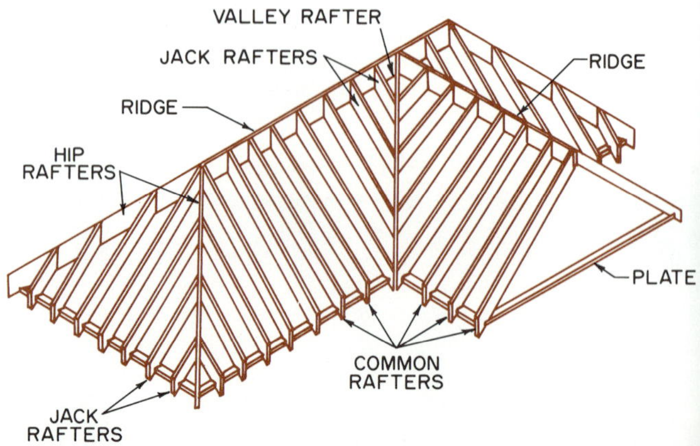

Gambrel roof

Gable roof

Hip roof

Mansard roof

Marshall Erdman and Associates, Incorporated

Flat roof

Shed roof

Fig. 43-10

The roof is an important part of the building's design.

Roof Framing Terms

Roof construction has its own vocabulary. Three of the most important terms are *rise, run,* and *span.* These measurements are shown in Fig. 43-12. The term *slope* is also important. A roof with a very slight slope rises gently. A steeply sloped roof rises sharply. Builders must know the slope, rise, and run of the roof to be built. The length of rafters needed to build the roof depends on the slope and the total run.

If you know the rise and the run of a roof, you can determine its slope. Divide the rise by the run. Slope is expressed as the ratio of rise to run. It is written as a fraction.

$$\frac{\text{Rise}}{\text{Run}} = \text{Slope}$$

Fig. 43-11

A typical roof truss for a small house. The web members and gussets add to the strength of the unit.

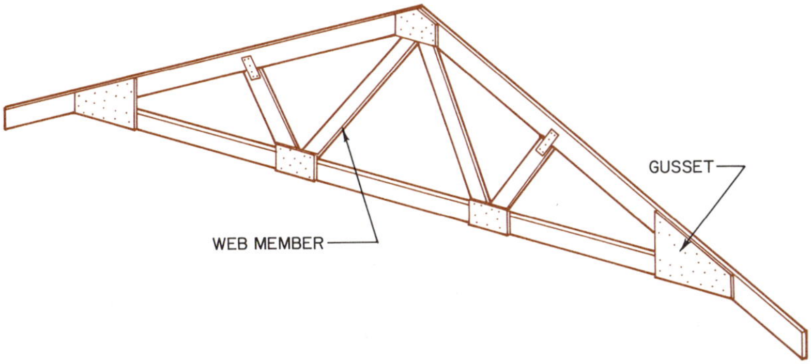

Fig. 43-12

For a simple gable roof, the run is one half the span.

Covering Walls and Roofs

On all kinds of walls other than masonry walls, *sheathing* is applied to cover the exterior. Sheathing may be lumber. But usually it is panels of plywood, gypsum, or insulating board. See Fig. 43-13. These materials make walls strong. In precut and prefabricated construction, sheathing makes the frame rigid. This permits easy transportation.

Another type of sheathing now in use is foam plastic sheathing. It is not very strong. However, it provides good insulation.

Fig. 43-13
Sheathing is used to cover wood-framed walls. Plywood sheathing has been placed on the corners of this structure to add more strength.

Fig. 43-14
Installing plywood sheathing over a roof frame.

Northwood Mills, Limited

Sheathing is also used to form a surface under the roofing. Sheathing used on the roof is called *decking*. Plywood is the most common roof sheathing. See Fig. 43-14. Shingles or other roofing materials are applied over the sheathing.

Installing Doors, Windows, and Louvers

Earlier in construction, window and door openings were left in the skeleton. After sheathing is applied, standard door and window frames are installed. See Fig. 43-15.

Also, the space between the roof and ceiling of a house must be *ventilated*. Vents or louvers are installed in the gable, in the *soffit* (bottom of roof overhang), or in the roof. See Fig. 43-16. Vents allow air to circulate through the space between the ceiling and roof. Vents in exterior walls allow air to circulate in the *crawl spaces* below floors. The circulation lets moisture and unwanted heat escape.

Finish Siding and Roofing

Finish siding is placed on top of wall sheathing. It resists moisture. This helps the structure shed water. Finish siding also provides insulation and makes the house look better. See Fig. 43-17.

Many materials are used for finish siding. Fiberboard, aluminum, enameled steel, and plastic are common. Masonry veneers (such as brick and stone) are also used. In the future, new siding materials may be made. Recycled materials may be used. This will cut costs. Also, exterior walls may be made with materials that provide more insulation.

Wood or asphalt shingles are common roofing materials. See Fig. 43-18. Slate, tile, and metal are others.

The *cornice* is the part of the house where the roof and sidewalls meet. On most houses, the roof overhangs (sticks out) from the wall. One common way to finish the overhang is with a boxed cornice. See Fig. 43-19.

Summary

The skeleton of a residence rests upon the foundation. The skeleton includes a wall system and a roof system. Openings must be left in the walls for doors, windows, and special vents.

The skeleton is covered with a "skin." Many different materials may make up this outer surface. It protects the inside from the weather. It also adds beauty to the structure.

Fig. 43-15
Installing standard window frames.

Fig. 43-16
Installing a vent in the soffit of a home.

Magnum Photos, Incorporated

Fig. 43-17
Finish siding is applied over a base of sheathing.

Fig. 43-18
Shingles are a popular finishing material for sloped roofs. Notice how the shingles overlap. This helps water to run off.

Fig. 43-19
A boxed cornice.

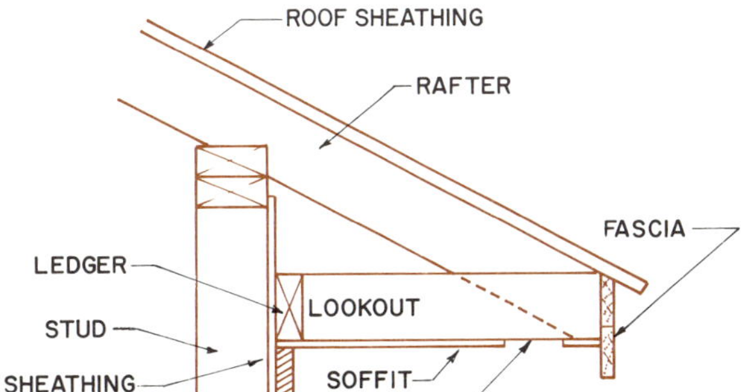

Terms to Know

cavity
component systems
cornice
crawl spaces
decking
finish siding
flat roof
footings
foundation
gable roof
gambrel roof
hip roof
insulation
load-bearing
louvers
mansard roof
masonry walls
non-load-bearing
panelized walls
post-and-beam wall system
rafter
rise
roof system
run
sheathing
shed roof
skeleton
"skin"
slope
soffit
span
stud system
stud wall
superstructure
trusses
veneer
ventilated
vents
wall panels
wall system

Think About It!

1. How do the walls of your school differ from those of your home?
2. Do the seasons and the weather affect roof design where you live?
3. How do you think the exterior walls of homes will be covered thirty years from now? Why?

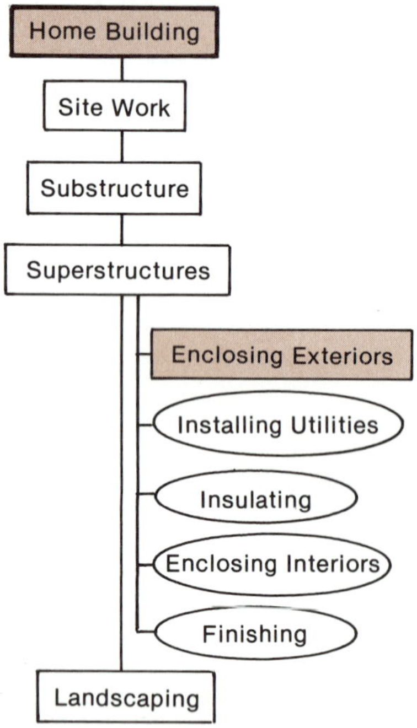

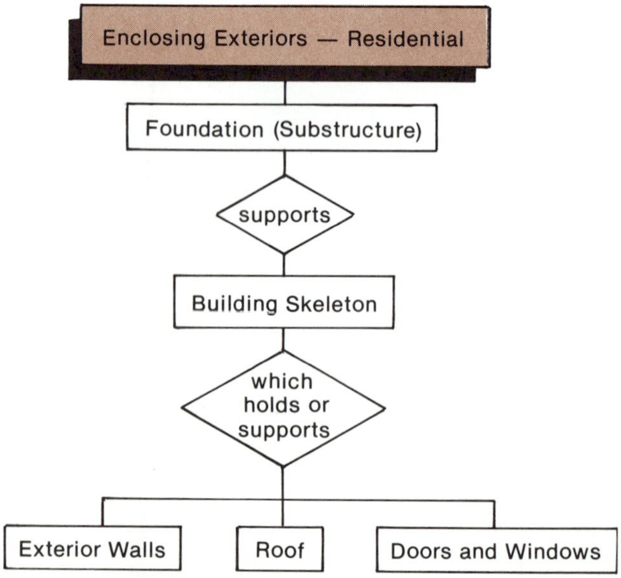

Reading 44

Enclosing Exteriors — Non-Residential

Open and Enclosed Structures

Most production practices apply to both *open* and *enclosed* structures. Some kinds of structures are usually open (left uncovered). Dams, roads, and towers used for utilities are examples. Most buildings, however, are wholly or partly enclosed.

This reading concerns enclosed *non-residential* structures. These are buildings, other than homes, that have a framed or solid wall superstructure. Some of the practices described here are used for residential structures, also.

Why a Structure Is Enclosed

The framework or walls of a building must be covered. This is done by constructing an *exterior* (outside) "skin." *Exterior walls* are part of the skin. The skin also includes the *roof*. See Fig. 44-1.

The skin protects a structure by keeping out sun, rain, snow, and air pollutants. It improves the structure's appearance. It also helps keep temperatures constant (even) inside a structure.

Enclosing a structure includes insulating, too. *Insulation* is installed in the roof and walls. It serves several purposes. The main purpose is to conserve energy. Insulation reduces the transfer of heat. In cold weather, it keeps the heat inside. In warm weather, it keeps the heat outside.

People Who Enclose a Structure

An architect usually decides what an exterior will look like. This person selects the materials and methods to be used to enclose a building. Selections are based upon function, economy (cost and quality), and appearance. Architects work with structural engineers. These engineers decide how the exterior skin and the framework should go together.

A general contractor is responsible for enclosing

Fig. 44-1

The roof portion of this school's exterior skin is in the shape of a dome. The shape provides more space inside the structure.

Bethlehem Steel Corporation

Fig. 44-2
Workers for general contractors or specialty contractors install exterior walls. These workers are installing glass panels.

Fig. 44-3
All the space underneath a flat roof is usable. This is one reason many office buildings have flat roofs.

the structure. The contractor's crew may do the work. Specialty subcontractors, such as masons, also do work on exteriors. See Fig. 44-2.

Many different types of skilled workers enclose a structure. Among them are carpenters, masons, and sheet metal workers. Roofers, glaziers (glass installers), and ironworkers also help enclose the structure. In fact, persons from all eighteen organized craft groups may become involved.

Roofs

A roof is a watertight skin that covers the top of a building. It may be *pitched* (sloped) or *flat*.

A pitched roof allows moisture to run off quickly. In cold climates, heavy snow cannot collect on it. A pitched roof may be used for appearance.

Some non-residential buildings have pitched roofs. But flat roofs are common on non-residential buildings. See Fig. 44-3. The space below the roof is completely usable. It is very important to get as much space as possible for the money spent. Flat roofs may be slightly sloped or slanted. This allows water to drain better.

Roofs are usually installed before the exterior walls are started. However, for some high-rise buildings, the exterior walls are built first. Then the structure is "topped out" with the roof.

Roof System

A *roof system* is composed of two main parts.

- Framework
- Roof deck

The *framework* supports the roof deck. It may consist of rafters, beams, and trusses. These are made of wood, steel, or concrete. The support for some roofs is built of reinforced concrete slabs. These slabs may be strong enough to support the roofing without any framework.

The *roof deck* is laid on top of the supporting framework. The deck may serve several purposes. It makes some frames more rigid. It supports the weight of rain or snow. On flat roofs, it also supports equipment, such as air conditioners, mounted on the roof. The deck may be made of steel, concrete, wood, or composition material. It must be covered with waterproof material.

Roofs often cover large areas. They are exposed to wide temperature changes. For these reasons, roof decks may be constructed with *expansion joints*. These joints allow the roof to expand (when heated) or shrink (when cooled) without cracking.

Insulation can be built directly into the roof covering. It is often placed right on top of the decking.

The most common roof for flat, non-residential structures is the *built-up roof*. See Fig. 44-4. In a built-up roof, several materials are combined. See Figs. 44-5 through 44-8. Specialty subcontractors apply the roofing materials. Built-up roofs are often guaranteed to last ten or more years.

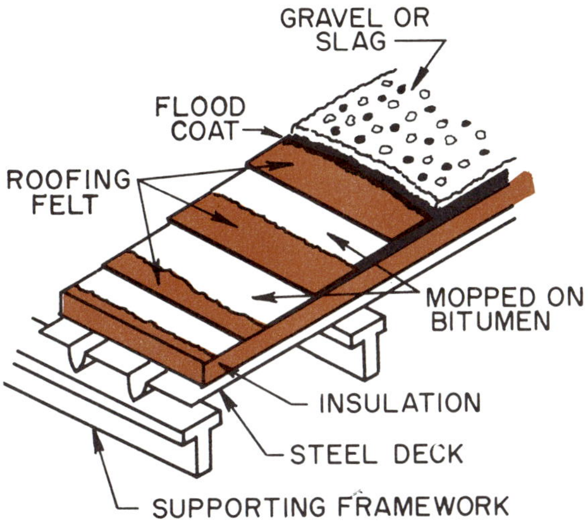

Fig. 44-4
In a built-up roof, several materials are combined to form a waterproof seal. Roofers apply the materials one layer at a time.

Fig. 44-5
Steel roof decking is laid over the framework.

Fig. 44-6
Insulation is laid on the decking.

Fig. 44-7
Roofing felt is placed over a layer of bitumen, a type of asphalt. Additional layers of these two materials are applied alternately.

Fig. 44-8
A thicker layer of bitumen and, finally, a layer of gravel are spread over the surface. This completes the built-up roof.

Libbey-Owens-Ford Glass Company

Fig. 44-9
Many modern structures use curtain wall construction. The glass in the building does not support the roof. Instead, it covers the framework. The framework supports the entire structure.

Exterior Walls

There are two types of exterior walls.

- Load-bearing
- Non-load-bearing

Many structures have *load-bearing* walls. A one-story warehouse is an example. The entire load of the structure is carried by the walls. There is no supporting framework inside.

Most tall structures have *non-load-bearing* exterior walls. That is, the framework carries the load of the structure. In framed structures, non-load-bearing walls are called *curtain walls*. See Fig. 44-9. Curtain walls always include a surfacing material. Frames and load-bearing walls within the structure may not. For example, in warehouses and factories, only the exterior walls are enclosed.

Many different kinds of materials are used to cover exterior walls. Masonry materials such as brick, stone, and marble are common. Wood siding is used. Panels of wood, plastic, glass, brick, stone, and aluminum are common coverings. See Fig. 44-10. Steel, enameled steel, and concrete form exterior surfaces.

One type of panel construction is called *tilt-up*. Concrete slabs are made in forms on the ground. When the slabs harden, they are lifted into place and fastened.

Fig. 44-10
Exterior panels may be made of many materials. The Gateway Arch in St. Louis is covered with metal panels.

Fig. 44-11
Windows serve many useful purposes.

A. Display windows allow people to see items in a store.

Sorg's Jewelers

B. Business transactions may take place through window openings.

Triple Treat, Incorporated

Exterior walls are fastened in position in many ways. They may be held by *mechanical fasteners*. Some materials are *cemented* in place. Others are *welded*. They must be installed according to the plans of architects and engineers.

Windows

Windows often make up a large portion of a building's surface. They may be used for any of a number of purposes.

Windows can be used for ventilation. They may be opened to allow fresh air into the buildings.

If properly positioned, windows allow solar energy to warm the interior of a building. And they provide natural light inside.

Windows also allow people to see in or out of the building. See Fig. 44-11. For example, display windows allow people to see items within a store.

Windows can improve the appearance of a building. They should also be energy efficient.

Insulation

We must conserve energy. For this reason, exterior surfaces (roofs and exterior walls) must be insulated. See Fig. 44-12.

Insulation should be built into the roof and walls. Insulation reduces *heat transfer*. It helps maintain constant temperatures inside the structure. It retards (slows) the passage of noise and moisture. Insulation also slows the spread of fire.

Rigid or *loose fill* insulation may be placed inside walls. *Batt* or *blanket* insulation is very common on the inside of exterior walls. Rigid or blanket insulation may be fastened in spaces under the floor and roof. Some rigid insulation is installed in the form of foam.

Fig. 44-12
Loose fill insulation is poured into the spaces in concrete block.

Portland Cement Association

Sealing a Building

When materials are joined together, they form a joint. Joints can allow moisture and air to enter a building. Unwanted moisture and air must be kept out of buildings. Structures must be sealed. Sealing a structure makes the skin of the exterior tight.

To seal a building, *flashing* is used where roofs join with exterior walls. It is made of strips of sheet metal. *Caulking*, a plastic putty, is used to seal cracks. *Weather stripping* is placed around windows and doors. It is made of rubber, metal, plastic, or felt.

Summary

The exteriors of framed or solid wall structures are enclosed by exterior skin. This skin consists of a roof and exterior walls. The roof and walls keep out sun, moisture, and pollutants. A roof system has a deck and a supporting framework.

Exterior walls may be load-bearing or non-load-bearing. Load-bearing walls support the structure. Non-load-bearing walls are also called curtain walls. A framework, not the walls, supports the load of the structure.

A wide variety of building materials are used for exteriors. Contractors install these according to plans and specifications of architects and engineers. Many skilled craftspersons work for the general contractor. Specialty subcontractors are also involved in enclosing exteriors. Buildings must be insulated and sealed.

Terms to Know

batt
blanket
built-up roof
caulking
cemented
curtain walls
enclosed
expansion joints
exterior
exterior walls
flashing
flat
framework
heat transfer
insulation
load bearing
loose fill
mechanical fasteners
non-load-bearing
non-residential
open
pitched
rigid
roof
roof deck
roof system
tilt-up
weather stripping
welded

Think About It!

1. Why is the exterior surface of a framed building sometimes called the "skin"?
2. What kinds of exterior materials enclose your school?
3. In your part of the country, are there more flat or pitched roofs? Why?

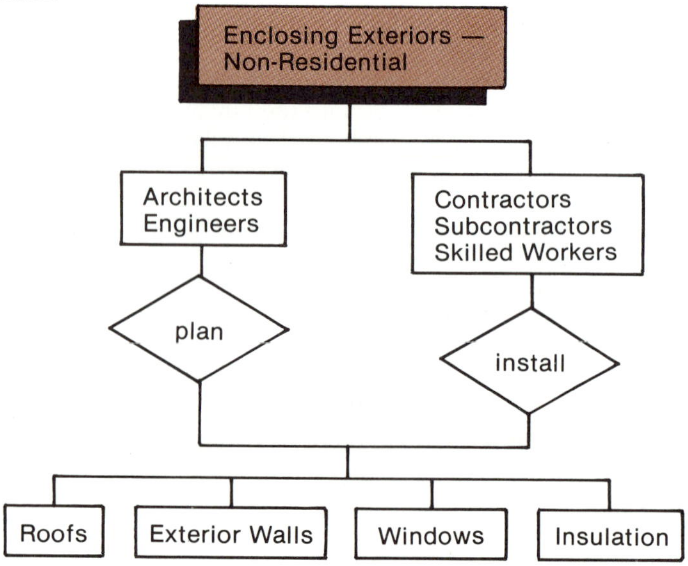

Reading 45

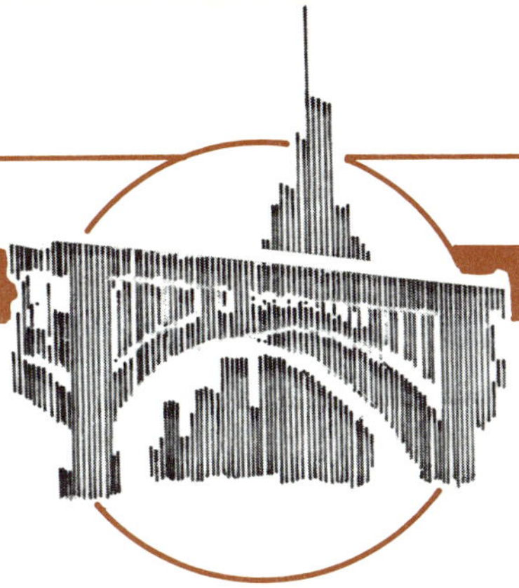

Insulating

Advantages of Insulating

Insulating saves energy. This is very important. It saves fuel. And it lowers operating costs. Insulating also improves safety and comfort. It can even reduce servicing costs.

This reading tells why *insulation* is needed and how it works. It also tells who installs it and what materials are used as insulation.

Reasons for Insulating

Insulation is used in a building for many reasons.

- Slows the flow of heat.
- Retards (slows) the spread of fire.
- Absorbs noise.
- Serves as a vapor barrier. Shutting out moisture reduces rotting, rusting, and paint problems.

See Fig. 45-1.

Three Kinds of Heat Transfer

Insulation materials reduce the transfer of heat. This is because they do not *conduct* heat. That is, insulation does not allow heat to pass through. It helps prevent heat from flowing out of a building. It also keeps unwanted heat from entering. Understanding how heat flows will help you understand how insulation works.

Fig. 45-1
Insulation reduces heat transfer, noise, and moisture in a building. It also can slow the spread of fire.

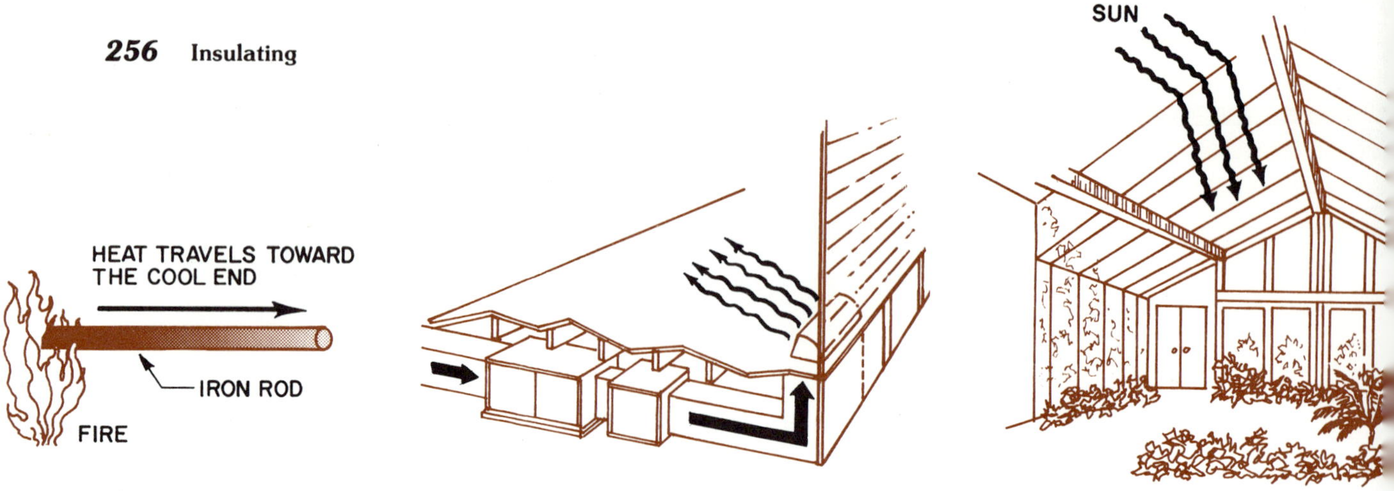

A. Conduction *B. Convection* *C. Radiation*

Fig. 45-2
Methods of heat transfer.

Heat may be transferred by

- conduction,
- convection, or
- radiation.

See Fig. 45-2.

Conduction of Heat

Heat energy can travel through a solid material. This happens, for example, when a metal rod is heated at one end. The other end soon becomes warm. The metal does not move. Only the heat energy moves through the metal. This process is called heat transfer by *conduction*. Look again at A of Fig. 45-2.

Heat travels through materials at different rates. The rate depends upon the insulating quality of the material. Most metals conduct heat very fast. Glass, stone, and clay conduct heat more slowly. Heat also moves slowly through many plastics. Suppose you have a metal cup and one made of plastic. Imagine filling them with hot water. Which one is more likely to feel hot first?

There are ways to slow conduction. One is to use a material that has been processed to become *porous* (fluffy). For example, solid glass conducts heat rapidly. It is a poor heat insulator. But glass can be made into a good insulating material. It is spun into thin fibers. This material is called fiberglass. It is made into a fluffy pad or *batt*. See Fig. 45-3. Much air is trapped in fiberglass batts. Heat cannot travel by conduction through the trapped air.

Convection of Heat

Air may be heated. If moved, it carries the heat. Air in a tropical region becomes warm. If the warmed air moves northward, it is a warm breeze. This process is called heat transfer by *convection*. A warm-air heating system works in much the same way. The air is heated. Then it is moved slowly around the building. The heat is conveyed (carried) with the air. This moving air is called a convection current. The heat is slowly passed off to objects in the current. See again B of Fig. 45-2.

Fig. 45-3
Fiberglass insulation is very porous. It slows the conduction of heat.
Owens-Corning Fiberglas Corporation

Fig. 45-4
The designer planned large windows for the sunny side of this home. The sun's heat will pass through the glass, helping to heat the home.

Radiation of Heat

Heat energy can be transferred in another way. This is by *radiation*. Look again at C in Fig. 45-2. Radiation is very different from conduction and convection. The sun's heat warms the earth. To do this, it travels millions of miles. It passes mostly through a vacuum (empty space). The heat does not come by conduction or convection.

Heat rays (radiant heat energy) are not absorbed as they pass through air. But when they strike a solid object, several things can happen. It depends on the kind of surface they strike. A shiny surface *reflects* (sends back) most of the heat rays. An object painted white reflects well, too. A dark object *absorbs* much of the radiant energy. The object becomes heated. Dark cars parked in the sun will get hotter than light cars. And a brown structure absorbs more radiant heat than a white one. But radiant heat can be kept out. Insulation that reflects the heat can be used. To let radiant heat in, transparent glass can be used. It lets most of the heat rays pass right through. (You can feel the sun's warmth through a closed window.) See Fig. 45-4.

Insulating Materials

Insulating materials are made from mineral, vegetable, or metallic products. Mineral and vegetable types work on the same principle (rule). The materials have thousands of tiny pockets. The pockets trap air. This keeps heat from passing into or out of the building. Metallic types reflect heat.

Mineral Insulation Materials

Mineral insulation is made from earth materials. These include sand, glass, mica, and certain plastics. Mineral insulation is made in several forms.

- Loose fill. This consists of loose particles.
- Batt. It is also called blanket insulation. Batt insulation is manufactured in standard widths. It is flexible. It is fitted into spaces, usually between studs.
- Rigid. This consists of panels made of mineral products.
- Plastic foam. This is the same as the material commonly used for cups for hot drinks.

Vegetable Insulation Materials

Insulation made of *vegetable* products includes many materials. Cork, wood, and plant stalks are examples. They are made into panels called fiberboard or insulation board.

Metallic Insulation Materials

Metallic insulation is a shiny foil. It is applied to insulation panels. It also may be glued to the back of gypsum board. Then the insulation is called foil-backed drywall.

Installing Insulation

Loose fill insulation can be poured into spaces in the walls. Sometimes it is blown into place by air pressure. Also, it may be mixed with concrete and spread over an area. See Fig. 45-5.

Fig. 45-5
A. *Installing insulation in an attic.*

B. *Placing a mixture of concrete and insulation on a flat surface.*

Zonolite Company

Batt or blanket insulation fits between studs and joists. See Fig. 45-6. Batts may be glued to a single sheet of multilayer Kraft® paper. This paper is like the kind used for grocery bags. Batts also may be entirely wrapped. Wrapped insulation has a *flange* or flap along the edges. The batt is fastened by nailing or stapling the flange to framing members. Wrapped insulation also may be used with masonry.

Rigid insulation, such as fiberboard, is nailed to the outside walls. See Fig. 45-7. It is then covered with wood siding, brick, or stone.

Plastic foam is made into panels. See Fig. 45-8. These are installed most often in masonry walls.

Plastic can also be injected as a fluid into an inside space. This fluid then foams and becomes a solid. Some plastic foams are harmful to your health. They give off dangerous fumes (gas). Foam insulation must be chosen carefully.

Metallic insulation is installed on the outside of the structure. The shiny surface is turned toward the outside. See Fig. 45-9. That way it reflects the sun's heat.

Reducing Further Heat Loss

The joints of buildings may be sealed to keep out air. This reduces *heat loss* by convection. Joints must be completely sealed with caulking. (Caulking

Fig. 45-7
Rigid insulation is nailed to outside walls.

Zonolite Company

Fig. 45-8
Plastic foam insulation can be installed between masonry walls.

Fig. 45-6
Batt insulation can be purchased in standard widths. It fits neatly between the studs.

Zonolite Company

Fig. 45-9
The shiny metal foil coating on this insulation material will reflect radiant heat.

Fig. 45-10
Weather stripping is placed around doors and windows.

is a plastic putty.) This is especially important around window and door frames. *Weather stripping* helps keep warm air from leaking out through windows and doors. See Fig. 45-10.

Storm doors and storm windows also help insulate. There is a dead air space between storm and regular doors and windows. The dead air reduces heat transfer. Multi-paned glass works the same way. Dead air between panes reduces heat transfer. See Fig. 45-11.

People Who Select and Install Insulation

Architects decide on the type of insulation to be used. They know the insulating features of most materials. On large structures, mechanical engineers may help decide. They figure the heat loss per hour. And they figure the *capacity* of heating and cooling units. This must be greater than the amount of hot or cool air lost. But using large units increases costs. And units that are too small are inefficient. They also cost more to repair and do not last well. Units should be just large enough to handle the job well. Smaller units can be used if buildings are well insulated.

Insulation is most often installed by the general contractor. Sometimes this work is subcontracted. The subcontractors specialize in insulating.

Skilled workers install insulation. They may be carpenters, insulators, masons, sheet metal workers, or roofers. Painters often do caulking work to seal joints.

Summary

Insulation is used for many reasons. It restricts heat flow and slows the spread of fire. Insulation deadens noise. It keeps out moisture. And insulation lowers heating and cooling costs. It does this by

Andersen Corporation

Fig. 45-11
This is a cross section of a multi-paned window. Air is trapped between the panes of glass. This reduces heat transfer.

helping to keep heat in or out. Insulation makes buildings more comfortable. Perhaps most important, it conserves energy.

Insulation may be vegetable, mineral, or metallic. Common forms of insulation are loose fill, flexible batt, and rigid panel.

To reduce heat loss even more, cracks and openings are sealed. This is done with caulking and weather stripping.

Terms to Know

absorbs	insulation
batt	metallic
capacity	mineral
conduct	porous
conduction	radiation
convection	reflects
flange	vegetable
heat loss	weather stripping
heat rays	

Think About It!

1. Suppose your home caught fire. Would it be better to have insulation in the walls and ceilings or not to have it? Why?
2. Is your home insulated? How could you find out?

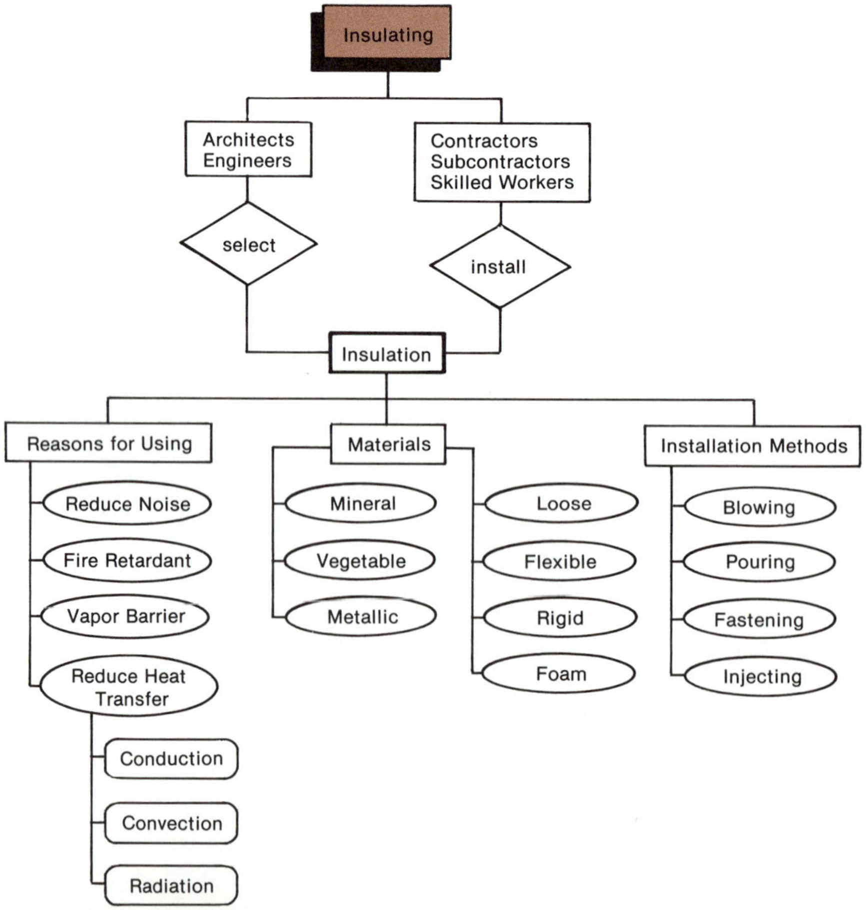

Reading 46

Enclosing Interiors — Residential

Interior of Structure

The skeleton of the structure has been put up. The exterior of the skeleton has been covered. The interior of the structure is protected from the weather. Now construction workers can begin their work inside.

A house has three basic kinds of inside surfaces — floors, ceilings, and walls. This reading will tell how these surfaces are enclosed.

Order of Inside Work

Inside work is done in a special order. First, *subfloors* are put down. This is done when the skeleton is constructed. See Fig. 46-1. Subfloors add strength to the skeleton. They give workers a surface to walk on.

Next, all utilities (such as electricity and plumbing) are *roughed in*. See Fig. 46-2. They are prepared for connection to outside utility lines. All ducts for the *climate control system* (heating, cooling, ventilating) are installed. Waterlines and waste disposal lines are installed. Pipes are placed in the walls, partitions, and through the floor and ceiling framing. If the subfloor is concrete, pipes are installed before the subfloor is put in place. Electrical wiring is installed through studs and joists. Look again at Fig. 46-2. Telephone lines and outlets are installed. In some panelized walls, utility systems are already installed. This is done at a factory. The panels are assembled on the site.

Next, the rough ceiling and wall materials are applied. The finish ceiling and wall finishes are put over these rough materials. Then, the finish floor is laid down. The last step will be finishing. This is described in a later reading.

Fig. 46-1
The subfloor is nailed to the top of the floor joists. This provides a walking surface.

Fig. 46-2
Wiring is run inside walls. Holes are drilled through studs to run the wire.

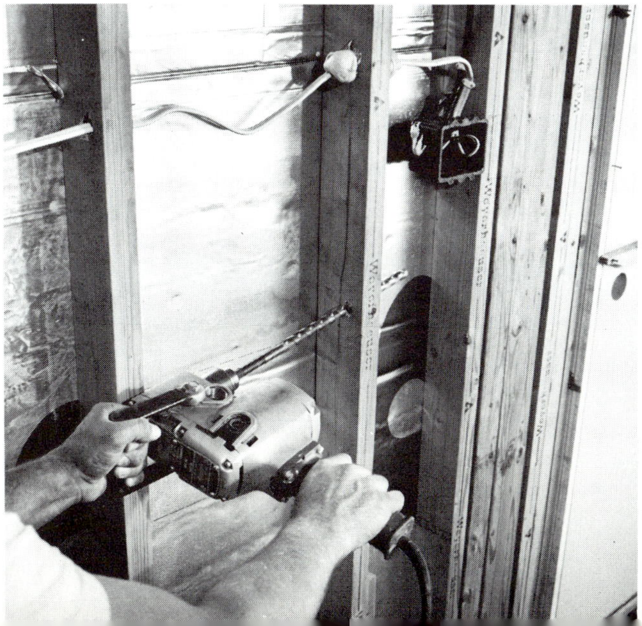

Installing Subfloors

A house skeleton may have a concrete floor. See Fig. 46-3. Concrete floors are usually only on the first or ground floor. In many houses, wooden floor joists are part of the skeleton frame. *Floor systems* made of wood or metal floor joists are above grade (ground level). They are on the first or ground floor. Look again at Fig. 46-1. Upper floors are almost always made with wood or metal joists.

Concrete slab-on-earth (grade) floors are excellent subfloors. Nothing needs to be done to prepare them for finish flooring.

Wood subfloors are nailed to floor joists before the exterior sidewalls are put up. The subflooring makes the skeleton strong. Plywood and other manufactured sheet materials are generally used for subflooring. They are placed at right angles to the joists.

Other subfloor systems may be made of planks and beams. One new type of subfloor is made of panels put together at a factory. The panels are called *stressed skin* or *sandwich panels*. Often, some utility systems have been built into the panels.

Installing Ceilings and Walls

Open spaces in ceilings and walls are filled with insulation material. This is done before the ceilings and walls are enclosed. See Fig. 46-4. Insulation helps keep the temperature inside from changing when the temperature outside changes. It also reduces noise. Some types of insulation have a vapor barrier to keep moisture out.

Fig. 46-3
Concrete slab floor construction.

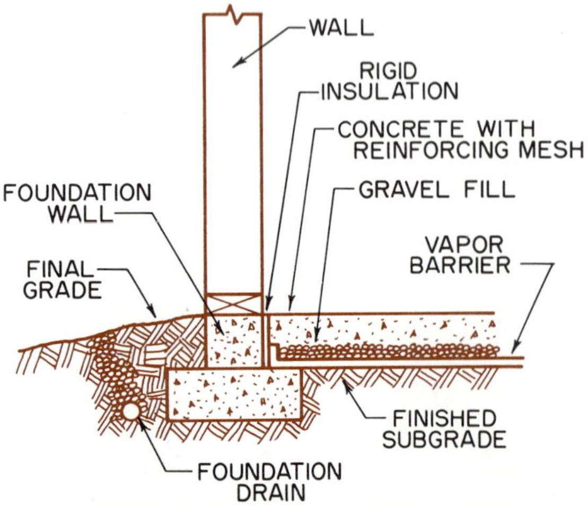

Fig. 46-4
Batt insulation is installed in the outside walls of structures. The shiny covering on the outside walls and the plastic sheeting over the installed insulation help keep out moisture.

Not long ago, *lath and plaster* were the most common materials used to enclose ceilings and walls. Some lath and plaster work is still done today. Rock lath or metal lath is used as a base. Then the *plasterer* puts a first coat of plaster over the base. This first coat is a *scratch coat*. A *brown coat* of plaster is put over the scratch coat. When a scratch coat is not used, the brown coat may be applied directly to the wallboard. After the brown coat has dried, a finish coat is applied. See Fig. 46-5. Applying lath and plaster is difficult. It can be done only by a highly skilled plasterer.

Today, most ceilings and walls are covered with sheets of *plasterboard* (also known as *drywall*). Plasterboard sheets add strength to the structure. There are two ways to apply plasterboard sheets to the ceilings or walls. One way is to nail them on. See Fig. 46-6. Another way is to use *adhesives*.

After plasterboard is put up, the joints must be covered. A *filler* is placed in the joints and in depressions (dents) around nails. Joints are then taped with a special tape and covered with another coat of filler. See Fig. 46-7. After this has dried, a final coat of filler is put on. All filled surfaces are then abraded (sanded). This makes them flat and smooth.

There are other materials used to enclose walls. Special plywood panels may be used. Several kinds of hardboard are available today. Plastic sheet *laminates* may be used in kitchens or bathrooms. Plastic, steel, ceramic or glass tile may be applied as a finish wall surface.

Fig. 46-5
Plastering.

A. *Spreading a final coat of plaster.*

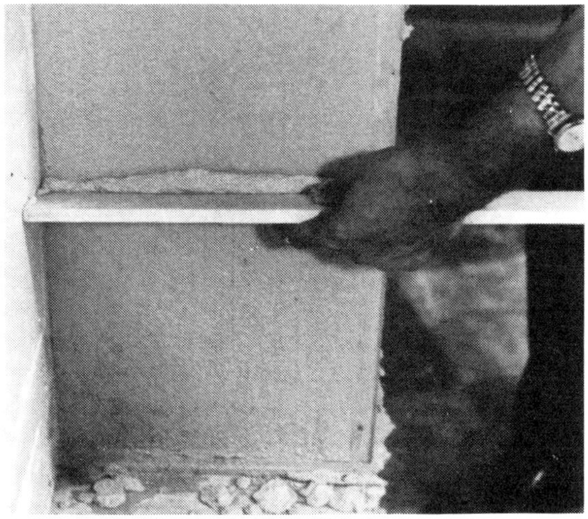

B. *Using a straightedge to level the plaster surface.*

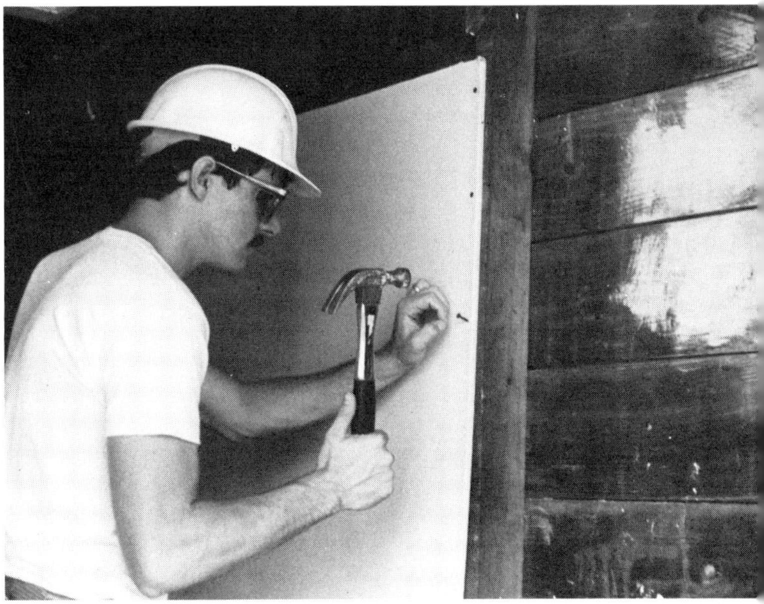

Fig. 46-6
Plasterboard is nailed to the studs.

Fig. 46-7
This worker is putting filler in the joints between plasterboard panels. The joints will then be covered with tape and more filler will be applied. Notice the stilts the person uses to reach high places.

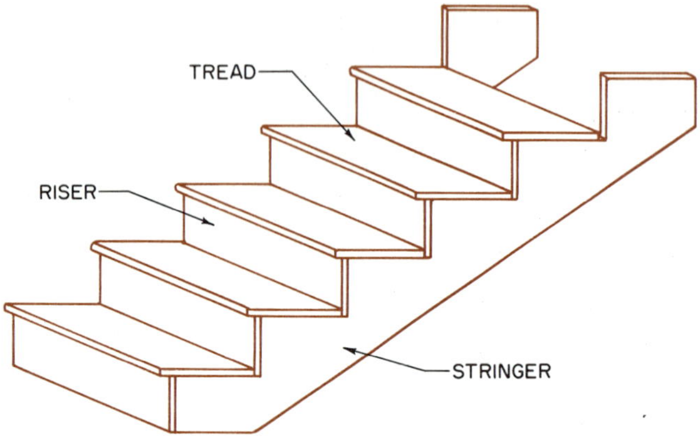

Fig. 46-8
Stairways are often made from stock parts.

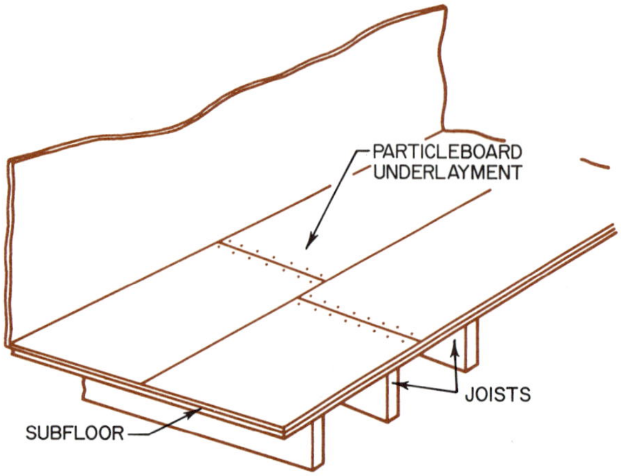

Fig. 46-9
Finish floors are usually installed over underlayment.

Fig. 46-10
Some types of finish flooring are applied with special adhesives. These carpet tiles are being installed directly over the concrete slab floor.

Other Inside Work

Stair framing and door openings were roughed in as parts of the skeleton of the house. After walls and ceilings are enclosed, stairways and doorways can be completed. This should be done before finish floors are put down.

Most stairways in homes are straight. Sometimes, they may be spiral (winding) or angular (with a sharp turn). Most staircase parts are *stock items*. They are made in a factory and bought in units. See Fig. 46-8.

Interior doors and frames may also be purchased as stock items. These units must be carefully installed. They need to be *plumb* (vertical) and square. Door frames are put into the rough openings of partitions. Then they are nailed in place.

Installing Finish Floors

After other inside surfaces are enclosed, finish flooring can be installed. This is done last so the floor surface will not be damaged.

There are many different kinds of finish floors. Wood, linoleum, ceramic tile, flagstone, cork, vinyl, and carpeting are just a few. Usually, an *underlayment* must be put over the subfloor before finish floors are applied. Underlayment may be hardboard, plywood, or particleboard. See Fig. 46-9.

Each finish floor material is installed in a special way. Some are attached by special fasteners. Others require adhesives. See Fig. 46-10.

After the finish floor is installed, only finishing is still left to do. *Trimming* will be done to cover all rough edges. Then, final painting, decorating, and cleanup will complete the process.

Summary

The skeleton of a house encloses the living area. The skeleton must be covered with other materials. First, the exterior is covered. Then the inside must be enclosed.

The inside of a home is enclosed at different times. First, subfloors are put down. This is done when the floor skeleton is built. Next, the walls and ceilings are put up. All utilities are roughed into the ceilings and walls. Next, insulation materials are put in place. Then the ceilings and walls are enclosed. Finally, finish flooring is installed.

Terms to Know

adhesives
brown coat
climate control system
drywall
filler
floor systems
inside work
laminates
lath and plaster
plasterboard
plasterer
plumb
roughed in
sandwich panels
scratch coat
stock items
stressed skin panels
subfloors
trimming
underlayment

Think About It!

1. Are the subfloors in your home wood frame or concrete? Are most subfloors in your neighborhood made like yours?
2. What interior work, if any, can be done during or before exterior work and the roughing in of utilities? Why?

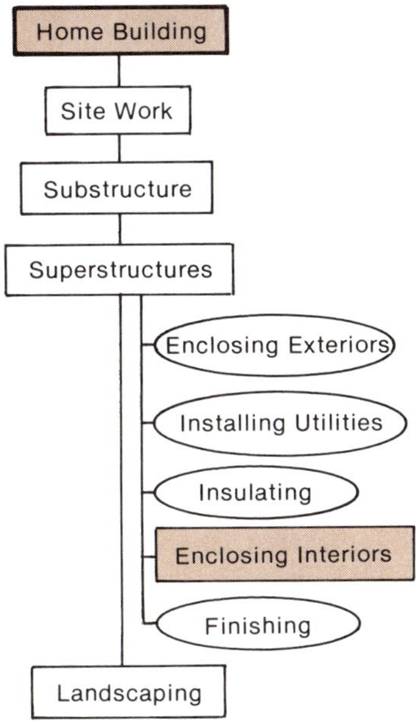

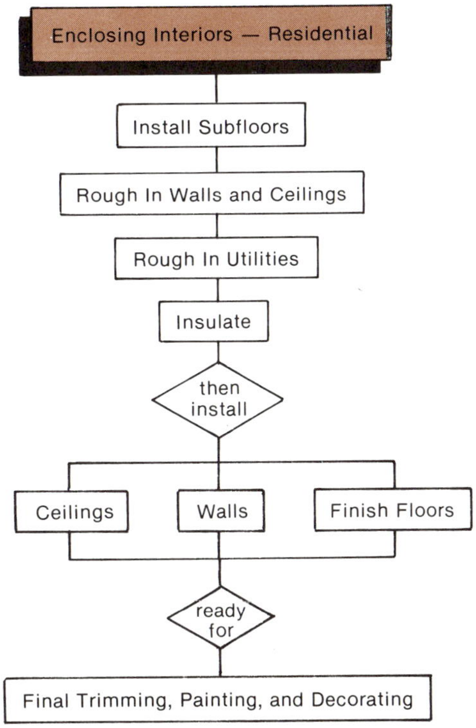

Reading 47

Enclosing Interiors — Non-Residential

Enclosing Interiors of Buildings

A roof and exterior walls cover the structure. Insulation is put in the exterior walls. These materials make up the exterior skin that protects the structure. The skin will also protect inside work. Once the skin is completed, the *interior* is *enclosed*.

This sequence is followed for both residential and non-residential structures. Some of the practices are common to both types of construction. But many practices used in enclosing commercial buildings are seldom used in building homes. In this reading you will learn how the surfaces of non-residential buildings are enclosed.

Why Interiors Are Enclosed

There are three principal surfaces inside a structure. These are ceilings, walls, and floors. They enclose interiors.

Interiors are enclosed to

- hide structural members and utilities.
- help control the building's temperature.
- help control sound.
- provide fire protection (if proper materials are used).
- add beauty.

Ceilings, walls, and floors also improve the building's *environment*. They add attractive textures, colors, and decorative details. See Fig. 47-1.

Interior enclosing work varies according to the *function* (planned use) of the building. Office buildings, stores, and schools are usually completely finished inside. Interiors of other structures, such as factories and warehouses, are not completely finished. Sometimes they are not even enclosed.

People Who Enclose Interiors

Architects plan the general function and appearance of interiors. Engineers may assist with certain details. But the surface materials are generally specified by architects. The architect of the building works closely with the owner.

Often a general contractor is responsible for getting the interior work done. In some cases, specialty subcontractors do parts of the interior work. See Fig. 47-2. Contractors employ many skilled workers. Among them are carpenters, masons, tile setters, ironworkers, and glaziers (glass workers). Others might include plasterers, cement finishers, and carpet layers. All these workers use special tools and skills to finish interiors.

Fig. 47-1
The materials used to enclose the interior of this bank provide attractive and functional surfaces.

Fig. 47-2
Drywall may be installed by a specialty subcontractor.

Homasote Company

A. Roof decking forms the ceiling of this recreation room.

Ceco Corporation

B. The ceiling of this library is the underside of its concrete roof.

Fig. 47-3
In some roofing systems, the same materials form both the roof and the ceilings.

Ceilings

Ceilings are the top surfaces of rooms. They limit the height of an enclosed space. Some ceilings are formed by the roofing system. See Fig. 47-3. But many ceilings are installed. They are put in just below the roof or below the floor above.

Ceiling materials are often applied to the underside of a floor. Different materials may be used.

- Lath and plaster
- Drywall
- Tiles

In a lath and plaster ceiling, metal lath (screen) is nailed to the underside of the floor joists. Then several coats of plaster are smoothed over the lath. See Fig. 47-4. For a drywall ceiling, sheets of gypsum board are nailed to the bottom of the floor joists.

Ceiling tiles may be applied in several ways. They may be attached to a surface such as drywall. *Mastic* (a pasty adhesive) is used to fasten them in place. Sometimes ceiling tiles are cemented to the underside of concrete flooring. Tiles may also be attached to the floor joists. This is done by nailing *furring strips* (narrow boards) to the joists. See Fig. 47-5. Tiles are then stapled or clipped to the furring strips.

Often the ceiling is made of *acoustical* (sound-deadening) materials. Some types of tiles deaden sound.

Often many pipes, wires, and ducts need to be covered. A *suspended ceiling* may be used for this reason. This is also called a false ceiling. It reduces the height of the room. Wire hangers are attached

Fig. 47-4
Plastering on lath is one of the oldest techniques of installing ceilings.

Fig. 47-5
Before ceiling tile can be installed, furring strips must be nailed to the joists.

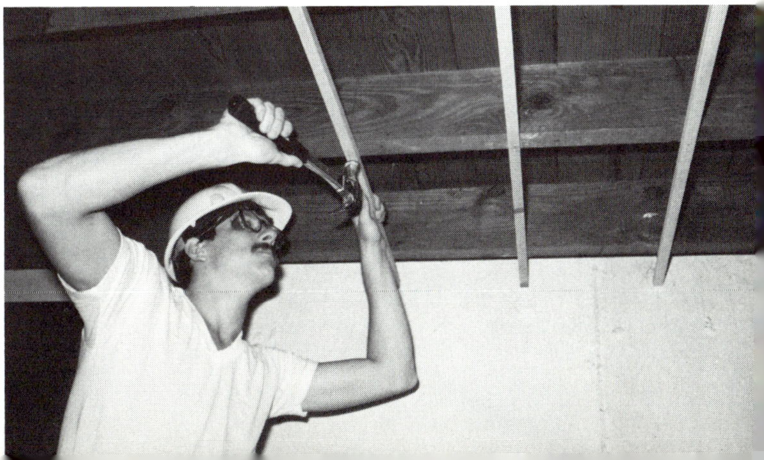

Fig. 47-6
Installing a suspended ceiling. A grid of supports is hung from wire hangers.

Fig. 47-7
Ceiling panels rest on the supports. Notice the built-in lights.

to overhead construction. A *grid* of supports is fastened to the hangers. See Fig. 47-6. Then ceiling panels are placed in the grid openings. See Fig. 47-7. A suspended ceiling may have built-in lights. These serve to light the space in the room below. Suspended ceilings are often installed after the walls are completed.

Walls

Interior *walls* are vertical surfaces that enclose rooms, hallways, and stairways. These include

- interior surfaces of exterior walls and
- partitions.

Fig. 47-8
Many types of materials are used for walls. These walls are made of glass. Armstrong Cork Company

Interior walls are usually installed before finished floors are put down.

Many materials are used to cover interior wall surfaces. Among these are wood, metal, glass, gypsum board, and plaster. See Fig. 47-8. Masonry surfaces of brick, tile, concrete, and marble are also common.

Interior spaces are divided by *partitions*. Partitions may be fixed or movable. Or, they may be room dividers.

Some *fixed partitions* carry part of the load of the roof or floor above them. They are called *load-bearing partitions*. They are not often used in modern concrete or steel-framed buildings. They limit the way the space can be divided.

Movable partitions are designed so that they can be moved if necessary. They are used to allow room sizes to be changed. Often they are made of panels of metal or wood. They may be made of wood or metal studs covered with gypsum board. Some are sets of hinged, rigid panels. Others are flexible panels that fold like an accordion. See Fig. 47-9.

Room dividers go only partway up from the floor to the ceiling. See Fig. 47-10. They may be partial walls, screens, planters, or counters.

Floors

The bottom surface of a room is called the *floor*. Rough floors or subfloors are installed as early as possible in construction. In concrete reinforced buildings, floor slabs are part of the framework. Subfloors provide walking surfaces for workers. And they serve as a place to store construction materials.

Fig. 47-9
Movable partitions are convenient to use when meeting rooms of various sizes are needed.

A "finish" layer is installed on top of a subfloor. Often finish floors are not laid until all other finishing is complete. Splashed paint or tracked-in dirt could mar them.

The traffic in non-residential buildings is usually heavy. Therefore, the flooring material selected must be able to withstand hard use. Wood floors are not common in commercial buildings. They are too difficult to maintain under heavy use. A number of other materials are used instead.

Concrete floors are common in commercial structures. They are installed over concrete slabs.

Fig. 47-10
Room dividers are often used to divide office space. They can be rearranged if space needs to be changed.

This is done by spreading a cement and sand mixture on top of the slab. The concrete is troweled (smoothed) when applied. When dry, the concrete may be painted.

Ceramic tile is set in a bed of mortar. The mortar is laid over the subfloor. The joints between tiles are filled with mortar, too. Ceramic tile floors are very attractive. They also take hard wear. Stone floors are laid in the same way as ceramic tile.

Resilient flooring is often used. It comes in rolls or as tiles. See Fig. 47-11. It is applied over a concrete or wood subfloor. This is done with an adhesive. Resilient flooring "gives" a little when walked on. Then it returns to its original shape. It is more comfortable to walk on than hard floors such as concrete.

A *terrazzo* floor is much like a concrete floor. A layer of mortar is put on a concrete slab or wood subfloor. Stone chips are embedded (firmly set) in the mortar. The surface of the topping is then given a high polish. See Fig. 47-12.

Fig. 47-11
Installing floor tile. A layer of adhesive is spread on the floor before the tile are placed.

Fig. 47-12
A terrazzo floor is good looking and very durable. It is made up of several layers of material.

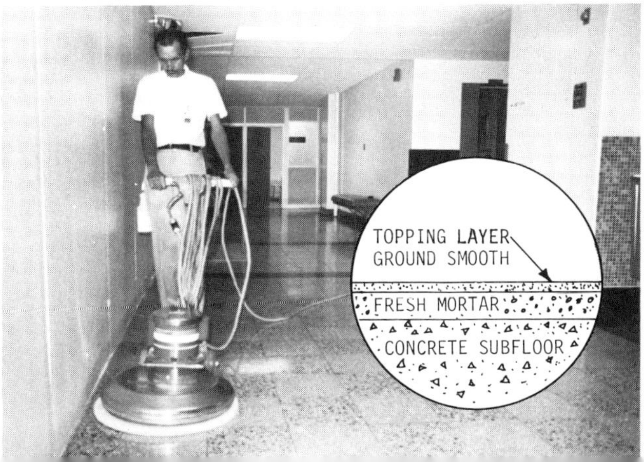

Esco Corporation

Fig. 47-13
Manufacturing plants often have metal flooring. It is strong, fireproof, and does not block the view of activities taking place.

Metal flooring is mainly for special uses. For example, metal floors are used in factories. And stair treads in fire escapes are made of metal. See Fig. 47-13.

Carpeting is used as a floor covering in many buildings. It is made of natural or *synthetic* (human-made) fibers. Carpeting may be installed over a concrete or wood subfloor. It wears well and is soft underfoot. It also helps reduce sound transfer (noise).

Summary

The interior of a structure is enclosed after the exterior is enclosed. Interiors are made up of ceilings, walls, and floors.

Ceilings limit the height of enclosed spaces. They may also conceal (hide) overhead framing and utility systems. They help control sound and temperature. And they add beauty.

The inside surfaces of exteriors are interior walls. Partitions are interior walls, too. Partitions may be fixed or movable. Or, they may be room dividers. Partition walls divide the floor space of a building.

Floor materials are selected to fit the activities in a building. Finish flooring is installed after most other interior work is done.

The design and method of installing interiors are very important. They make a building function as planned. They help provide a pleasant and safe interior environment.

Terms to Know

acoustical
carpeting
ceilings
ceramic tile
enclosed
environment
fixed partitions
floor
function
furring strips
grid
interior
load-bearing partitions
mastic
movable partitions
partitions
resilient
room dividers
suspended ceiling
synthetic
terrazzo
walls

Think About It!

1. Is there a difference between the interior walls of your school classrooms and the interior walls of your home? If so, what is the major difference?
2. Can you name a room or a building in which the ceiling is part of the roofing system?
3. Identify the type or types of flooring materials used in your school.

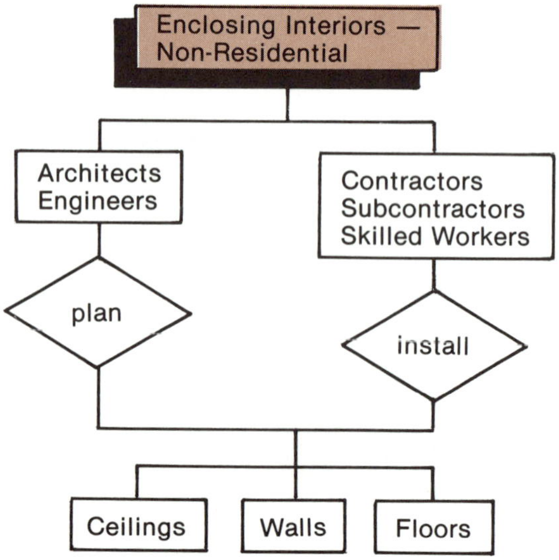

Reading 48

Finishing the Project – Residential

Finishing Work

After the main parts of a home are built, some important work remains. This work includes *trimming, painting,* and *decorating*. It also includes *installing accessories*.

Trimming is done near the end of construction. Rough edges and openings are covered. See Fig. 48-1. Because trim work will be seen, it must be neat. It must be done with care. Trimming requires special skill.

Painting and decorating are done for several reasons. They protect indoor and outdoor surfaces. They also improve the appearance of surfaces. Painting and decorating may be done to keep things clean. Sometimes painting and decorating are done for one of these reasons. Sometimes they are done for all three reasons.

Everyone knows what painting is. But what is decorating? In this reading, decorating will mean ways of protecting and beautifying surfaces, other than by painting. Decorating may include designing and *furnishing* the interior of a building. *Interior decorating* means designing and furnishing. Interior decorating is usually not considered construction work. It is an applied art.

Interior decorators usually work on their own. See Fig. 48-2. They may work for a large department store or painting and decorating firm. People hire interior decorators to help them decorate their homes. Interior decorators tell the contractor what kinds and colors of finishes will go in various rooms. They may also help the family select rugs, drapes, furniture, and other furnishings. Many families do their own decorating. They tell the contractor what finishes they want in their homes. Then they decide on furnishings to go in the rooms.

Fig. 48-1
Trim is used to cover rough edges.

Fig. 48-2
An interior decorator studies a room carefully before selecting colors and materials to decorate it.

Trimming

Most trimming is finish carpentry. Trimming covers rough edges and openings. See Fig. 48-3. It is done on both the outside and inside of the structure.

On the exterior, trim is used where it will improve appearances or seal out the weather. For example, it is put on windows, doors, cornices, and corners. Sometimes a tighter seal is required. *Caulking* (a putty-like substance) provides a tight seal. Caulking is applied with a special tool that looks like a gun.

On the interior, trimming is used to cover and seal in much the same way as outside trimming. See Fig. 48-4. Inside trimming is always in view. It must be done well to look nice. There are many interesting kinds of trimming.

Moldings are a type of trimming. They are commonly used to cover the edges where the wall meets the ceiling or floor. There are many types of moldings. See Fig. 48-5. Moldings are costly. They must be installed carefully. The type of molding used may be chosen by the owner. Or, it may be selected by an interior decorator, contractor, carpenter, or supplier.

Carpenters install the trim if it is wood. Metal trim is installed by metalworkers. Masons install stone trim.

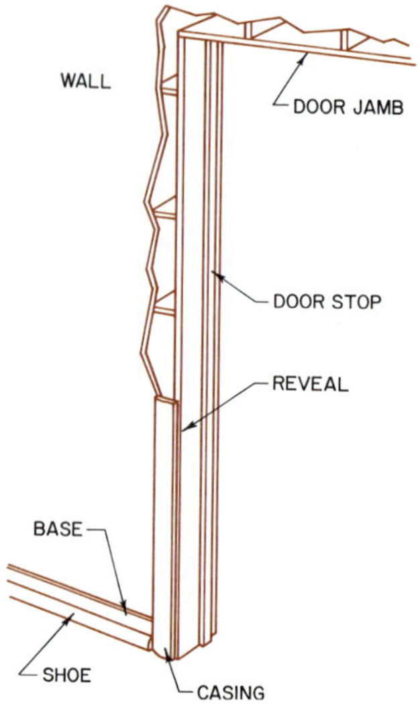

Fig. 48-3
Trim is put around the rough door opening. Notice the door stop. It keeps the door from swinging too far.

A. Untrimmed floor

B. Trimmed floor

Fig. 48-4
The rough edges between floors and walls are trimmed with molding.

Fig. 48-5
Many types of moldings are available. They are sold in building supply stores.

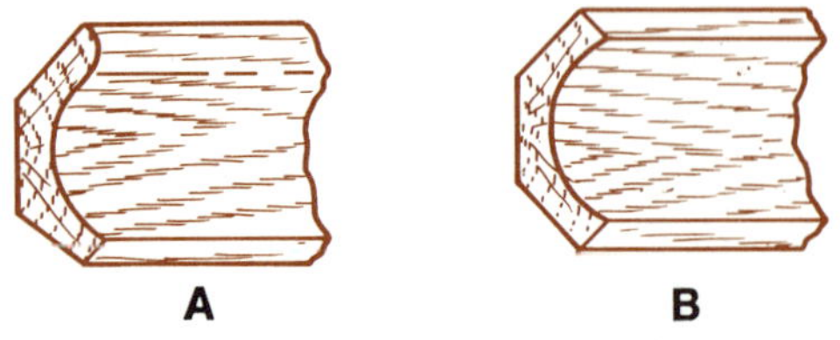

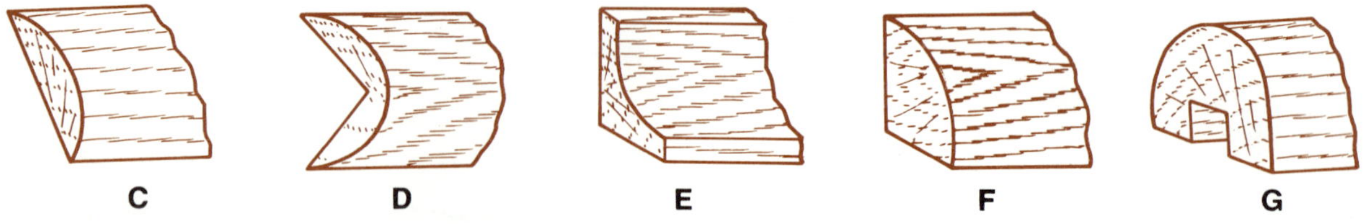

A. ***Crown.*** *Covers wall and ceiling edges. Also used for picture frames or mantel trim.*
B. ***Cove.*** *Used as trim between wall and ceiling edges or inside corners.*
C. ***Half Round.*** *Used to cover joints between two panels.*
D. ***Corner Bead.*** *Covers wallboard or paneling joints on outside corners.*
E. ***Scoop.*** *Covers gaps between baseboard and floor.*
F. ***Quarter Round.*** *Like scoop molding, used to cover gaps between baseboard and floor.*
G. ***Edge.*** *Used to frame room divider paneling.*

Painting

Both inside and outside surfaces of the home must be protected. Paint is the most common protective coating. See Fig. 48-6. All paints have three main parts. The *base* (such as oil or latex) is one part. This is sometimes called the *vehicle*. Another part is the color, sometimes called *pigment*. The third part is a *thinner*. The thinner makes the paint flow easily. Some paints also have a drier. This substance causes the paint to dry quickly.

The appearance of a paint after it dries is called its finish. There are three main paint finishes. One is called *flat* (dull). Another is called *gloss* (shiny). The third is *semigloss* (between dull and shiny).

Flat finishes are used to give a soft (not shiny) appearance. Flat finishes do not wash or wear as well as the other finishes. But they are used most often. They look best in most settings. Glossy finishes resist wear and are most easily cleaned. Semigloss paints have some of the advantages of gloss and flat paints.

It is important to choose the right paint for the job. Many things about the paint must be considered.

- Durability. (How long the paint will last)
- Cost of application.
- Appearance.
- Ability to withstand wear. (Especially important for floor coverings)
- Fire resistance.
- Resistance to air pollutants.
- Amount of surface preparation needed.
- Water resistance.
- Ease of cleaning.

An important part of painting is preparing the surface. Wood surfaces need more preparation than any other surface. Painters fill the cracks, knots, and nailholes with filler. Then they abrade the surface smooth. Joints and nail dents in drywall surfaces should already be covered. No further preparation is needed.

Before they start, painters protect surfaces that are not to be painted. This makes cleanup easier. Inside, they spread coverings — usually cloths or plastic sheets. These "drop cloths" are put on the floor and over window sills and other built-in features. If necessary, painters use drop cloths outside, too. For example, they might cover walks or driveways with drop cloths. See Fig. 48-7.

Copper Development Association, Incorporated

Fig. 48-6
Painted walls provide a plain background for patterned furniture and decorations.

Fig. 48-7
This painter is using drop cloths to keep paint off areas not to be painted. The cloths are easy to place and to remove. And they make cleaning up much easier.

H. Armstrong Roberts

Painters use *scaffolds* and ladders to reach high places. See Fig. 48-8. Scaffolding gives painters a platform on which to stand. It also gives them a place to set paint cans and equipment. Some scaffolds are upright metal frames. Planks are laid across the frames at the desired height. Some scaffolds are platforms suspended by ropes. For small jobs, ladders are more commonly used. They are easier to handle and to move than scaffolds.

H. Armstrong Roberts

Fig. 48-8
Using a scaffold saves time for painters. These workers can paint several window frames before moving their equipment.

Fig. 48-9
Strips of wallpaper must be measured carefully. It is important to match the pattern exactly from strip to strip.

Stauffer Chemical Company

Painters may apply paint by brushing, rolling, or spraying. Each method has advantages. The best method to use depends on three factors.

- The kind of work surface.
- The amount of detail in the work.
- The kind of paint used.

Spraying is seldom used for interior house painting. It leaves too much mess. Brushes and rollers are both commonly used for interior painting. Trim and small areas are usually painted with a brush. Large areas are most often painted with a roller. Flat finishes tend to roll on better than glossy ones.

Stains and clear finishes are similar to paints. They are usually applied to woodwork. Stains and clear finishes let the wood show through. Stains change the natural color of the wood. Clear finishes protect the wood and improve its appearance without changing its color. Varnish and lacquer are examples of clear finishes.

Decorating

Wallpaper is a common decorating material. See Fig. 48-9. Many modern "wallpapers" contain no paper at all. They may be made of cloth or plastic. But they are still called wallpaper.

Wallpaper comes in rolls. The rolls are cut into strips to fit a wall or ceiling. Paste is spread on the back of each strip. Then the wallpaper is applied to the wall. Some wallpapers come prepasted from the factory. At the site, the prepasted surface is moistened to make it stick to the wall.

Paperhangers are people who apply wallpaper. They roll it out and cut it to length. Then they apply water or paste to the strips of wallpaper and put them in place. Look again at Fig. 48-9. Finally, paperhangers brush the wallpaper. This smooths it and pushes out air bubbles.

It was once common to use wallpaper to cover all the walls in a room. And many rooms were wallpapered. Ceilings were often wallpapered, too. Now wallpaper is more often used to cover only one or two walls of a room.

Wallpapers are available in a large variety of colors and patterns. Homeowners can choose designs that suit their individual tastes. Wallpapers cost more per square foot than paints. But they can add color and style that paints cannot.

Fig. 48-10
Paneling costs more than paint or wallpaper, yet it usually outlasts them both. Panels are often made to look like boards.

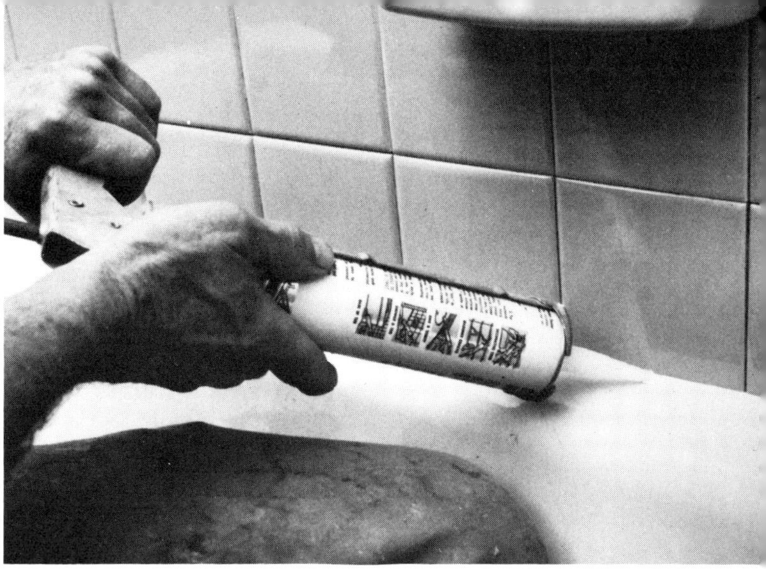

Fig. 48-11
Grouting prevents moisture from loosening the tile.

Rigid sheet materials are often called *paneling*. See Fig. 48-10. They are used to protect and decorate walls. In bathrooms and kitchens, all walls may be covered with some type of paneling. There are fewer patterns of paneling than there are of wallpaper. But panels are available in many different finishes and patterns.

Paneling is usually installed by carpenters. The most common panel size is ¼″ × 4′ × 8′. Because panels are rigid, they must be cut very carefully. They must fit the area they are to cover. Tiny mistakes show clearly, especially if trim is not used on the joints.

Tongue-and-grooved boards also are used to panel walls. They are about three times as thick as sheet paneling. They are also much more expensive. For this reason, paneling is seldom done with boards.

Tile is another common decorating material. It is used mainly on bathroom and kitchen walls and on entrance hall floors. It decorates and protects surfaces that come in contact with water.

Most tiles are ceramic. They come in a variety of shapes, sizes, and colors. Special decorative tiles may be used for trimming. Plastic tiles also are available. They do not wear as well as ceramic tiles.

Tile setters are people who apply tiles. First, they spread an adhesive on the wall or floor. Then they place the tile on the adhesive and press it in place. The adhesive *sets* (hardens). Next, a plaster-like material called *grouting* is used to fill the spaces between tiles. Grouting seals the joints. See Fig. 48-11. It also adds to the decorative effect. Grouting may be white. Or, it may be a color that goes well with the tile color.

Installing Accessories

Installing accessories is the last work done in a home. It is done just before final cleaning. It includes such work as putting in light fixtures and curtain rods. Look at Fig. 48-12. Like other finishing, this work does not add to the strength of the house. *Accessories* add usefulness and beauty. They should be selected and installed carefully. This work will affect the family's feelings about the home.

Fig. 48-12
Installing a sink faucet is a finishing task.

Summary

Trimming, painting and decorating, and installing accessories are types of finishing work. They are done after the main construction is complete. Finishing work must be done carefully. It is important to protect the work that has already been done.

Finishing work is important. It requires much care and skill. It affects the way people feel about their homes.

Terms to Know

accessories
base
caulking
decorating
flat
furnishing
gloss
grouting
installing accessories
interior decorating
moldings
painting
paneling
paperhangers
pigment
scaffolds
semigloss
thinner
tongue-and-grooved
trimming
vehicle

Think About It!

1. Pick a room in your home or apartment. List the finishing work that was done in that room. Include all tasks of trimming, painting and decorating, and installing accessories.
2. Suggest finishing tasks that could have been done as alternatives to each task on your list. Give reasons for your choices. For example, you might say, "Light yellow paint instead of dark blue wallpaper would make this room look larger." Or, you might say, "Lamps would give a softer light than ceiling lights."

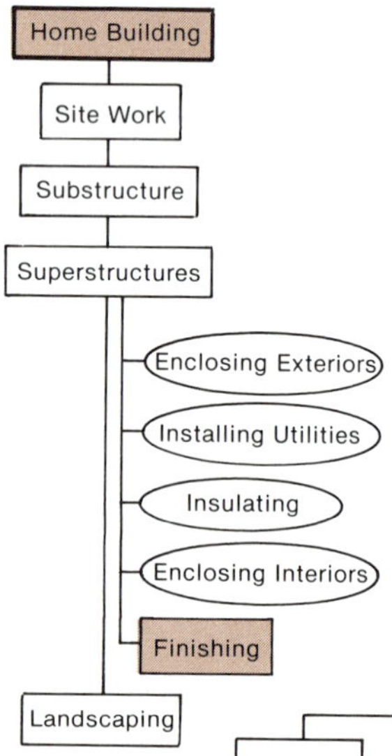

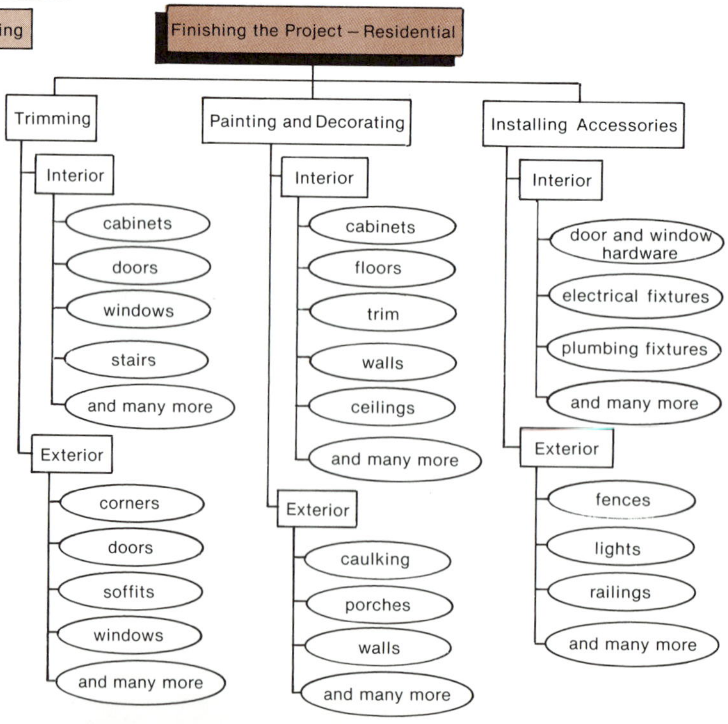

Reading 49

Finishing the Project – Non-Residential

Finishing Practices

Completing a non-residential structure involves many *finishing* practices. Some of them are the same as those used to complete residences. These are discussed in another reading. But some practices used in finishing non-residential structures are seldom used in home building.

All non-residential buildings need finishing work. Schools, stores, and factories are examples of non-residential buildings. See Fig. 49-1. And finishing is just as important in projects that are not buildings. These include highways, dams, and bridges. See Fig. 49-2. Finishing is usually the work that is most noticed. It must be done accurately and carefully.

The non-residential finishing practices that are the same as residential finishing practices are

- trimming.
- painting and decorating.
- installing accessories.
- completing the site.

Landscaping may also be considered non-residential finishing. It is treated separately in another reading.

Fig. 49-1
Every day hundreds of people walk the stairs of this social center. The steps must withstand the wear, and the railings must remain secure. Finishing materials and techniques are very important.

Fig. 49-2
This highway project required much finishing work. Signs, railings, lights, and reflective markers were installed.

Trimming

Trimming is done to cover edges, joints, and openings in buildings. See Fig. 49-3. Little trimming is done on projects other than buildings. Trim should look attractive. And it should be *durable* (long-lasting). It should be easily and quickly installed.

A variety of materials are available for trim. Most people prefer the appearance of wood trim. But wood trim has disadvantages. It scratches and chips easily. And it is costly to buy and *maintain*. Also, it requires time and skill to install woodwork. For these reasons, wood trim is seldom used in non-residential buildings. Other trimming materials are commonly used. These materials include ceramics, masonry, metal, and plastic.

Ceramics and masonry are used because of their *durability*. They may cost much more to buy and install. However, it costs little to maintain them. Over many years of heavy use, their cost per year is small.

Plastic and metal trim are usually the cheapest to buy and install. And they are easy to install. They may be clipped or snapped in place. But they may not wear well. Still, the cost of replacing them may be reasonable.

Painting and Decorating

The *painting and decorating* of non-residential buildings is done in much the same manner as for residences. But there may be some differences in the painting techniques used.

Large buildings have many more types of surfaces to finish than do homes. They may require many different kinds of finishing materials. Often, interior decorators help select these materials.

Many types of paint may be needed. They serve different purposes. Paint's appearance is always important. But in commercial buildings (businesses), durability and the speed of application are most important. Paint manufacturers can suggest the best type of paint to use.

Methods of applying paint in commercial buildings often are different from those used in residences. In commercial buildings, finishes are often sprayed. This is not often done in homes. And larger *scaffolding* and other equipment may be used. Ceilings may be higher. Wall areas may be larger. The larger equipment helps workers reach these surfaces.

In special painting operations, specialized equipment is used. See Fig. 49-4. Painting markings on highways is an example of a special painting operation. Applying tennis court surfaces is another.

Installing Accessories

Installing accessories is the next step in finishing the project. Accessories include a wide variety of items. See Fig. 49-5. *Fixtures* are accessories. Plumbing and electrical fixtures are major accessories installed in commercial structures. Heating,

Fig. 49-3
Trim around edges, joints, and openings gives a room a finished look.

Fig. 49-4
Painting is not limited to buildings. The lines painted on highways serve as safety or informational markers.

TRW, Incorporated

cooling, and ventilating fixtures are also used. Hardware, handrails, and communications fixtures are other accessories put in business structures.

Accessories are installed in most projects. Even a highway needs railings, lights, and signs.

Furnishings differ from accessories or fixtures. Fixtures are permanently attached to structures. They usually stay inside the structure when it is sold. But furnishings are movable. They are seldom sold with a structure. They are most often moved to a new location. See Fig. 49-6.

Usually, accessories are installed after the structure is enclosed and can be locked. They are usually not installed until painting and decorating have been done. But sometimes fixtures must be installed as the structure is built. For example, a large boiler or large electrical items may not fit through doors or windows. They must be put in before the structure is enclosed. See Fig. 49-7.

Completing the Site

One of the last tasks in finishing the project is *completing the site*. See Fig. 49-8. This means adding constructed items such as sidewalks. Parking lots and lights are other examples. This work usually is completed before final landscaping.

Fig. 49-5
The fencing around this spillway was installed for safety.

Fig. 49-6
Furnishings are usually moved when a building is sold.

Fig. 49-7
This large electrical accessory is a transformer. It is being installed in a factory that is under construction.
Carolina Power and Light Company

Fig. 49-8
Paving an entryway may be one of the last tasks in finishing a building project. This task is part of completing the site.

People Who Do Finishing Work

Finishing work is planned by architects and engineers. General or specialty contractors may contract to do finish work. Nearly all types of construction workers are involved. Most often, finishing is the last work done on the site by a particular trade. For example, electricians may rough in lines early. Later they return and install fixtures. Finally, they make connections and test their work. Plumbers, sheet metal workers, masons, and other workers may follow the same procedure.

Summary

Finishing work is important on all construction projects. Non-residential finishing includes trimming, painting and decorating, installing accessories, and completing the site. See Fig. 49-9.

Non-residential finishing materials and practices may be the same as those used on residential projects. But different materials, equipment, and techniques are needed for non-residential projects.

Terms to Know

completing the site
durability
durable
finishing
fixtures
furnishings
installing accessories
landscaping
maintain
painting and decorating
scaffolding
trimming

Think About It!

1. Name several kinds of finishing work.
2. In what ways are residential and non-residential finishing alike? How are they different?
3. What is the major difference between fixtures and furnishings?

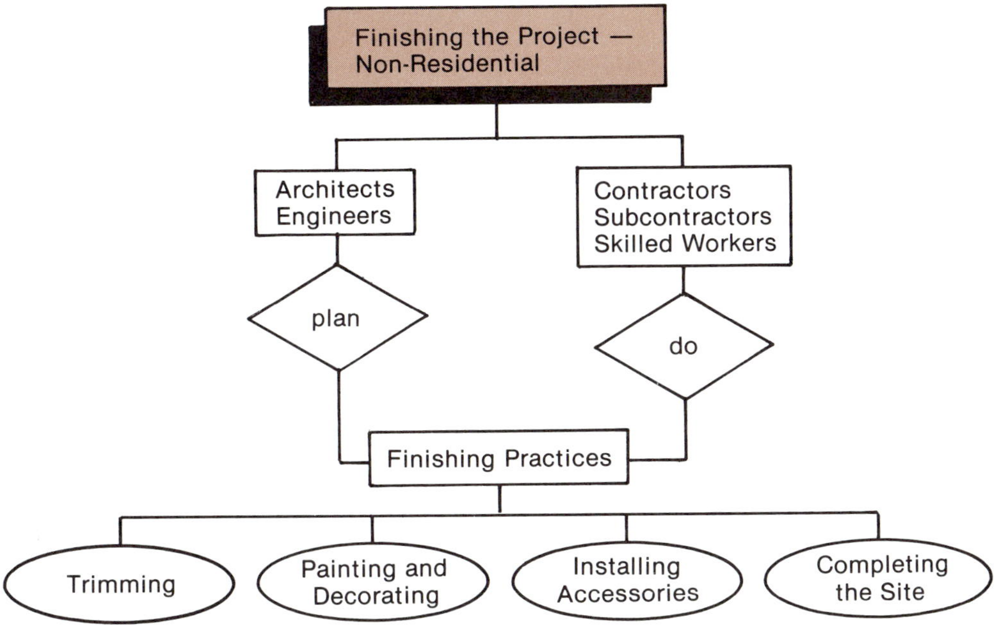

Fig. 49-9
Finishing work can be seen. This adds to its importance.

Making Inspections

Why Inspections Are Made

One way to assure proper construction is by making inspections. A dictionary definition of *inspect* is "to look at carefully." It also means "to examine officially." Usually the person who does this "official looking" is called an *inspector*. Inspecting may be a person's full-time job. Sometimes it is only a part-time duty. People other than inspectors may inspect work, too. For example, supervisors monitor ongoing work. Inspecting is also known as *monitoring*.

Inspecting is part of the management function of controlling. Controlling includes directing, monitoring, reporting, and correcting. Inspecting is done to make sure construction work meets minimum requirements. See Fig. 50-1.

When Inspections Are Made

Inspections begin when drawings are made of the project. And building materials are inspected before they are shipped by suppliers. Suppliers want to be sure their products meet standards. Inspections of the work continue at various stages of construction. Before the owner accepts the project, a final inspection is made.

The facts learned from inspections are reported to managers. See Fig. 50-2. Any needed corrections are made. The work is reinspected. Then inspectors make another report.

Fig. 50-1
Inspecting is an important part of controlling a construction project.

Fig. 50-2
The safety of the many people traveling on a highway may depend on how well the highway is constructed. This is an inspection form for a state highway building project. Managers stay informed to make sure the job is done right.

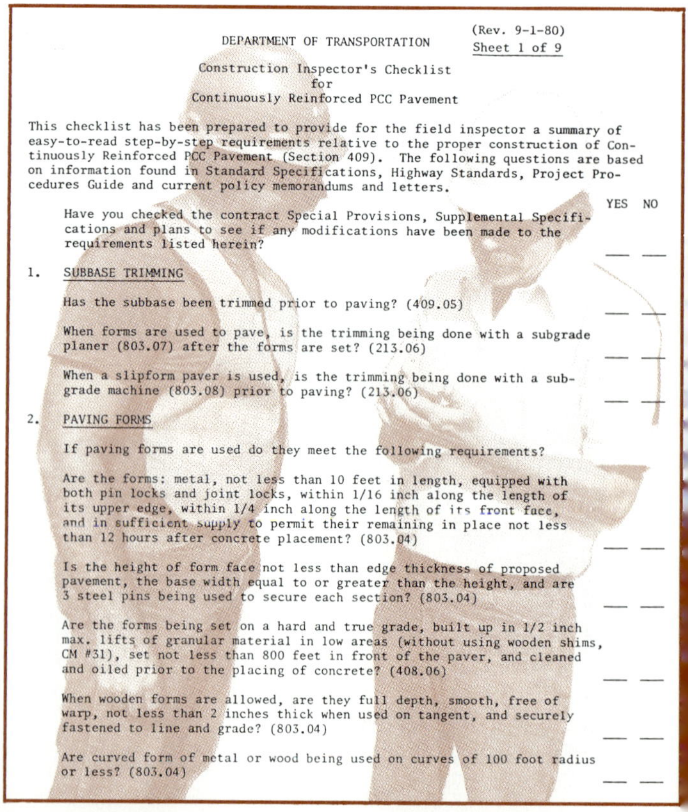

DEPARTMENT OF TRANSPORTATION (Rev. 9-1-80) Sheet 1 of 9

Construction Inspector's Checklist
for
Continuously Reinforced PCC Pavement

This checklist has been prepared to provide for the field inspector a summary of easy-to-read step-by-step requirements relative to the proper construction of Continuously Reinforced PCC Pavement (Section 409). The following questions are based on information found in Standard Specifications, Highway Standards, Project Procedures Guide and current policy memorandums and letters.

YES NO

Have you checked the contract Special Provisions, Supplemental Specifications and plans to see if any modifications have been made to the requirements listed herein? ___ ___

1. SUBBASE TRIMMING

Has the subbase been trimmed prior to paving? (409.05) ___ ___

When forms are used to pave, is the trimming being done with a subgrade planer (803.07) after the forms are set? (213.06) ___ ___

When a slipform paver is used, is the trimming being done with a subgrade machine (803.08) prior to paving? (213.06) ___ ___

2. PAVING FORMS

If paving forms are used do they meet the following requirements?

Are the forms: metal, not less than 10 feet in length, equipped with both pin locks and joint locks, within 1/16 inch along the length of its upper edge, within 1/4 inch along the length of its front face, and in sufficient supply to permit their remaining in place not less than 12 hours after concrete placement? (803.04) ___ ___

Is the height of form face not less than edge thickness of proposed pavement, the base width equal to or greater than the height, and are 3 steel pins being used to secure each section? (803.04) ___ ___

Are the forms being set on a hard and true grade, built up in 1/2 inch max. lifts of granular material in low areas (without using wooden shims, CM #31), set not less than 800 feet in front of the paver, and cleaned and oiled prior to the placing of concrete? (408.06) ___ ___

When wooden forms are allowed, are they full depth, smooth, free of warp, not less than 2 inches thick when used on tangent, and securely fastened to line and grade? (803.04) ___ ___

Are curved form of metal or wood being used on curves of 100 foot radius or less? (803.04) ___ ___

People Who Make Inspections

Inspection is vital (necessary) to the success of a project. It requires inspectors who are completely familiar with what is being inspected. Often, inspectors have a trade background. For instance, an electrical inspector may be an experienced electrician. Chief inspectors may have a background in architecture or engineering. Many different kinds of inspectors may become involved in a construction project. See Fig. 50-3. Some of them are not employed by contractors.

Designers and Building Officials

Architects and engineers carefully inspect their plans. These plans must be given to state, county, or city building departments. Department officials go over the plans very carefully. They make sure the plans follow *local building codes*.

Construction Workers

The contractor's or subcontractor's employees inspect to achieve *quality control*. They want to make sure the job is done in a professional manner. For example, ironworkers assembling structural steel will check to see that the bolts are tight. A supervisor will make sure that beams are level and columns are plumb. A superintendent will constantly watch all job details. All this is done to make sure a high-quality project is built.

Owners' Inspectors

Owners may have inspections made by persons they employ. Owners want to be sure they are getting what they are paying for.

Government Employees

On-the-job inspections also are made by government employees. These inspections usually take place to protect public health or safety. For example, *city building inspectors* may monitor construction projects. They inspect structural supports, wiring, plumbing, and heating systems. They check them to make sure local building codes have been followed. These inspectors also ensure that the plans are followed as approved by the building department.

Lending Agencies

Money is usually lent to pay for construction. The *lending agencies* often send out inspectors to look at the work. They do this to protect their interest in lending the money.

Insurance Companies

An *insurance company* often sends inspectors to a construction site. It may do this for two reasons. The company may be providing construction insurance for the contractor. Or, it may be insuring a part of the structure. For example, the company may be asked to provide boiler insurance. The insurance goes into effect when the building is completed. But the insurance company usually insists on inspecting the installation of the equipment it insures.

What Is Inspected

Inspectors usually have three major concerns when making inspections.

- Materials used.
- Methods used.
- Quality of work done.

Fig. 50-3
Work on a construction project is inspected carefully and often.

A. Before actual construction begins, designers inspect the plans.

B. Inspectors and the workers themselves check work during construction. Here an inspector uses a special wrench to check a bolted connection.

J. Walter Thompson Company

Each of these factors is important in itself. If all three meet inspection standards, the project will last for many years.

Materials

To pass inspection, materials must meet or exceed (go beyond) the specifications. Sometimes materials are inspected before they reach the job. This is done either by a manufacturer's representative or by a member of one of the inspecting organizations. For example, lumber is graded and marked at the mill. Suppose it is stamped as "No. 2 Common." The inspector merely has to check this marking against the specifications.

Fig. 50-4
Materials may be checked in a laboratory. Inspectors must make sure the proper materials are used.

Fig. 50-5
Inspectors check to be sure materials are installed correctly.

Sometimes the inspecting organization runs extensive (very complete) laboratory tests on materials. See Fig. 50-4. The inspector rejects defective (faulty) materials. Then the contractor must replace them with good quality materials.

Methods

The proper methods must be used to prepare or install materials. Inspectors watch workers on the job. They check to see that workers are preparing materials properly. See Fig. 50-5. The right number or amounts of materials must be used. Also, inspectors check to see that sizes are correct according to specifications. For example, before concrete is placed, inspectors check the reinforcing steel. They make sure it meets the requirements of the plans and specifications. Suppose the requirements read as follows:

> All reinforcing steel shall be wire-brushed or otherwise cleaned to remove all loose rust and scale . . . Vertical reinforcement shall be No. 6 (3/4" diameter) bars spaced on 12" centers . . .

If the inspector finds the steel has been improperly prepared, defects are reported. The steel may not be clean enough. It may be the wrong size. Bars may be spaced too far apart. Corrections must be made. Otherwise the structure may fail.

Quality

A third concern of inspectors is work quality. See Fig. 50-6. They examine a job while it is in progress. An example is placing concrete. Cement and aggregates may separate. Or, concrete may be too wet or too dry. A watchful inspector will see that work is done correctly.

Inspectors also examine completed work. An installed waterline is an example. Inspectors apply pressure to find out whether the line leaks. Such checks tell whether any final corrections need to be made. A well-inspected project will meet high standards.

Summary

Inspection is a necessary part of construction. It is an ongoing process. It starts at the beginning of the designing process. It continues until the contractor completes the work.

Employees of the contractor are constantly inspecting their own work to control its quality. Inspectors may be hired by the owner. They ensure

Fig. 50-6
These inspectors are using an air meter to check the quality of a sample of concrete.

that the job is built according to the contract. State, county, or city government inspectors also inspect construction. They are concerned about public health and safety.

Inspection is important to all concerned. It helps contractors stay in business. It helps workers keep their jobs. It helps owners get what they pay for.

Terms to Know

city building inspectors	lending agencies
inspect	local building codes
inspector	monitoring
insurance company	quality control

Think About It!

1. Why do inspections need to be made at the site?
2. When are inspections made?
3. How do local building codes and project specifications relate to inspecting?

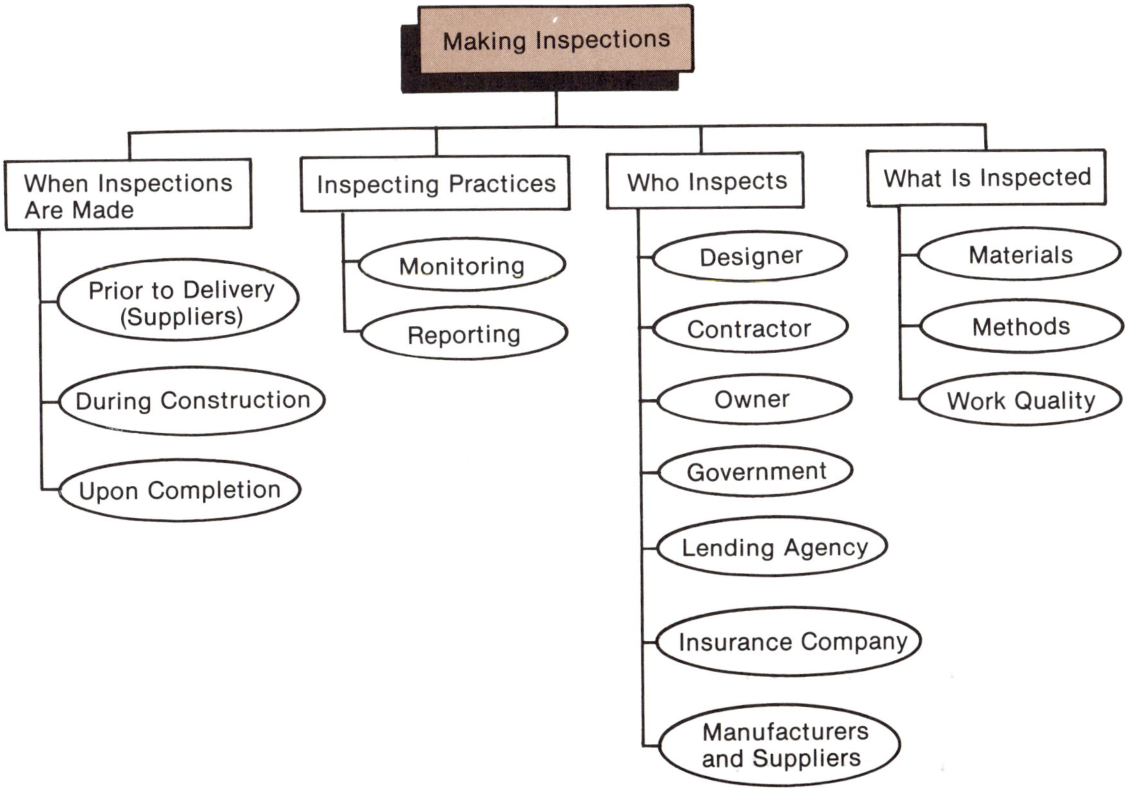

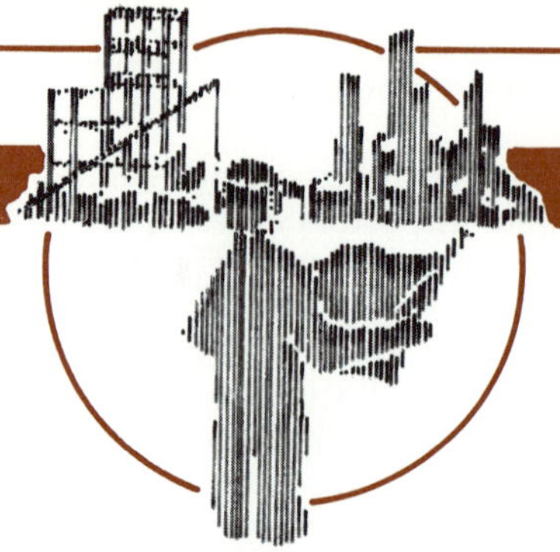

Reading 51

Landscaping

The Purpose of Landscaping

Landscaping is changing natural scenery. It completes a project. Landscaping includes everything from planning the site to cleaning it. It is the art of making sites beautiful. When it is done well, it makes sites more useful and more valuable, too.

Sites often need to be changed. There may be something about the site that makes it an unsuitable building setting. Good landscaping makes even the finest structure look better.

The Landscape Plan

A *landscape architect* makes the *landscape plan*. See Fig. 51-1. A landscape plan shows the kinds of shrubs, trees, hedges, and ground cover to be used. It also shows where they are to be located and where the structures are to be.

A *landscape contractor* may be hired to develop the plan. Most often the plan is developed without charge. This is done to get the landscaping contract. The property owner usually helps make the plan.

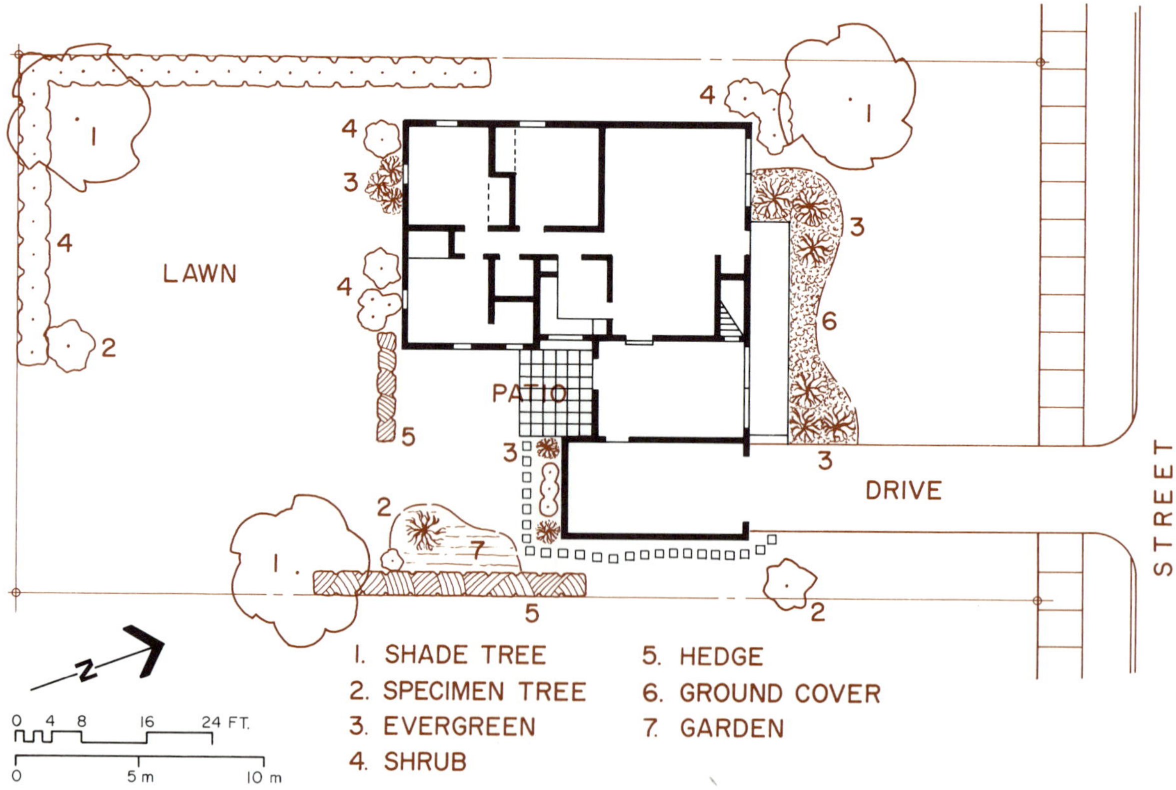

Fig. 51-1
A landscape plan.

Fig. 51-2
Building accesses and features are built before the final shaping of the earth.

Building Accesses and Features

Before the ground is shaped and finished, *building accesses* and *features* are made. See Fig. 51-2. Building accesses include driveways, walkways, and parking areas. Features include patios, fences, walls, benches, pools, lights, sculptures or statues, and plant boxes.

Accesses and features are usually added to the site after the foundation has been *backfilled*. Backfilling is replacing the soil that was dug out to make the foundation.

Shaping and Finishing the Earth

The next step in landscaping is shaping the land surface as shown on the plan. This is important for good drainage. If the ground is too steep, the soil will wash away. If the ground is flat, water will gather in puddles. If either of these things happens, plants may die.

First, the earth is *rough graded* to give the needed slope. Stones are taken from the site. Then the lot is *finish graded* (smoothed).

Preparing the Soil for Planting

Topsoil is the upper layer of earth. It is the richest layer. Plants grow in topsoil.

When the foundation was dug, topsoil was set aside. Now the topsoil is put back over the site. If more topsoil is needed, the contractor has it hauled in.

The topsoil can be put back in two ways. It may be done by hand with a wheelbarrow, shovel, and rake. Or, it may be done by machine. See Fig. 51-3. The machine method is used if the area is very large.

After the topsoil is put back, the ground is *cultivated* (broken into fine pieces). This levels out the soil. The next step is to rake the plot (lot). After the soil has been raked, chemicals may be added to enrich it for good plant growth.

Planting Trees and Shrubs

At this point, trees and shrubs are planted. See Fig. 51-4. Following the plan, the gardener marks the locations of the trees and shrubs with stakes. Holes are dug. Then the trees and shrubs are planted. Trees may need to be supported against the wind until their roots can hold them up. Sometimes the soil around trees and shrubs is not rich enough to help them grow. Then fertile (rich) soil is brought in. It is placed around the trees and shrubs, on top of the soil that is already there. If none of the original soil is fertile, it is removed. It is replaced with rich soil.

Laying Lawns

The lawn is the most important part of the landscape. Usually, lawns are planted. But in some parts of the country, rocks may be used for lawns instead of grass or other plants.

Planted lawns control surface water. They give the site a pleasing look. Sodding and seeding are the two methods commonly used for laying planted lawns.

Sodding gives an almost instant lawn. Strips of sod are laid out on the prepared ground surface. See Fig. 51-5. They are placed close together and rolled or tamped (packed by tapping) in place. The sod is watered regularly to ensure root growth.

Seeding is less expensive than sodding. But it takes longer to get a lawn this way. First, the seed is spread evenly. This may be done by hand or by machine. The seed is then raked into the soil. Next, the lawn is rolled. This compacts (presses down) the soil to keep the seed from blowing or washing

Fig. 51-3
Topsoil is spread over the site before planting is done.

Fig. 51-4
Trees may be planted to make the site more attractive.

The Daily Pantagraph

The Daily Pantagraph

A. *Strips of rolled grass sod are delivered by truck to the site.*

B. *The sod is spread over the topsoil. With careful watering, it will soon take root to form a permanent lawn.*

Fig. 51-5
Sodding is a quick way to start a lawn.

away. The ground may be covered with straw or burlap. The covering protects the tender plants from the hot, direct sunshine. It also reduces erosion. Finally, the ground is watered. This helps the seeds begin to grow.

Planting Flowers

Gardeners prepare flower beds by adding topsoil or peat moss to them. Then they plant flowers in the locations marked on the landscape plan.

Multiphase Landscaping

It is best to complete landscaping soon after construction. However, the cost may be too great for the homeowners to pay all at once. If so, the landscaping operations may be *phased*. This means that the landscaping is done in steps. The steps are done over a period of time. The lawn and large trees should be put in during the first year. Shrubs and flowers may be added in later years.

Cleaning

During construction, the site is kept as clean as possible. Most *debris* (such as boxes or scraps of materials) is burned or hauled away. After landscaping, temporary facilities, fences, and equipment are removed. Any remaining debris is hauled away.

Summary

Landscaping the site is the last step in constructing. It begins with a landscape plan. Landscaping includes these jobs:

- Building accesses and features.
- Shaping and finishing the earth.
- Preparing the soil for planting.
- Planting trees and shrubs.
- Laying the lawn.
- Planting flowers.
- Removing equipment and cleaning up.

No matter how well a project is designed, a well-landscaped site will make it more pleasing to look at. See Fig. 51-6.

Fig. 51-6
Good landscaping makes a well-designed structure even more attractive.

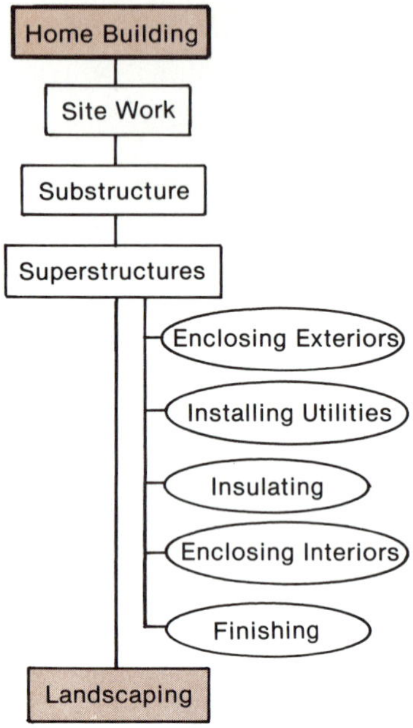

Terms to Know

backfilled	landscape plan
building accesses	landscaping
cultivated	phased
debris	rough graded
features	seeding
finish graded	sodding
landscape architect	topsoil
landscape contractor	

Think About It!

1. Why is landscaping done?
2. Why is landscaping done after the structure is completed?
3. How would you decide whether to sod or seed a lawn?

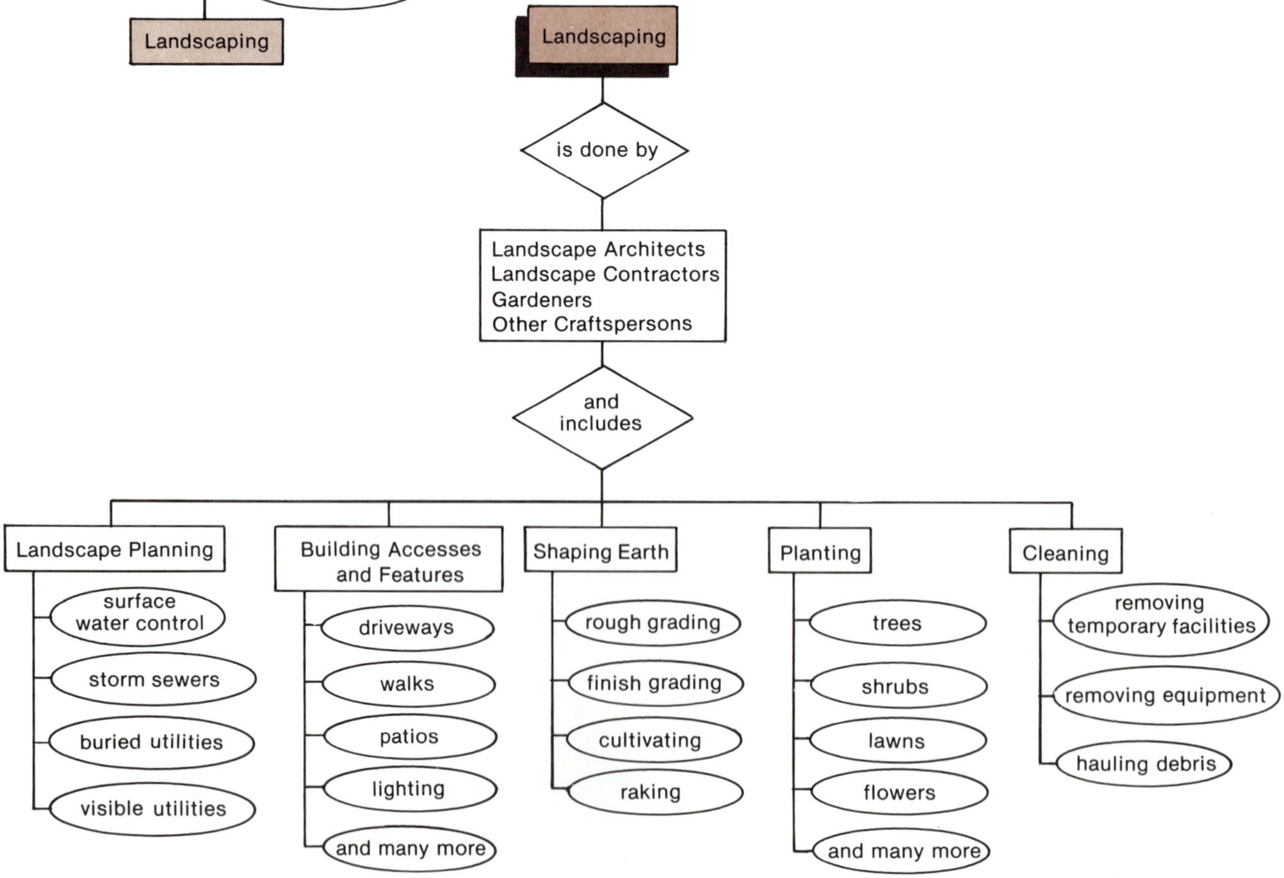

Reading 52

Transferring the Project

Requirements for Transfer

Work has been completed at the job site. Now the project is *transferred* from the contractor to the owner.

An owner has a contract. It may be with the architect or engineer. It may be with the general contractor. This reading describes how such contracts are completed. It tells how the owner takes over the project.

When the project is complete, everyone must be satisfied. The owner expects good quality work. The contractor expects to be paid. Therefore, several things must be done before the owner starts using the structure.

- An inspection is made.
- Bills are paid in full, and releases are signed.
- Service warranties and equipment operating manuals are assembled and given to the owner.
- The final payment is made to the contractor.

Inspection

The completed project is carefully *inspected*. See Fig. 52-1. This is done by inspectors. They are representatives of the contractor and of the architect or engineer. Sometimes representatives of the owner inspect, too. They look at every part of the job in detail. They check to make sure the work has been done according to the construction contract. This contract includes the plans and specifications. All mechanical and electrical equipment is tested. Everything must operate properly.

Sometimes, the inspectors find *defects* (faults). They make a list of these. See Fig. 52-2. The list is called a *punch list*. Most defects are very small. A smudge on a wall is a defect. Some defects are major. For example, a highway may be uneven.

A punch list may have only a few items. Or it may have none. On a large project, there may be several hundred items.

Soiltest, Incorporated

Fig. 52-1
A road roughness indicator is used when making a final inspection of highways.

Fig. 52-2
Defects in construction are recorded on a punch list. If these holes are visible inside a finished building, they are defects. They are then noted on the punch list, and the contractor must fill them.

Fig. 52-3
Was the defect corrected? The owner checks to be sure. The architect or engineer also makes a final check of the structure. All corrections must be made and approved.

The finished punch list is given to the contractor. Correction of the defects is started immediately. The contractor's employees may do the work. However, some defects may be the subcontractor's fault. These are corrected by the subcontractor's employees.

After the corrections are made, they must be approved. See Fig. 52-3. This is done by the architect or engineer and by the owner. Then the job itself needs no more work.

Before the owner makes final payment, however, some paperwork remains. This includes an exchange of papers between contractor and owner. The papers are

- releases.
- warranties.

Releases

When no complaints have been made, *releases* may be signed. This is done to show that all people involved with the project have been paid.

Payment must be made for all work done. But sometimes, for some reason, payment is not made. The unpaid person may take *legal action* to obtain the payment due. He or she makes a complaint of nonpayment. When payment is made, this person signs a release. To release means to let go. Signing the release shows that the person is willing to release (let go of) the complaint of nonpayment. There are two types of releases.

- Release of claims
- Release of lien

Release of Claims

A contractor may have a *claim* on a construction contract. A claim is a request for extra money. Suppose construction is to begin. The contractor checks the soil on which the structure will stand. The soil has more rock than shown on the plans. Rock is more expensive to excavate than ordinary earth. The original contract allowed a certain amount of money for excavation. But the contractor feels more money is due for this extra work. The contractor presents a claim to the architect or engineer. The architect or engineer gives the claim to the owner. Sometimes the claim is not settled before the project is done. Such a claim is *outstanding* (not yet decided).

The contractor and architect or engineer try to *settle* claims. To settle means to reach an agreement. Sometimes no agreement can be made. Then the contractor, architect or engineer, and owner decide how to settle the claim. This is done through *arbitration* or legal action. See Fig. 52-4.

When the contractor is satisfied, the claims are settled. He or she signs a *release of claims*. This releases all claims against the owner. The contractor may not present any new claims to the owner.

Fig. 52-4
When a claim cannot be settled through discussion, legal action may be taken.

Release of Lien

Another type of release is a *release of lien*. A *lien* is a charge against property for money due. The money may be owed to a skilled worker, subcontractor, contractor, or materials supplier. The debt stands (remains) because the owner is not satisfied with the work or materials. The person who has not been paid goes to court. A *mechanic's lien* is obtained. See Fig. 52-5. The property becomes the *security* for the money due. This means that, if necessary, the property can be sold to pay the debt.

Suppose, for example, that a paving subcontractor surfaces a parking lot for a motel. But the owner does not pay the general contractor for the work.

Mechanic's Lien Notice
FOR LABOR AND MATERIAL

To

L. Curtis

OWNER OF

485 Washington

Hastings

Form 63 – MECHANICS' LIEN - Notice to Owner for Labor or Material — Perfection Legal Forms & Printing Co., Rockford, Ill.

To (1) L. Curtis

You Are Hereby Notified That (the undersigned) have been employed by (2) Haas Building Company to furnish

(3) labor and materials to build 20' x 24' reinforced concrete slab

under 10-6-80 contract with you on your property described as follows: (4)

Situated in the State of -----, County of Clark, and Township of Woodbury, and further described as follows:
Being Lot Numbered 46 of Meadow Estates No. 2 as the same is numbered and delineated upon the recorded plat thereof, of record in Plat Book 16, pages 4 through 12, inclusive, Recorder's Office, Clark County.
Grantor obtained title in Deed Book 232, page 190.
Premises also known as 485 Washington Road, Hastings, -----.

and that there (5) was due the undersigned on the 29th day of January 19 82

therefor the sum of One thousand five hundred ---------- Dollars,

and that (undersigned) will hold the buildings and your interest in the grounds, liable for the amount that

(6) is due the undersigned on account thereof.

DATED at 4 p.m. this 14th day of March A. D. 19 82

Robertson Concrete Co.
721 Dillard Drive, City

B. J. Robertson

(1) Insert owner's name here.
(2) Insert name of employer.
(3) Here state whether labor or material and the nature of the demand.
(4) Here state the building and where situated in general terms.
(5) "Was" or "will become."
(6) "Is" or "will become."
Note.—Make in duplicate, preserve copy and note date of service thereon.

— Lien laws may vary from state to state.

Fig. 52-5
If the owner does not pay for work done, a worker can obtain a lien against the property.

Therefore, the general contractor does not pay the paving subcontractor. The subcontractor goes to a local court of law. The subcontractor obtains a lien against the whole motel property. This property is the security. If payment is not made, the subcontractor can have the motel property sold. The money from the sale will be used to pay the subcontractor. Actually, such a forced sale of property rarely happens. The owner and general contractor usually pay the subcontractor.

The job is finished. The owner normally requires a complete release of lien from the contractor. This release is on behalf of the contractor, subcontractors, suppliers, and skilled workers. The release of lien is an *assurance* to the owner. The property is not security for any debt. All people who worked on the job have been paid.

Warranties

A *warranty* is a promise. There are two types of warranties.

- Contractor's warranty
- Suppliers' warranties

Contractors promise to allow no defects in the work. Suppliers promise to allow no defects in the materials. There is usually a time limit on warranties.

The Contractor's Warranty

Most construction contracts include one-year warranties. The contractor will fix any defects in the structure for one year. Perhaps the paint starts to peel after nine months. The contractor must return and make the repairs.

Contractors prize their *reputations* highly. Sometimes, at no charge, they correct defects after the warranty time is up.

Suppliers' Warranties

Suppliers often warrant (provide a warranty on) their products for one year or longer. See Fig. 52-6. An air conditioner manufacturer, for example, may warrant the air conditioner's compressor unit. The manufacturer may promise that the compressor unit will work for five years.

When construction is completed, the contractor gives the owner two things. First, there are the suppliers' warranty certificates. Second, the owner receives manuals and operating instructions for installed equipment. Examples of installed equipment are water heaters, stoves, dishwashers, and air conditioners.

Final Payment

All defects of the project are corrected. The owner obtains the needed releases of claims and liens from the contractor. The owner has all warranties and operating manuals. Now the owner formally accepts the project. The owner pays the contractor all amounts due under contract.

The contract is now considered *closed* (ended). See Fig. 52-7. The owner takes over all responsibility for the completed project. The contractor is released from responsibility except for warranties. A contract between the owner and the architect or engineer would be closed in a similar way.

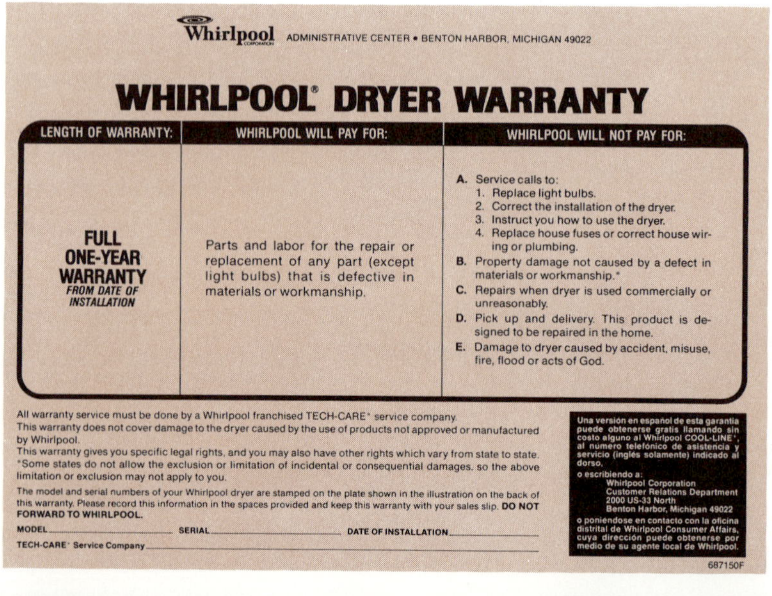

Whirlpool CORPORATION ADMINISTRATIVE CENTER • BENTON HARBOR, MICHIGAN 49022

WHIRLPOOL® DRYER WARRANTY

LENGTH OF WARRANTY:	WHIRLPOOL WILL PAY FOR:	WHIRLPOOL WILL NOT PAY FOR:
FULL ONE-YEAR WARRANTY *FROM DATE OF INSTALLATION*	Parts and labor for the repair or replacement of any part (except light bulbs) that is defective in materials or workmanship.	**A.** Service calls to: 1. Replace light bulbs. 2. Correct the installation of the dryer. 3. Instruct you how to use the dryer. 4. Replace house fuses or correct house wiring or plumbing. **B.** Property damage not caused by a defect in materials or workmanship.* **C.** Repairs when dryer is used commercially or unreasonably. **D.** Pick up and delivery. This product is designed to be repaired in the home. **E.** Damage to dryer caused by accident, misuse, fire, flood or acts of God.

All warranty service must be done by a Whirlpool franchised TECH-CARE® service company.
This warranty does not cover damage to the dryer caused by the use of products not approved or manufactured by Whirlpool.
This warranty gives you specific legal rights, and you may also have other rights which vary from state to state.
*Some states do not allow the exclusion or limitation of incidental or consequential damages, so the above limitation or exclusion may not apply to you.

The model and serial numbers of your Whirlpool dryer are stamped on the plate shown in the illustration on the back of this warranty. Please record this information in the spaces provided and keep this warranty with your sales slip. **DO NOT FORWARD TO WHIRLPOOL.**

MODEL ________ **SERIAL** ________ **DATE OF INSTALLATION** ________

TECH-CARE® Service Company ________

Una versión en español de esta garantia puede obtenerse gratis llamando sin costo alguno al Whirlpool COOL-LINE®, al número telefónico de asistencia y servicio (inglés solamente) indicado al dorso,
o escribiendo a:
Whirlpool Corporation
Customer Relations Department
2000 US-33 North
Benton Harbor, Michigan 49022
o poniendose en contacto con la oficina distrital de Whirlpool Consumer Affairs, cuya dirección puede obtenerse por medio de su agente local de Whirlpool.

687150F

Fig. 52-6
The contractor gives the owner warranties and operating instructions for installed equipment.

Reprinted with permission of Whirlpool Corporation.

Fig. 52-7
The contract is closed when the final payment is made.

Think About It!

1. When is a project closed? What does this mean to the contractor? to the owner?
2. What might be some examples of punch list items for a new highway? a new house?

Summary

When construction is completed, an inspection is made. The contractor corrects any defects that are found. Then the project is transferred to the owner. This involves the release of claims and liens. It also involves providing warranties. All operating and servicing manuals are given to the owner. Then the owner makes the final payment. The contract is closed. The owner takes over responsibility for the project.

Terms to Know

arbitration
assurance
claim
closed
defects
inspected
legal action
lien
mechanic's lien
outstanding
punch list
releases
release of claims
release of lien
reputations
security
settle
transferred
warranty

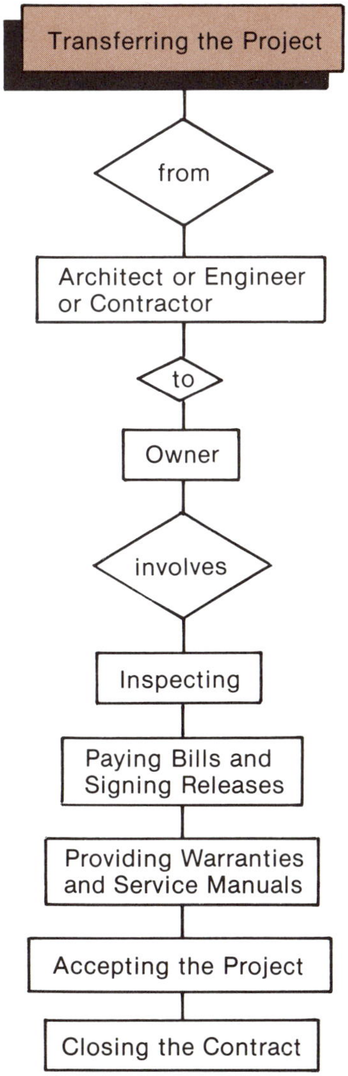

Reading 53

Servicing Property

Servicing Activities

Processing in construction is producing constructed works. After processing, much work must continue to be done. This work assures that projects continue to be useful. These activities are called *servicing* or *postprocessing*. (The prefix "post" means "after.") This reading is about servicing. Servicing consists of five activities.

- Operating
- Maintaining
- Repairing
- Altering
- Installing

Your school building is an example of a constructed work that is serviced. Someone keeps the heating plant operating. When custodians clean the school, they are maintaining it. Broken windows or floor tiles are replaced. This is repairing. Altering may be adding a partition in a classroom. Installing may be putting in a new public address system.

Fig. 53-1
About one-third of the total money spent on construction is for servicing existing projects.

Santa Fe Railway

Servicing Costs

Servicing is not a minor expense. Servicing costs are about half as much as the cost of all new construction. See Fig. 53-1. (Notice that 33% is half of 66%.) During times of economic troubles, people have less money to spend. New construction often is affected first. Fewer new projects are started. A larger proportion (part) of money spent on construction is spent on servicing.

People Who Do the Servicing

Owners are responsible for servicing. They may do their own servicing. Or, they may hire others to do the work. People who do this work are service workers.

Service workers are sometimes hired on a permanent basis. For example, cities hire operators for sewage treatment plants. And many large companies hire their own service workers. Sometimes building owners hire services from (contract with) a company. For instance, an industrial firm may contract with a protective agency. Employees of the agency police the industrial plant. Other agencies provide *char* (housekeeping) *services*. These include such tasks as sweeping and polishing floors and cleaning windows.

Some servicing work requires much knowledge. Many skills and materials are the same as those needed for construction. Workers from every trade and craft are employed in servicing. But some service workers may have little knowledge of construction. Custodians and building operating personnel may be among these.

Illinois State University

A. *This worker is monitoring a steam heating unit. The water temperature and pressure must be checked periodically twenty-four hours a day.*

B. *A closed-circuit TV system enables one person to watch many areas at the same time.*

Fig. 53-2
Operating is an important part of servicing a structure.

Operating

Operating a project involves two things.

- Equipment must be kept running properly.
- The project must be protected.

See Fig. 53-2.

Mechanical and electrical equipment may have automatic controls. Operators may not need to be present at all times. However, they must check the operation of equipment at certain set periods. Some facilities require operators twenty-four hours a day. One example is a central heating plant for a skyscraper. Others might be a dam and locks on a river.

Large or expensive property needs security protection. Service workers called guards help prevent theft and vandalism (damage to property). They also sound alarms if there is a fire or an equipment failure. Automatic alarm systems and closed-circuit television are widely used. They have reduced the number of guards needed.

Maintaining

Maintaining is taking care of property. See Fig. 53-3. Some maintenance, such as housekeeping, needs to be done often and regularly. Other maintenance needs to be done periodically (at longer stated periods). This type of maintenance includes, for example, changing filters on a ducted heating or cooling system.

Often there is a regular schedule of *periodic maintenance*. Filters may be changed every two months. Painting may be done every four years. The grass may be mowed every two weeks. Exterior masonry walls may be sandblasted (cleaned) every twenty-five years.

Some maintenance is done on a "spot" basis. This means it is done whenever the need arises. Changing burned-out light bulbs is an example of this type of maintenance.

Repairing

Repairing is fixing. In time, most structures are damaged or begin to show wear and tear. There are many reasons for this. Weather or air pollution may cause damage. A bad choice of materials or poor work may cause the structure to decay. Overloading, improper use, or neglect may damage the structure. And materials *deteriorate*. This is natural. It happens with all structures.

Fig. 53-3
Mowing grass along a highway is an example of maintaining property.

Fig. 53-4 Bacon and Van Buskirk Glass Company

Replacing even a large pane of glass is a fairly minor repair. Minor repairs should be made promptly. Otherwise, additional repairs may become necessary.

Fig. 53-5 Northwood Mills, Limited

Remodeling may be more expensive than new construction. But it may be the best way to meet changing needs.

If repairs are done when needed, they prolong the life of a structure. They also increase the value and efficiency of the structure.

Using Construction Practices in Repairing

Some new construction practices are used in making repairs. Repairs are done on a smaller scale. Of course, the whole structure is not rebuilt. For example, a hole in a wall may be repaired. Or a section of a road may be torn out and replaced.

Repairs may be quite minor. Replacing a broken window is an example. See Fig. 53-4. But some repairs are major. They cost a great deal of money. For example, the lead roof of an old church was repaired. Then it was given a plastic coating. The total cost was $200,000. On another building, pieces of the limestone facing were falling into the street. The entire exterior masonry had to be *repointed*. This means the mortar joints were repaired. The cost was $50,000.

Replacing Equipment

Replacing equipment is another form of repairing. Suppose a water heater stops working ten years after being installed. The old heater may be replaced by a new one. Sometimes only one part needs replacing. Perhaps only the temperature control device on the water heater needs to be replaced.

Altering

Altering (remodeling) means changing the structural form of a project. See Fig. 53-5. The usual reason for making alterations is to meet new needs. For example, a two-lane highway may be too crowded. It may be widened into four lanes to carry more traffic.

When a structure is altered, it may be made larger or smaller. Sometimes part of a structure is defective. If so, just that part is removed.

Unit costs (costs per square meter or foot) for alterations are usually quite high. It may cost more to remodel than to replace.

Installing

Installing usually involves placing new equipment in a structure. A simple example might be the installation of an air compressor.

Placing insulation in an old structure also is installing. It is an example of installing something other than equipment.

Installing is sometimes done by manufacturing employees. Sometimes it is done by a construction crew. They often work together. The contract for a large installation usually states who will do installing.

Installing is often done to modernize old construction. See Fig. 53-6. Installing new equipment is important. When it is done, old structures sometimes serve in place of new ones.

Summary

Servicing or postprocessing takes place after the project is completed. It includes operating, maintaining, repairing, altering, and installing.

Postprocessing is done by owners or service workers. Some servicing requires construction knowledge and skills. When a structure is properly serviced, it serves its purpose better.

Fig. 53-6
Installing new equipment in an old structure can improve its appearance as well as its usefulness.

Terms to Know

- altering
- char services
- deteriorate
- installing
- maintaining
- operating
- periodic maintenance
- postprocessing
- repairing
- repointed
- servicing

Think About It!

1. How does postprocessing differ from processing? Why is postprocessing important in construction?
2. Is there a street, highway, or bridge in your community that needs servicing? List several servicing activities that should be carried out to improve its condition.

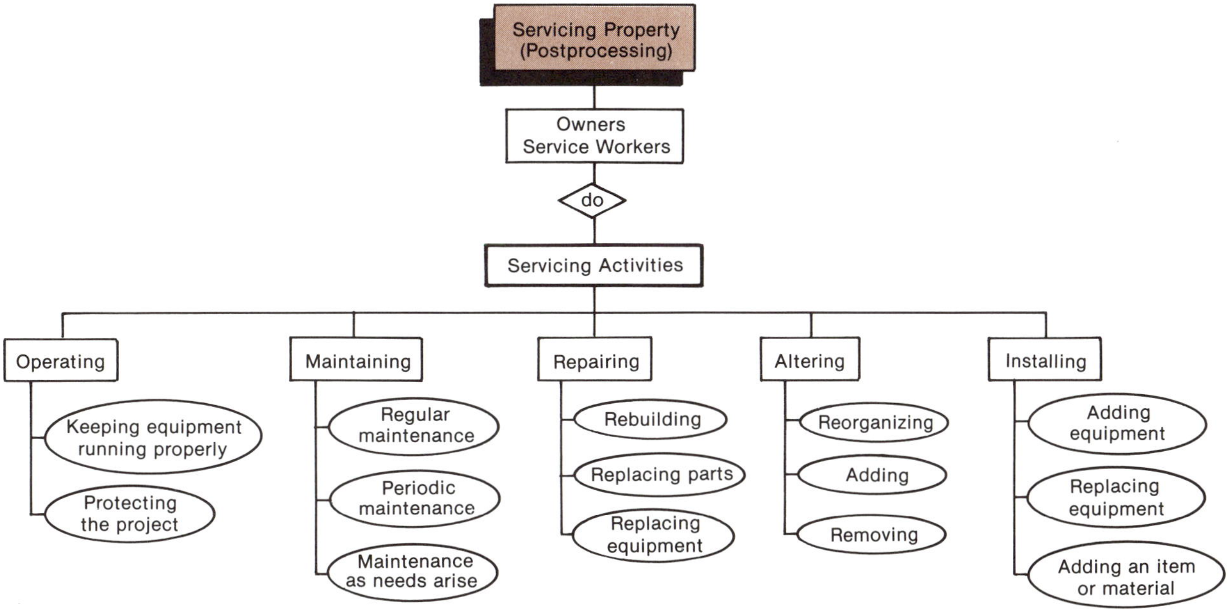

Reading 54

Building Dams

Purposes of Dams

Dams are placed across a stream or river to block the flow of water. The primary purpose of a dam is to control water.

Small upstream dams are important. They keep soil from washing away. They control floods and supply water. However, most of this reading is devoted to larger projects.

Dams may be made of earth, concrete, masonry, steel, or wood. Earth dams and concrete dams are the most common.

This reading tells why and how dams are built. Besides the purposes of dams, you will learn about the

- materials used to build dams.
- people who build dams.
- planning and design of dams.
- features of dams.
- acquisition of land for dams.
- construction of dams.

The Reservoir

When a dam is built, the water level on the upstream side rises. This forms a *reservoir*. A reservoir holds or stores water. See Fig. 54-1. Water is released from the reservoir as needed.

Water in a storage reservoir has many uses. It is piped to cities and farms. On farms, water may be used for livestock. Or, it may be used to grow crops. It is run into the fields. The water's flow is controlled. Watering crops in this way is called *irrigating*. See Fig. 54-2.

Dammed water can be used to produce electricity. Fast moving water can cause objects to move. It has force or energy. Water falling from a high reservoir has energy. The energy is used to run generators in a power plant. Generators produce electricity. This type of power plant is called a *hydroelectric plant*. See Fig. 54-3.

The lake formed by a dam is sometimes used for *recreation*. People may swim, boat, ski, and fish.

Fig. 54-1
When a dam is built across a stream, the water level behind the dam rises. A reservoir is formed.

Lyle C. Axthelm — U.S. Department of the Interior

The lake is also used to *propagate* (increase the number of) fish. Areas around the lake are often saved for wildlife.

The Outlets

Outlets are openings that allow water through the dam. *Outlet works* are gates. They regulate (control) the flow of water from the reservoir through the dam. The needed amount of water flow varies. It depends on the purpose to be served downstream.

The outlets of a dam can be used to keep a minimum water level downstream. For example, enough water may be supplied to allow boats to move freely. Or water may be needed to carry away *effluent* from sewage disposal plants.

In times of little rainfall, water levels in streams drop. Then the outlets are opened to allow more water to flow downstream.

After heavy rains, less water is allowed to flow. Controlling the flow keeps the stream from rising

Fig. 54-2
In dry areas, a dam may provide water for irrigation. Irrigation makes this land more productive and more valuable.

A.G. D'Alessandro — U.S. Department of the Interior

Fig. 54-3
In a hydroelectric plant (lower right of photo), the energy of falling water is changed into electric energy.

Freelance Photographer's Guild

Fig. 54-4
A dam might have prevented this destructive flood.

above its banks. This prevents the flooding of farms or cities. Regulating the height to which downstream water can rise is called *flood controlling*. See Fig. 54-4.

Controlling the flow of water also prevents erosion. *Erosion* is the washing away of soil. In a long, hard rain, much water runs off very quickly. It picks up and carries off bits of soil. This makes a stream very muddy. The stream rises. It flows faster. The swift current washes away the soil along the riverbank. It may wash away soil from under roadbeds or bridge substructures. This is called undercutting. But water can be held back by dams. This reduces erosion and protects structures.

Special and Multipurpose Dams

A special kind of dam is used for *navigation* (shipping). It is located on a river or canal. Such dams are used with structures called *locks*. Locks are enclosures that have gates at each end. The gates at one end are opened to allow boats to enter the lock. The lock is then filled or drained until it is the same level as the next water level. Then the gates at the other end are opened. In this way, boats can pass from one water level to another.

Many dams serve more than one purpose. They are called *multipurpose projects*. An example is Keystone Dam near Tulsa, Oklahoma. It regulates the flow of the Arkansas River. It prevents floods, permits navigation, and carries away sewage effluent. The dam also provides water to the city of Tulsa. The water is used for residential and non-residential purposes. The dam also produces hydroelectric power. A lake is formed by the dam. This lake is used for recreation, fish, and wildlife.

A dam may also serve a secondary purpose. It may provide a roadway across a river or stream. This makes a bridge unnecessary.

Materials

It is possible to make a dam of a single material. The material may be earth, concrete, masonry, steel, or wood. But most larger dams use a combination of materials. Keystone Dam is an example. It is made of earth, steel, and concrete. It also has small amounts of wood and masonry.

People Who Build Dams

Many different people may own or *initiate* (start) a dam and reservoir project. Farmers may build earth dams across small streams. This prevents soil erosion. It also provides water for livestock. Real estate developers may build dams to provide recreational lakes. Cities may build dams to store water for residential and non-residential use. Or, special districts may construct dams. A special district consists of several cities and counties. Its dams may be built for flood control and electric power. They also may be built for recreation or water supply.

Federal Agencies

The federal government has three agencies that build dams. One is the Soil Conservation Service of the Department of Agriculture. It builds small dams to prevent erosion and floods. Another is the Bureau of Reclamation of the Department of the Interior. It builds larger dams for irrigation. These are also used for flood control and electric power. The third is the U. S. Army Corps of Engineers. It builds large projects for flood control, navigation, and electric power.

Dams built by these three agencies may serve other purposes, too. They may be used for water supply, recreation, and fish and wildlife conservation.

Engineers

Dams are designed by civil engineers. See Fig. 54-5. Mechanical and electrical engineers assist them. Engineers supervise the work of the contractors. In addition, they work as managers to operate the project after completion.

Contractors

Dams are constructed by one or more general contractors. Usually there is more than one contract. Sometimes two or more contractors work on a large contract. This is called a *joint venture*. Most of the work is done by the general contractors' employees. Special work may be subcontracted. This includes landscaping and structural steel work.

Workers

A large dam project may involve building other structures. For example, buildings and highways may be needed. Many types of skilled workers are employed for these jobs. Operating engineers (equipment operators), ironworkers, and cement finishers are hired. Carpenters are needed, too. They construct concrete forms. Many laborers also work on these projects.

Planning and Designing a Dam

Planning a dam involves a great deal of time and work. Planners must show that the dam is the best solution to the problem. And the dam must benefit the environment. The benefits to be gained must be greater than the cost of the dam. Then the most desirable site must be found. And the problem of moving structures already on the site must be considered.

Feasibility Study

Planning a dam begins with a *feasibility study*. Planners will use the study to help determine whether the dam is practical. This may take a long time. Five to ten years is not unusual. Public hearings are held. People may voice their opinions for or against the project.

In a feasibility study, a preliminary site choice is made. Engineers prepare a very general design. Included with it is a cost estimate. This helps planners figure the approximate cost of the project. Five major features are considered.

- Condition of the soil.
- Amount of water to be stored.
- Cost of the real estate.
- Actual cost of the proposed dam.
- Relocation of other construction within the proposed dam area.

U.S. Army Corps of Engineers

Fig. 54-5
Dams are designed by civil engineers.

Effect on Environment

Planners must show that the dam will greatly improve the environment. A study is made. It describes the effects the dam will have. The document based on this study is called an *environmental impact statement*.

Economic Analysis

A complete economic analysis is made. It determines the yearly benefits and the yearly costs. The yearly benefits may include prevention of flood damage. And they may include the value of water supplied to the area. See again Fig. 54-2. The yearly costs include those of operating and maintaining the dam.

A *benefit-cost ratio* is also figured. To be a feasible project, the benefits must be greater than the costs. Suppose a dam is to be used only for water supply. The value of the water may be $120,000 per year. The yearly cost may be $100,000. The annual benefit-cost ratio is $120,000/$100,000. This can be simplified to 12 to 10. Since the benefits (12) are greater than the costs (10), this is a feasible project. What is obtained is more than what is paid.

The feasibility study is referred to a legislative body for approval. In the United States, federal projects are referred to Congress.

Site Selection

The site for a dam must be selected carefully. The most desirable spot is where a river valley narrows. See Fig. 54-6. The site should be downstream from a wider part of the valley. Soil conditions along the banks must be checked. The soil must be able to withstand the dam's weight and the water pressure. The soil beneath the dam should not be very porous. It must not permit water to leak under the dam. The availability of construction materials is also checked.

Relocation of Other Construction

Roads usually crisscross the area being considered for a dam. Sometimes railroads and utility lines cross the area. Homes and towns may be in the reservoir area. All of these must be moved or relocated. If not, they may be flooded by the reservoir waters.

Sometimes relocation costs are greater than the dam costs. For example, an interstate highway may need to be moved thirteen kilometers. Moving a highway means rebuilding it. This is expensive. Designing relocations usually takes more effort than designing the dam. And reaching agreements with property owners takes much effort.

Features of a Dam

Some parts of a dam are below grade. That is, they make up the substructure that holds the dam in place. The substructure prevents the dam from being swept downstream. The portion above grade is the superstructure.

Fig. 54-6

Where is a narrow spot in the river valley? Is there a wide spot upstream for the water to collect? People selecting the site for a dam must consider many factors. Then they must choose carefully.

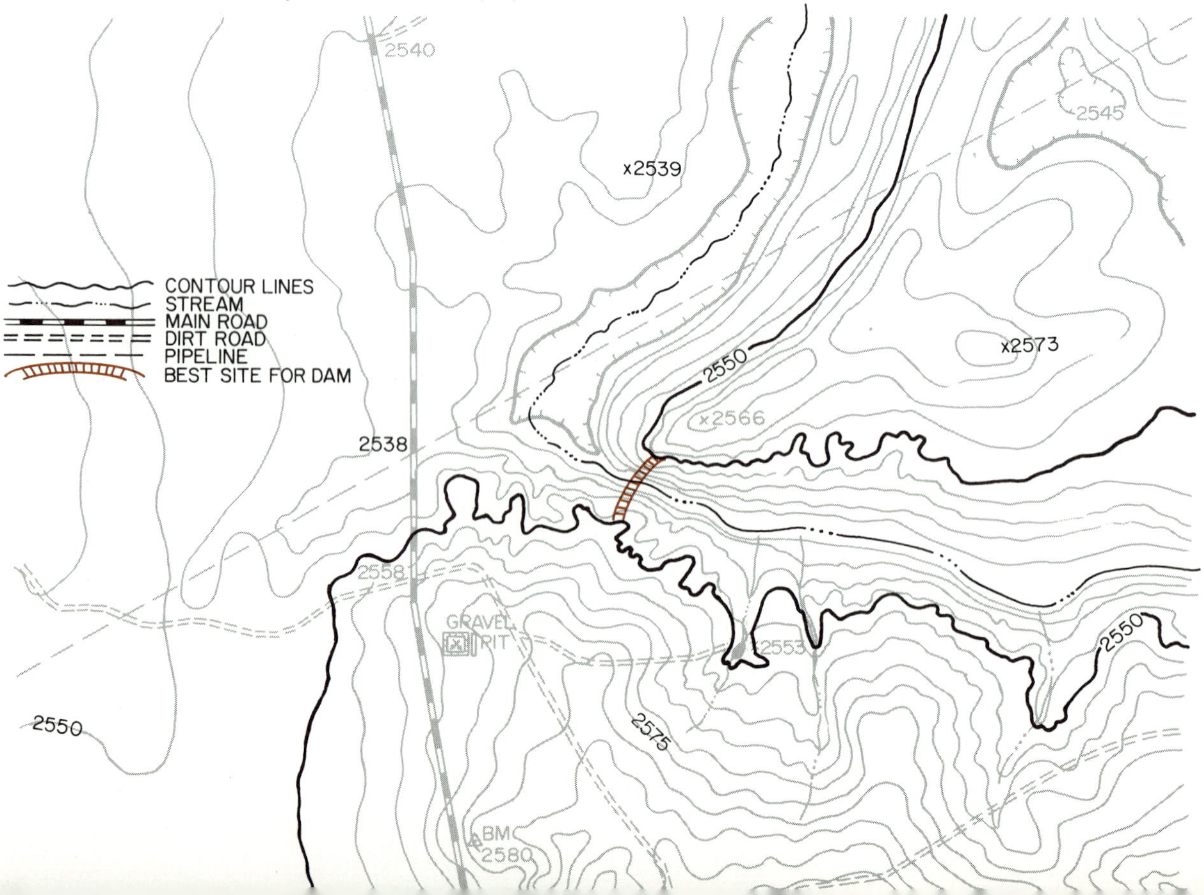

There are three other features of large dams.

- An earth embankment.
- The outlet works.
- A spillway.

See Figs. 54-7 and 54-8.

The purpose of an *embankment* is to hold back the water. This means that it does not transmit water.

As you know, the outlet works regulate water flow from the reservoir through the dam. Sometimes these gates are in tunnels. The outlet works may include a power plant.

The *spillway* releases water automatically. It is often lined with rock or concrete. This keeps it from eroding. The spillway releases water in an emergency. One example of such an emergency is a heavy rainfall. Without the spillway, the reservoir could overflow.

The spillway can be located away from the dam. It may be a tunnel through the earth around the dam. The tunnel discharges (lets out) water below the dam. The water runs out of the face of the canyon or valley wall.

E. E. Hertzog — U.S. Department of the Interior

Fig. 54-7
The embankment and the spillway can be easily identified on this dam.

Fig. 54-8
Water may run through large tubes built around the dam. A.E. Turner — U.S. Department of the Interior

Land Acquisition

Before construction can begin, land must be purchased for the dam and access roads. Other land is acquired in stages. With each stage, the land is farther upstream from the dam site. Land for relocations is purchased as needed. *Land acquisitions* may take a long time. For a major project, they may require two to seven years.

Constructing a Dam

There are many ways of constructing a dam. The way given here is typical for a large project. It is based on a time schedule of several contracts. Some things happen at the same time. As work is being done at the dam site, other work is being done on relocations.

Before work begins, two contracts are made. One is for access roads to the dam site. The other is for a construction field office.

Sometimes a site is far from a town. Then a small city must be built to house workers. This is expensive. It is avoided whenever possible.

The next contract is made for building the dam's *abutments* (ends). See Fig. 54-9A. And a contract may be signed for the *cofferdams*. See Fig. 54-9B.

Fig. 54-9
Constructing a dam.

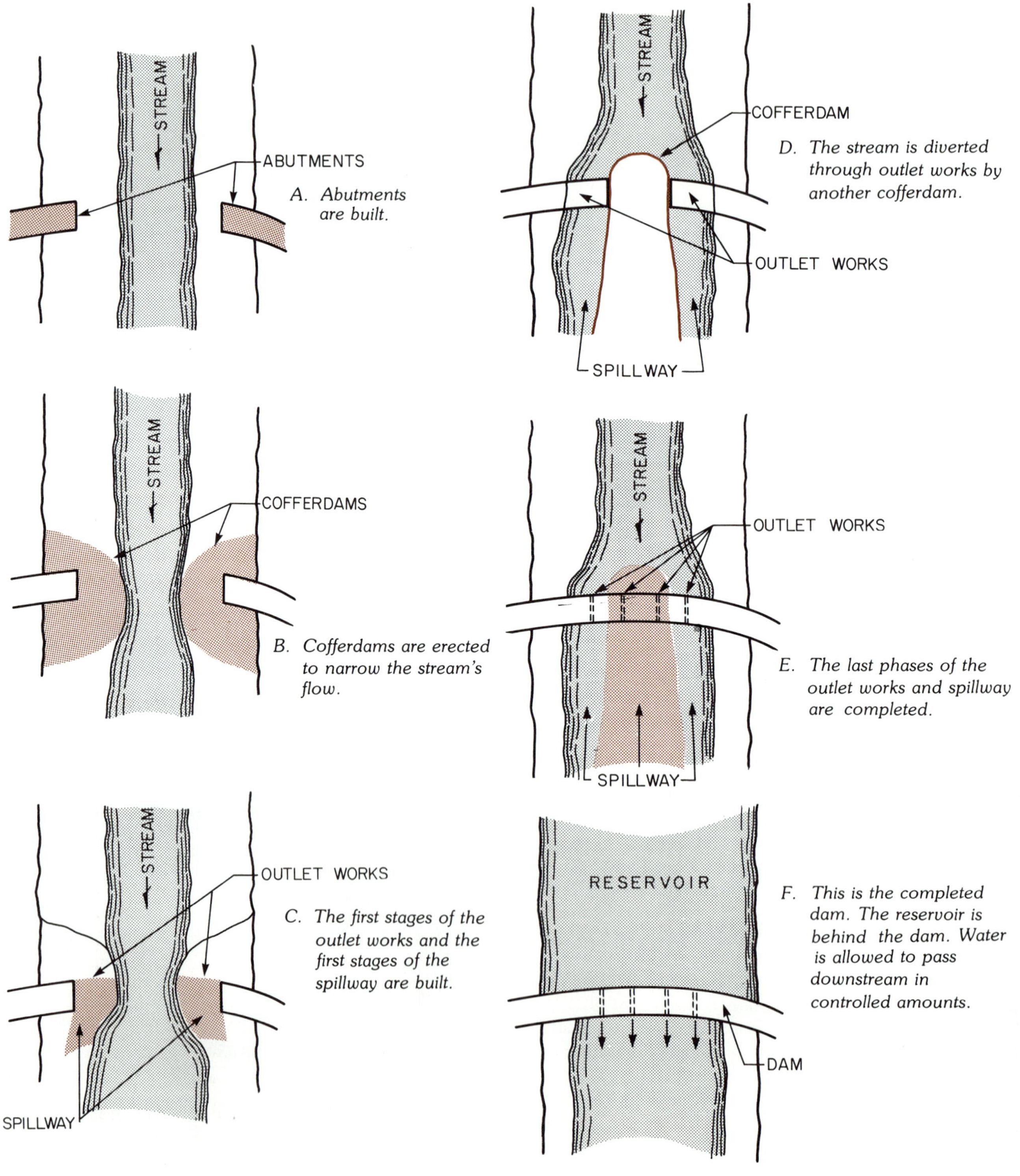

The first cofferdams enclose part of the embankment and the outlet works. The first stage of the embankment should be built on dry land. The outlet works must also be built on dry land. The cofferdams provide the dry land. The river continues to flow in the same channel. See Fig. 54-9C.

The outlet works are completed. The first stage of the embankment construction is well underway.

Next, another contract is *let* (issued) for removing the cofferdams. It provides for moving the river channel. Water will now pass through the outlet works. The same contract allows construction of the second stage of the embankment. For this, another cofferdam is often needed. See Fig. 54-9D. The spillway may be part of an embankment's first and second stages.

The embankment, spillway, and outlet works are completed. See Fig. 54-9E. Then the cofferdam is removed. Water behind the dam forms a reservoir. Water passes downstream through the outlet works and down the spillway. See. Fig. 54-9F.

Sometimes a power plant is provided. The last major contract is for construction of this plant.

Building a large dam takes a long time. In some cases it may require ten or more years. See Fig. 54-10.

Fig. 54-10
Building a large dam can be an awesome task.

H. S. Holmes — U.S. Department of the Interior

Summary

Dams are built for several reasons. The reasons may include water supply, flood control, navigation, and recreation. Dams are commonly built of several different materials. Large, multipurpose dams are a major undertaking. Completing the planning, design, and construction of these dams often takes years.

Terms to Know

abutments
benefit-cost ratio
cofferdams
effluent
embankment
environmental impact statement
erosion
feasibility study
flood controlling
hydroelectric plant
initiate
irrigating
joint venture
land acquisitions
let
locks
multipurpose projects
navigation
outlets
outlet works
propagate
recreation
reservoir
spillway

Think About It!

1. Is there a large dam in your region? If so, how does it affect your life?
2. What are some of the services provided by a multipurpose dam?

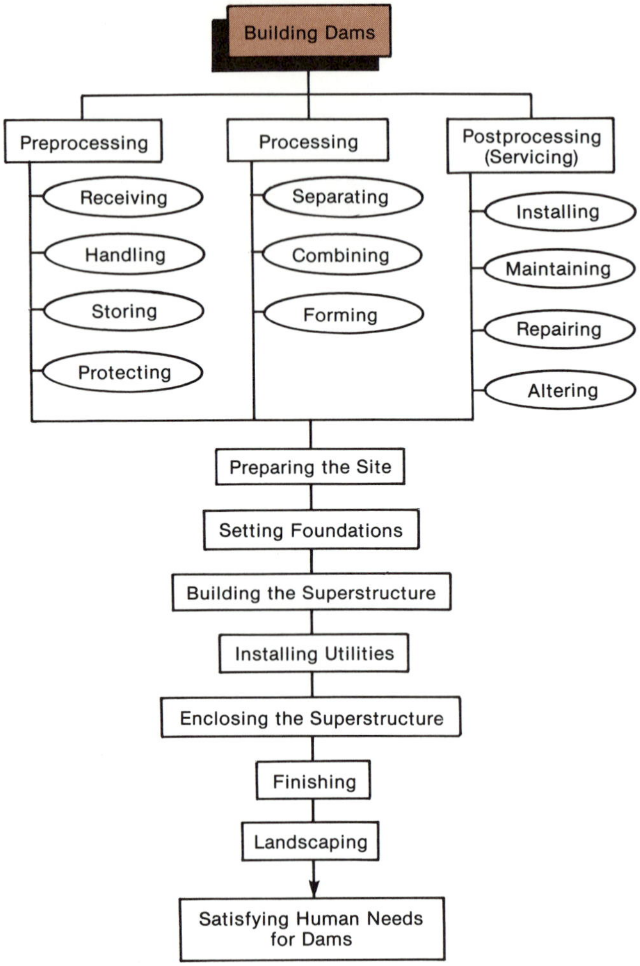

Reading 55

Bridge Building

Bridges

Bridges provide a way over obstacles such as rivers. Like other structures, bridges consist of two parts. Each has a substructure and a superstructure. The *abutments* (ends) are the substructure. Some bridges have *piers* between the abutments. Piers are also part of the substructure. The substructure supports the superstructure and the traffic load. See Fig. 55-1. The roadway is the superstructure. It also supports the traffic load.

Bridges may be *fixed* or *movable*. See Fig. 55-2. Fixed bridges are permanently anchored in place. Movable bridges lift, open, or swing out of the way to allow boats to pass.

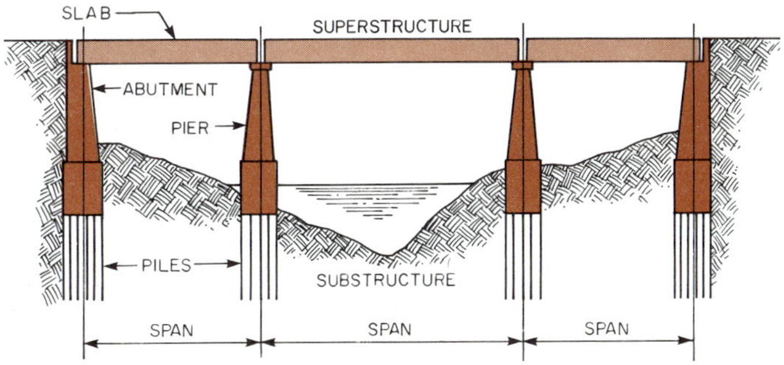

Fig. 55-1
A typical bridge structure.

Fig. 55-2

A. *Fixed bridge*

Marathon Oil Company

B. *Movable bridge*

Fig. 55-3
Types of bridges.

A. Through bridge *B. Deck bridge*

Two types of bridges are *through bridges* and *deck bridges*. A through bridge has a roadway between main beams or trusses. In a deck bridge, the roadway rests upon the main beams or trusses. See Fig. 55-3.

People Who Build Bridges

The owner of a bridge may be a division of government (federal, state, county, or city). A private company, such as a railroad, may own a bridge. And an individual may own a small bridge on private property.

Bridges are designed by civil engineers. Sometimes the engineers are assisted by architects. Manufacturers of steel and concrete products may also help design the bridge. Bridges must be designed carefully. They must safely support all traffic that will use the bridge.

Contracting

There are several methods of contracting for a bridge. The owner may award the whole job to one contractor. Or, separate contracts may be written for the substructure and superstructure. One contract may be subcontracted to a specialist in heavy concrete work. This company will build abutments and piers for the substructure. And another contract may go to a structural steel company. This specialist will build the superstructure. Subcontracts also may be let for such work as painting, lighting, and landscaping.

Many skilled workers are employed in bridge building. Some are ironworkers, operating engineers, and carpenters. Others are cement finishers and laborers. These are the main people involved.

Materials

Bridge substructures were once commonly made of stone masonry. Today, concrete is most often used. Sometimes the earth alone will not support this concrete. Then steel or concrete piles are used. See again Fig. 55-1. Superstructures may be made of steel, concrete, masonry, or wood. Materials are often combined. For example, a concrete roadway may be placed on steel girders. And the roadway and girders may rest on a substructure of concrete abutments.

Precast concrete beams are often used in bridges. Often these beams are prestressed. Prestressed beams contain steel wires that are stretched and tightened. A prestressed concrete beam is stronger than an ordinary reinforced concrete beam of the same size. The precast beams are made in a concrete products yard. They are hauled to the job.

Fixed Bridges

Fixed bridges are anchored in place. They do not move. They may extend across water, railroad tracks, valleys, or roads. There are several types of fixed bridges.

- Slab bridges
- Truss bridges
- Arch bridges
- Cantilever bridges
- Suspension bridges

Fig. 55-4
These metal beams will support a concrete slab bridge.

Slab Bridges

One of the simplest types of bridges is the reinforced concrete *slab* bridge. See again Fig. 55-1. Unsupported slabs can be used for short spans and light loads. A *span* is the distance between supports. On longer spans, *beams* are needed to support the slabs. See Fig. 55-4.

Truss Bridges

Trusses are often used in place of beams to support a roadway. A truss is a rigid framework. It is made by connecting long, straight pieces of wood, steel, or aluminum. The connected pieces form triangles. Examples of trusses are shown in Fig. 55-5.

Arch Bridges

Arch bridges can be used for spanning deep ravines. Arch bridges built long ago had arches made of stone masonry. Modern arches are made of steel or reinforced concrete. The roadway and arch are connected by bars called spandrels.

In a steel arch bridge, the spandrels are made of steel. See Fig. 55-6. This is called a *spandrel braced* steel arch.

Figure 55-7 shows an open spandrel reinforced concrete arch. Arches of reinforced concrete are very common. There are various ways of applying the roadway load to the arch. The roadway in Fig. 55-7 is supported by spandrel walls.

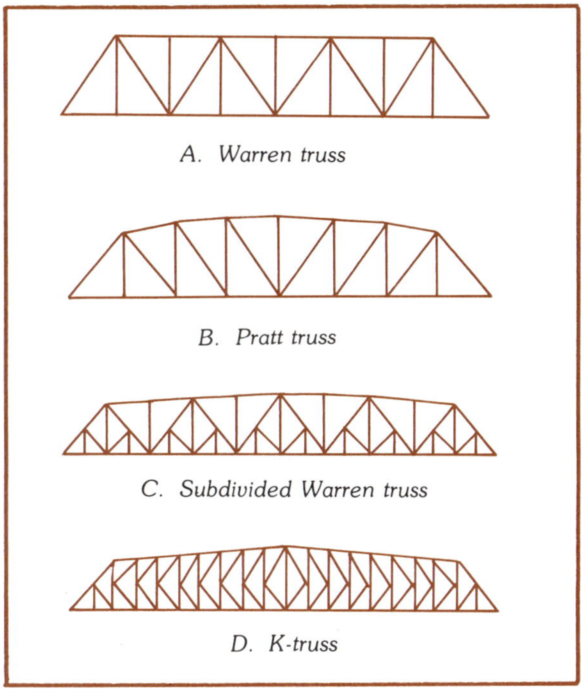

Fig. 55-5
Simple trusses.

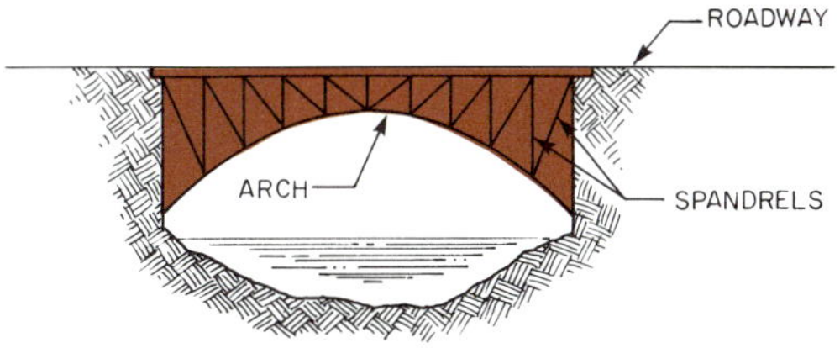

Fig. 55-6
A spandrel braced steel arch.

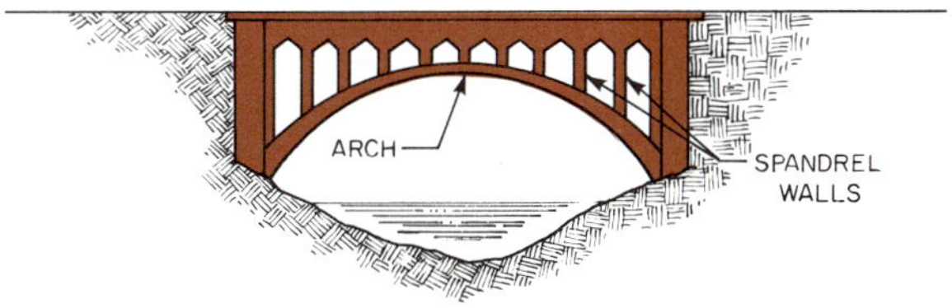

Fig. 55-7
An open spandrel reinforced concrete arch.

Fig. 55-8
Cantilever bridges.

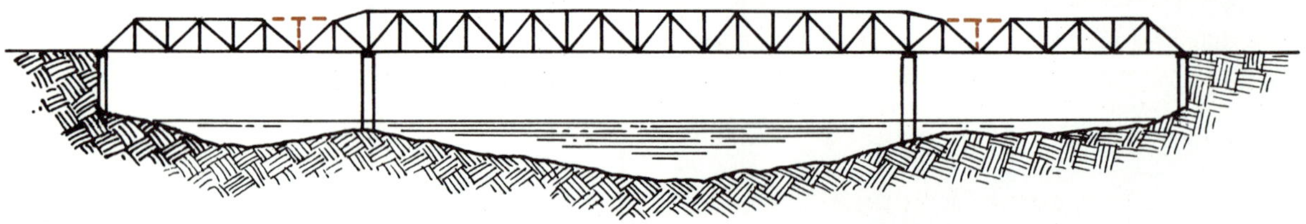

A. *Cantilever truss bridge with two suspended spans.*

B. *Cantilever bridge with one suspended span.*

Cantilever Bridges

The *cantilever* bridge is used for fairly long spans (up to 600 m). Cantilever bridges may be built of reinforced concrete or steel girders. But usually they are made of steel trusses. See Fig. 55-8.

Suspension Bridges

Suspension bridges are used for very long crossings. See Fig. 55-9. In suspension bridges, the load is carried mainly by cables. The cables are made of thousands of parallel, galvanized-steel wires. The cables pass over towers. They are anchored at each end of the bridge. The roadway is hung from the cables by "suspenders." Suspenders are smaller, vertical cables.

Suspension bridges tend to sag at the center. Trusses are put in to reduce this sagging. They stiffen a suspension bridge. And they spread its live load to several suspenders. Trusses also help support the roadway across the length of the bridge. In the United States the bridge with the longest span is the Verrazano-Narrows suspension bridge in New York. Its span is approximately 1300 m. The Golden Gate Bridge in San Francisco is the next longest. It is 1282 m, only about 18 m shorter.

Movable Bridges

Many ships that use waterways are tall. To allow a tall ship to pass, many high fixed bridges are built. But an alternative is to build a low bridge that is movable. Movable bridges are of three types.

- *Bascule* bridge
- *Lift* bridge
- *Swing* bridge

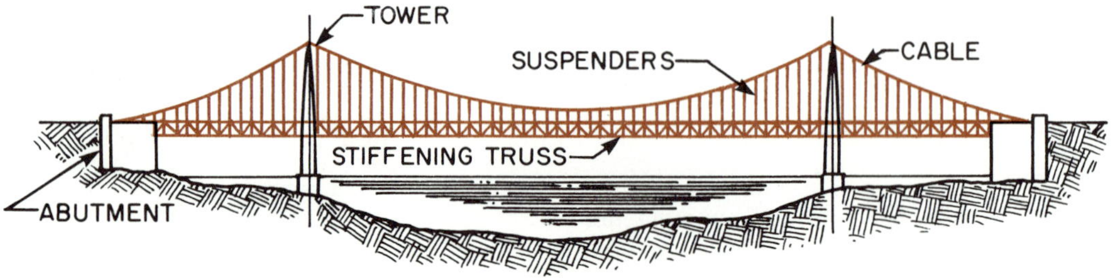

Fig. 55-9
A suspension bridge.

See Fig. 55-10. Notice how each type of bridge operates.

Movable bridges have a disadvantage. Opening bridges for boats and ships stops the land traffic. That is why high, fixed bridges are more common. Land and water traffic can move at the same time.

Temporary Bridges

Most temporary bridges are made of wood. Floating bridges are temporary bridges used under emergency conditions. They are made by tying together boats called *pontoons*. Timbers or steel cover the pontoons. They form the roadway of the floating bridge.

Planning and Design

The planning and design phases for a bridge are like those for other construction. The designer must make important decisions. What type of bridge is most suitable? Where should the bridge be located? What materials should be used for the superstructure? What safety features are needed? There are no set rules for these decisions. See Fig. 55-11.

Many more questions may arise. Should the bridge be fixed or movable? Should intermediate piers be placed? If so, how many? At what elevation (height) should the bridge be located? Should the bridge be a deck bridge or a through bridge? Should steel or reinforced concrete beams be used? Should an arch, cantilever, or suspension design be used?

Many factors affect the answers to such questions.

- Length of the obstacle to be spanned.
- Traffic load to be carried.
- Foundation conditions.
- Requirements for clearance below the bridge.
- Appearance.

Building the Substructure

Abutments support the ends of a bridge. They also serve as retaining walls. They support the earth that lies under the roadway near the bridge. This part of the road is called the approach.

Piers are intermediate supports. They prevent a bridge from sagging. Piers are used when the bridge's weight cannot be carried by one span. Refer back to Fig. 55-1.

Abutments and piers are usually made of reinforced concrete. They should be constructed so

Fig. 55-10
Movable bridges.

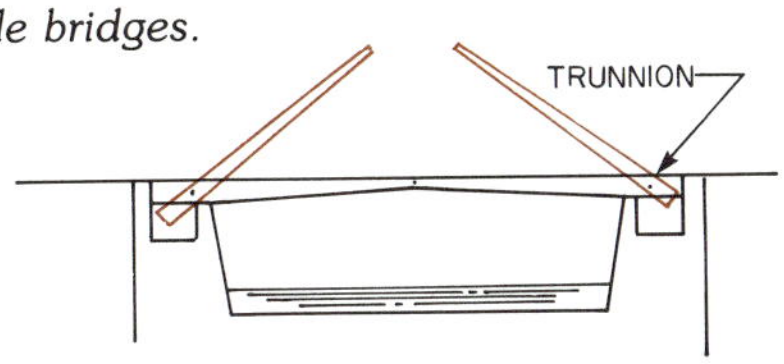

A. Bascule bridge

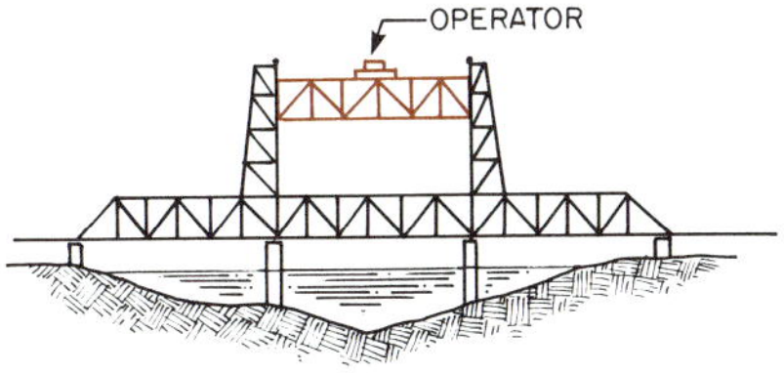

B. Vertical lift bridge

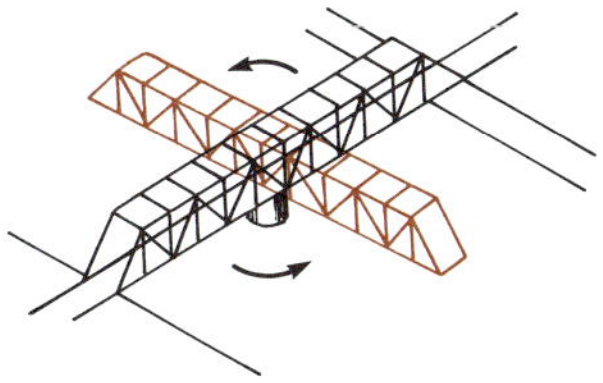

C. Swing bridge

Fig. 55-11
"Where should the bridge be located?" This question and others must be answered during the planning phase.

Hewlett-Packard Company

Fig. 55-12
A cofferdam may be constructed to enable workers to place piers under water.

Fig. 55-13
Clever methods must be used in bridge building. Materials and subunits may need to be transported to remote places. And builders must devise ways to put them in place.

they will settle as little as possible. They must rest on solid earth. The earth supports them. It also supports the superstructure and the traffic. When solid earth is very far underground, *piles* are used. Piles may be made of timber, steel, or concrete. They are driven down into the solid surface. Then the abutments and piers are placed on the piles. Together, the substructure parts are as strong as natural rock.

Most piers and abutments are built using common construction practices. Placing piers deep under water requires special methods. See Fig. 55-12. These methods include using barges and floating cranes. Divers may even be needed to place the piers. Sometimes temporary *cofferdams* are used. These are much like cofferdams used in building dams.

Building the Superstructure

Methods of building the superstructure vary with the type of bridge. Some bridges are made of steel and precast concrete. For these bridges, as much assembly work as possible is done off the site. For example, many pieces can be welded together at a shop. Assembled subunits (pieces) can be quite large. They must be transported and lifted into place. See Fig. 55-13. Pieces of a bridge are carried to the site by truck, rail, or barge. They are hoisted into place with a crane. The structure is welded or bolted together.

Sometimes concrete is put in place on the bridge itself. Then it is necessary to provide a large number of supports. These supports are called *falsework*. They hold the weight of the forms and the wet concrete. First, the supports are placed. Then the concrete roadway is formed and placed. Sometimes the weight of the bridge must be reduced. Then a steel grille (open metalwork) is used instead of a concrete roadway.

The final step involves placing fixtures. These may be safety items such as guardrails, lights, and signs. If the bridge is steel, it is painted after being erected.

Summary

Bridges are built to carry traffic over obstacles. Like other constructed works, they have two major parts. These are the substructure and the superstructure.

Bridges are either fixed or movable. Slab bridges and slab and beam bridges are examples of fixed bridges. Others are truss, arch, cantilever, and suspension bridges. Bascule, lift, and swing bridges are movable.

The safety of traffic crossing the bridge is important. Bridges must be designed and built carefully.

Terms to Know

abutments	piers
arch	piles
bascule	pontoons
beams	slab
cantilever	span
cofferdams	spandrel braced
deck bridges	suspension
falsework	swing
fixed	through bridges
lift	trusses
movable	

Think About It!

1. Can you move around your community with ease without crossing a bridge? If not, how many times do you cross a bridge each day?
2. List the types of bridges you have seen.

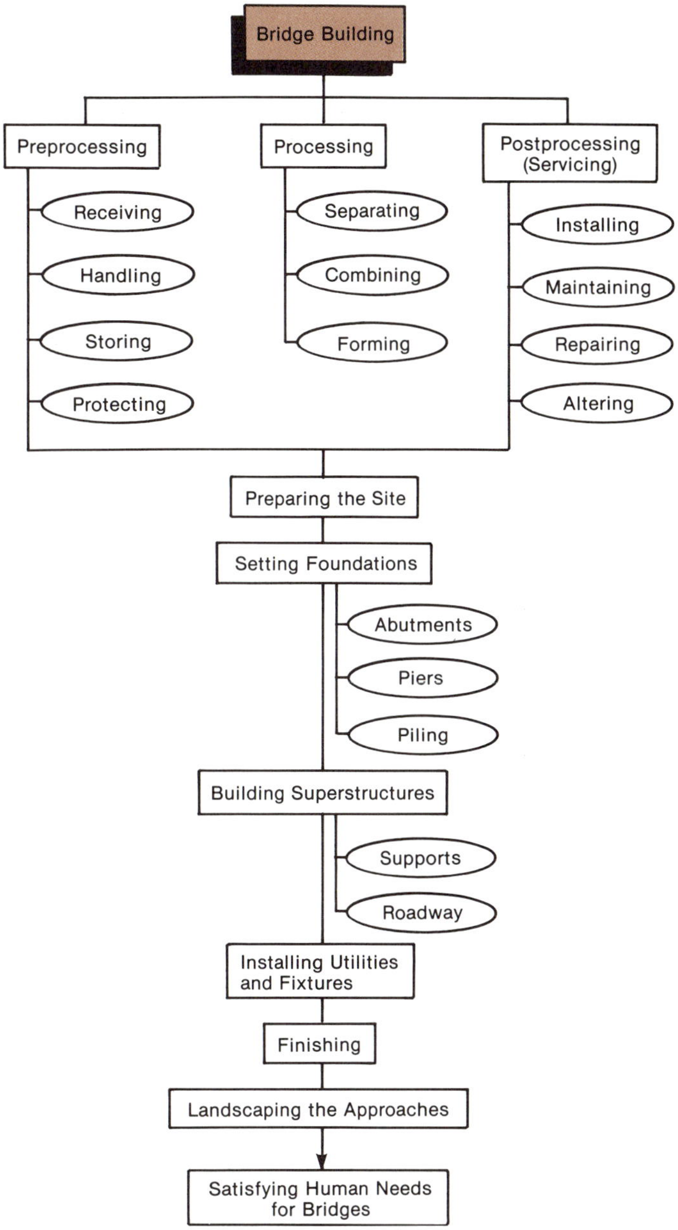

Reading 56

Road Building

U. S. Highway 54

This reading tells the story of a road. You will follow the project from its planning to its completion. The road is a section of U. S. Highway 54, located in central Kansas. It is about two and one-half kilometers long. It includes a bridge. See Fig. 56-1.

Most roads are built in much the same way as U. S. 54. Certain steps are followed. These include

- planning,
- contracting,
- construction, and
- inspection.

Planning

Planning for the road began soon after the Army Engineers made a feasibility study for a reservoir (place to store water). They had already determined that the reservoir site was to be near U. S. 54. But their study indicated that the reservoir water would flood the highway.

There were two possible solutions. The road, including the bridge, had to be relocated. Or, it had to be raised above the water. Next, a feasibility study was made for the road. It was made to find how and where the road should be built.

Feasibility Study for the Road

Engineers needed to know how high and wide to build the bridge. Boats were to be used on the reservoir. As part of the feasibility study, engineers gathered information about the size of the boats. Studying this information helped them determine the *clearance* needed under the bridge. Clearance is the space between the boat and bridge. Engineers also studied rainfall information. The amount of rainfall would affect the water level. These data (facts) helped engineers find the needed height and width of the bridge.

The road construction was not to stop traffic. Engineers considered several different methods.

- **Plan 1.** Move the road a mile* north.
  Findings. The route would be too long.
- **Plan 2.** Move the bridge 500 feet south.
  Findings. The soil and foundation conditions were poor in that location.
- **Plan 3.** Raise the height of the road by building an embankment. The embankment would be above the present road. A new bridge would also be built.

*The actual planning and construction of this project in Kansas was done using U.S. Customary measures.

Fig. 56-1
A section of Highway 54 was rebuilt. It included this bridge.

Findings. This would be too costly.

- **Plan 4.** Raise the bridge and embankment ten feet. Five old piers would be used. A sixth pier and seventh span would be added. This would extend the opening to the needed width.

Findings. This plan was most feasible. It would cost least. It would require the least construction time. And constructing the project would interfere little with traffic flow. See Fig. 56-2.

Funding

Plan 4 was chosen. Now it had to be funded. Money was provided on this schedule.

1st year — Planning and buying the land.
2nd year — Starting work on the road and bridge.
3rd year — Completing the road and bridge.

Designing and Engineering

Certain agreements were made between the Kansas Highway Commission and the Army Engineers. The Commission was to design the road and bridge. It would also award (present) the contracts. Both parties agreed to a *cost reimbursable contract*. This stated that the Commission would be paid for all costs. This included both planning and building costs.

Site work began. Surveyors made detailed maps of the site. Soil testers tested the soil. See Fig. 56-3. This helped determine foundation conditions for the new highway.

A high embankment was needed on approaches to the raised bridge. Questions had to be answered. How much should the bank slope? What kind of earth was available for landfill? What method would be used to place and roll the fill?

Other questions were asked, too. And decisions were made. The extra bridge span would be like the rest of the bridge. The superstructure would be made of steel girders. The pavement would be concrete.

A study of the road was made. It revealed the best type of surface material. A *flexible pavement* such as asphalt or blacktop was needed.

Ways to avoid blocking traffic during construction were studied. A temporary *bypass* detour (a way around the site) was the solution chosen. This would be north of the site. Its location would be far enough away not to interfere with construction.

The bypass road was designed. A bypass is temporary. Lower cost materials are used. This bypass would be a double-surfaced asphalt road. A layer of

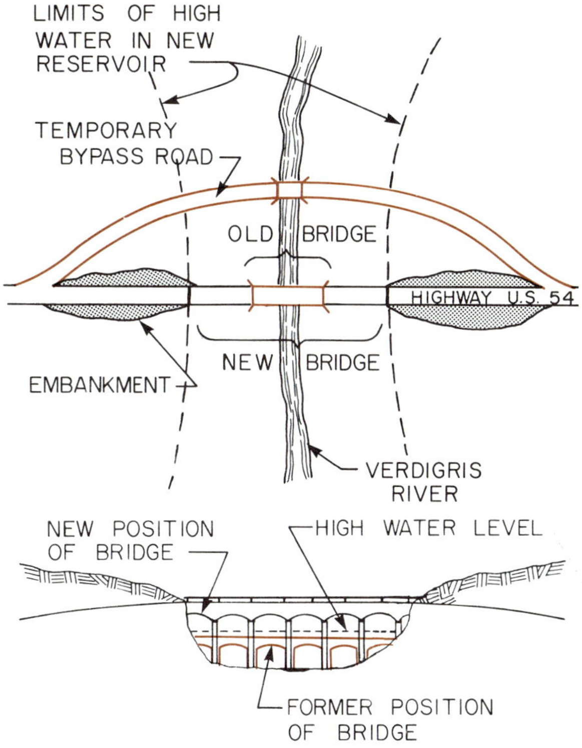

Fig. 56-2
This plan was selected as the best solution. The bridge and embankment would be raised ten feet.

Fig. 56-3
Before building a highway, many soil samples must be tested.

Soil Testing Services

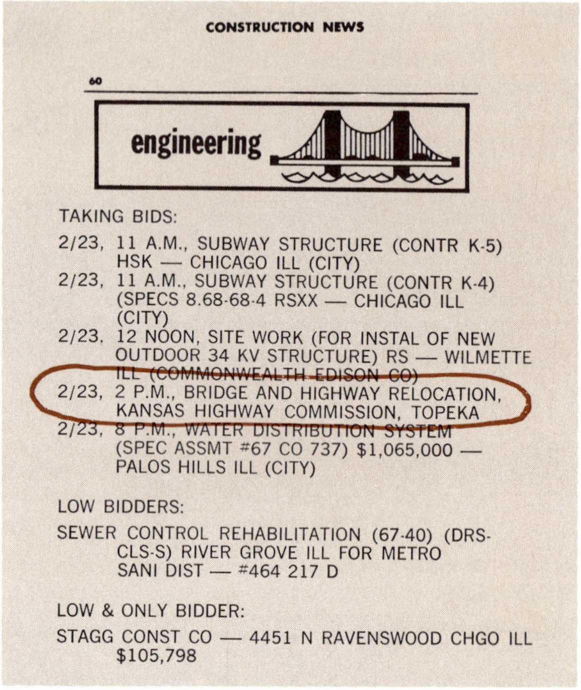

CONSTRUCTION NEWS

60

engineering

TAKING BIDS:

2/23, 11 A.M., SUBWAY STRUCTURE (CONTR K-5) HSK — CHICAGO ILL (CITY)

2/23, 11 A.M., SUBWAY STRUCTURE (CONTR K-4) (SPECS 8.68-68-4 RSXX — CHICAGO ILL (CITY)

2/23, 12 NOON, SITE WORK (FOR INSTAL OF NEW OUTDOOR 34 KV STRUCTURE) RS — WILMETTE ILL (COMMONWEALTH EDISON CO)

2/23, 2 P.M., BRIDGE AND HIGHWAY RELOCATION, KANSAS HIGHWAY COMMISSION, TOPEKA

2/23, 8 P.M., WATER DISTRIBUTION SYSTEM (SPEC ASSMT #67 CO 737) $1,065,000 — PALOS HILLS ILL (CITY)

LOW BIDDERS:

SEWER CONTROL REHABILITATION (67-40) (DRS-CLS-S) RIVER GROVE ILL FOR METRO SANI DIST — #464 217 D

LOW & ONLY BIDDER:

STAGG CONST CO — 4451 N RAVENSWOOD CHGO ILL $105,798

Fig. 56-4

An invitation to bid was placed in trade journals and newspapers.

rock is placed over a layer of oil. The bridge would be made of wood.

The first problems were solved. Next the Commission made detailed working drawings, specifications, and schedules. Some guidelines were needed. The Commission used the standard specifications of the Kansas Highway Department. It also used the rules of the U. S. Bureau of Public Roads.

Contracting

Final approval of the detailed plans and specifications was given. The bidding papers were prepared. The papers included standard rules for Kansas Highway contracts.

Now contractors were given an *invitation to bid* on the job. And the job was announced in state and regional trade publications. See Fig. 56-4. It was also announced in area newspapers.

Contractors were given thirty days to prepare their bids. During this time the Commission had an owner's estimate made. It was used later in checking the bids. On the set day, bids were received. Bidding opened.

The bids were carefully checked. Then a contract was let (issued). The successful bidder then gave a performance bond. This assured that work would be done as planned and specified. Soon after, the contractor was given a notice to begin work.

Construction

Some clearing had to be done. See Fig. 56-5. But the area around the site was mostly pasture. Not much clearing was necessary. The first task was site layout. A surveyor did this.

Before construction could begin, traffic had to be rerouted. Therefore, the temporary bridge and bypass road were built first. Piles were driven for the wood bridge. The beams and roadway were added. Grading and paving were done for the bypass road.

While this was going on, some work began for the permanent structure. The contractor's personnel made schedules. And they bought materials.

Fig. 56-5

A highway construction site must be cleared, excavated, and filled.

Fig. 56-6
A crane such as this was used to lift the bridge span into place.

Fig. 56-7
Paving the surface is the last step in building a road.

They also rented or bought equipment. They hired workers. Subcontracts were let. Detailed shop drawings were made.

Constructing the temporary bridge and road took three months. Traffic was *detoured* (sent around). Now work began on the bridge and embankment. These were the two main parts of the job.

The whole superstructure was lifted by hydraulic jacks and screw jacks. They lifted it three feet at a time. Then I-beams were put in place of the jacks. A *lift* (layer) of concrete two and one-half feet thick was placed around the I-beams. Four lifts this size raised the bridge ten feet.

The embankments were made high enough to meet the ends of the bridge. Then workers built new concrete abutments (end piers). They also built the new sixth pier. Meanwhile, steelwork for the new seventh span was being made in the shop.

Finally, the new span was lifted into place. See Fig. 56-6. The concrete roadway was built over it. Then new sections of guardrail were added.

During this time, the embankment was being built. The old road was torn up with a bulldozer and rooter (machine like a plow). The ground was excavated (dug) to the proper level. Next, the contractor began the fill. Tractor scrapers carried earth from nearby pits. Fill (earth) was placed in six-inch layers. Each layer was rolled by sheepsfoot rollers pulled by tractors. This packed the soil in the embankment according to specifications.

Riprap (broken stone) was placed on the sides of the embankment. This prevented rain or reservoir water from eroding (washing away) the embankment. Then seed was planted in the top of the embankment. The embankment was coated with straw. This helped protect the seed. Finally, the asphalt surface of the road was placed. See Fig. 56-7.

The road and bridge were finished. They were opened to traffic. See Fig. 56-8. Then the bypass was closed and its bridge was taken down. The bypass road was torn up.

Fig. 56-8

Finally, the road and bridge were reopened. Compare the raised bridge with the bridge as it was before construction was done (Fig. 56-1). Note where the old substructure ends. Then note the lifts that were added to raise the bridge.

Inspection

Throughout construction, Kansas Highway Commission inspectors were busy. They made sure that plans and specifications were followed exactly. This involved checking many things.

- Firmness of the fill.
- Quality and strength of the concrete.
- Placement of reinforcing steel.
- Alignment of all parts of the structure.

After construction was completed, the Army Engineers and Kansas Highway Commission made another inspection. It was very detailed. A list of faults was given to the contractor. The contractor made the repairs. Then the Commission accepted the work and gave the contractor the final payment.

Summary

We have followed the story of a road. It began with a feasibility study and a design. Contracting and land acquisition were the next steps. Construction began. Inspections were made. At completion of the work, another inspection was made. These are the usual steps followed in building a road.

Terms to Know

bypass
clearance
cost reimbursable contract
detoured
flexible pavement
invitation to bid
lift

Think About It!

1. How would your life be changed if there were few or no roads in your community?
2. What are the major steps in building a road?

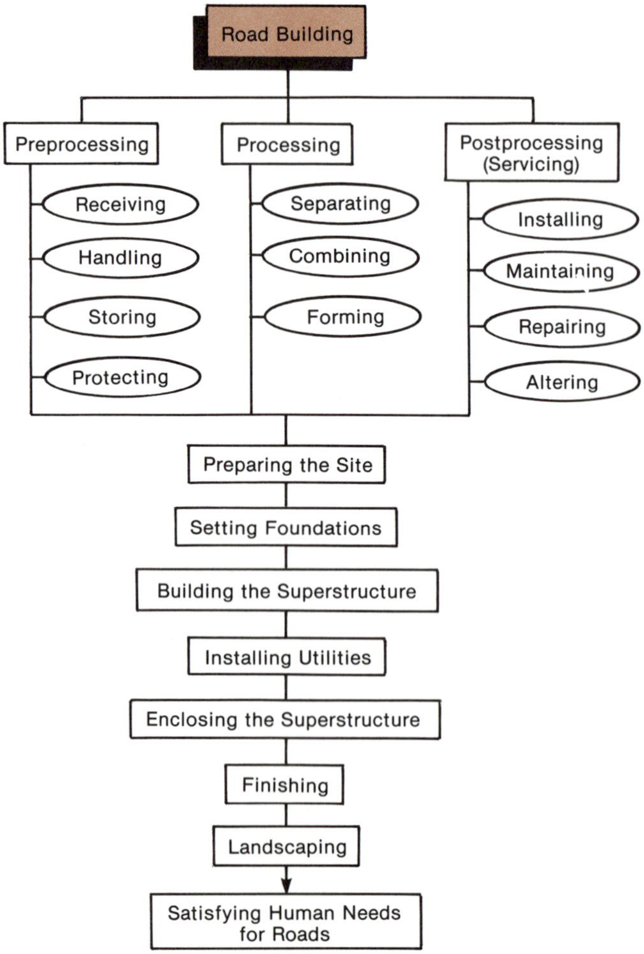

Reading 57

Building Skyscrapers

Skyscrapers

For centuries, a building's weight had to be carried by its walls. The taller buildings were, the thicker their walls were. Thin walls could not be made strong enough to support tall buildings. Thick walls cost too much. They also took up too much space. Thus, the height of buildings was limited. And there were no elevators. People did not want to walk up more than about six stories. This, too, limited the height of buildings.

Construction technology has removed these limitations. Within the last century, elevators and steel framing have been developed. It is now possible to build *skyscrapers*. See Fig. 57-1.

Skyscrapers have many uses. For example, they may be office buildings, apartments, or hotels. But all skyscrapers have one function in common. They make maximum use of costly land.

Bethlehem Steel Corporation

Fig. 57-1
These steelworkers are setting steel for a building that will "scrape the sky." Steel framing has helped make skyscrapers possible.

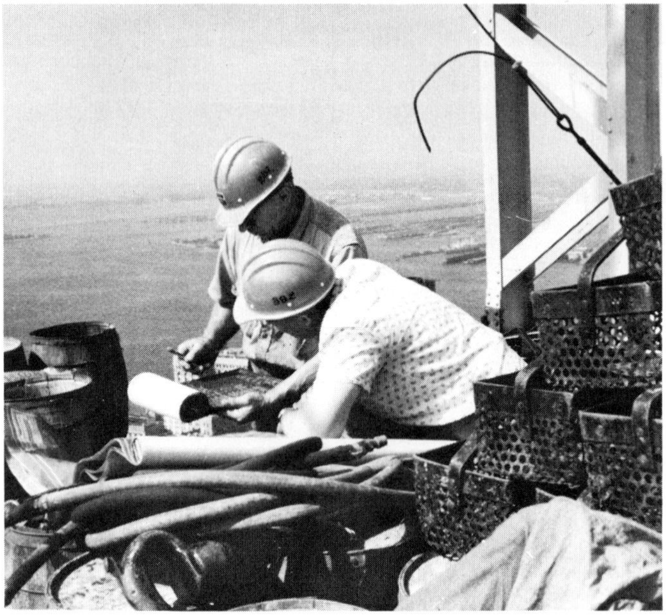

Bethlehem Steel Corporation

Fig. 57-2
A skyscraper is the product of the knowledge, skills, and hard work of many people.

Fig. 57-3
"How much office space is needed?" A feasibility study provided the answer to this question. Planners were able to determine just how large this building should be.

Patent Scaffolding Company

People Who Build Skyscrapers

Many different people may need a skyscraper to be built. Company owners may be project initiators. They may see a need for more downtown office space. A real estate developer may want to rent out downtown offices for a profit. Government employees may need office space.

A skyscraper is a major undertaking. Therefore, several architectural and engineering firms are usually hired to design it. These firms employ specialists. Foundations, framework, and utility systems of skyscrapers are complicated to design.

A general contractor is usually selected to build the skyscraper. The contractor generally employs many subcontractors. Subcontractors may do the mechanical or electrical work. Or, they may do other special jobs. The job of erecting the steel frame is usually subcontracted. And complicated foundation work is sometimes subcontracted.

It takes many kinds of construction workers to build a skyscraper. See Fig. 57-2. These include many laborers. They work for the contractor or the subcontractors.

Planning

The first step of planning to build a skyscraper is a feasibility study. See Fig. 57-3. The initiators consider whether there is a need for a skyscraper. They find out whether offices or apartments can be rented profitably. And they must find out whether a good site is available.

Buying the Land

When the right location for the skyscraper is found, the owner buys the land. Usually the purchase is made through a realtor.

It may take several years to buy the land. Often the site consists of several plots of ground. Each plot may be owned by a different person or group. Some plot owners may not want to sell. Others may ask very high prices for their land. If the buyer and seller cannot agree, the buyer may have to find another site.

Designing the Skyscraper

Designing a skyscraper takes a long time. Architects begin designing while land is being bought.

Foundations must be designed carefully. Skyscrapers place heavy loads on their foundations. The building's weight makes up much of this load. And the wind blowing against the building adds to the load. Sometimes the foundation must be designed to resist earthquakes. The design must provide for a firm foundation.

Careful attention is also given to the design of the superstructure. The weight of the floors, walls, and roof is tremendous. The strength to support this weight must be designed into the framework. And the building should have a pleasing appearance.

Utility systems also must be planned. Finally, consideration is given to architectural details. These details include exterior and interior walls and finishes.

All plans are detailed in drawings. Many, many drawings and specifications are needed for building a skyscraper. If all the drawings and specifications were put together, they might weigh over half a ton. If laid out flat, they would cover several acres.

Demolition

Very few skyscrapers are built on open fields or vacant lots. Usually, existing buildings must be demolished (torn down). *Demolition* is done with crowbars, cutting torches, bulldozers, explosives, or a wrecking ball. See Fig. 57-4. Sometimes several different types of wrecking equipment are used together.

Safety is very important in demolition work. Traffic is sent around the site. Shields are placed around and over equipment operators. Breathing masks are worn to protect against dust. Guards and signalers keep people away from dangerous areas. These and other steps may be taken to avoid accidents. Both the wrecking crew and the public must be protected.

Foundations

There are heavy loads on skyscraper foundations. Therefore, they are set on *bedrock* wherever possible. But sometimes bedrock is far below the earth's surface. Then piles are often used.

Steel sheet piles are usually driven around the perimeter (outside edge) of the site. This piling (pile structure) is tightly braced. The braced piling provides support for streets near the site. And it supports underground utilities, subways, and existing buildings. See Fig. 57-5. Sometimes very elaborate supports, called *underpinnings*, are needed. They give extra support to adjacent (nearby) structures. Steps must be taken to pump out any underground water.

Fig. 57-4
Often, existing structures are cleared to make room for a new skyscraper.

Fig. 57-5
Steel sheet piling is driven into the ground around a skyscraper building site. It supports streets and other structures next to the site.

Patent Scaffolding Company

Fig. 57-6
Recent developments have allowed reinforced concrete to be used in high-rise structures.

Fig. 57-7
A skyscraper building project may be in various stages of completion. Lower stories are enclosed and finished while upper stories are being built.

There are other foundation operations, too. These include excavating, placing footings or pilings, and placing underground exterior walls.

Framework of Superstructure

The framework of a skyscraper is made of steel. Steel framework has two major kinds of parts. These are *beams* (horizontal members) and *columns* (vertical members). The beams carry the weight of the walls and the floors. They also keep the columns from moving sideways. The columns carry the loads from the beams to the foundation.

Steel framing members are usually welded or bolted together. Rivets are not often used. As much of the framework as possible is *prefabricated*. That is, pieces and subassemblies are made in the shop. Then they are joined together on the site.

Members are lifted into place with a *traveling crane*. Lifting may be done from the ground with a *tower crane*. Or, it may be done from the top of the construction. Then a *derrick* is used. The derrick can be moved up two stories at once.

The framework is usually erected two stories at a time. Members are hoisted into place and fastened by temporary bolts. Then the beams are made level. The columns are made *plumb*. Next, the permanent connections are made. The process is repeated, two stories at a time. This goes on until the framework is *topped out*.

Framework is not always needed. The development of reinforced concrete has made it possible to erect high-rise buildings that have *load-bearing walls*. See Fig. 57-6.

Floors and Walls

After the frame is put up for a few stories, the contractor begins to enclose the interior. All floor systems (except the top covering) are installed. And the exterior walls are installed. See Fig. 57-7. Normally, this work is done one story at a time. It is begun at the bottom. However, one story need not be entirely completed before another is started.

Utilities

Utilities are usually installed right after the structural framework is completed. The installation of utilities involves many tasks. Plumbing, ductwork, and electrical and communications lines are installed. Elevators are put in.

Finishing the Interior

Finishing is done to each story after it is enclosed. This includes insulating, installing partitions, finishing floors, trimming, and painting. Accessories and fixtures are installed last.

Patent Scaffolding Company

Fig. 57-8

Elevators lift load after load of materials, equipment, and workers to where they are needed on a skyscraper project. So busy are these elevators that their use may be scheduled months in advance.

Coordinating the Work

Building a skyscraper requires teamwork. Working space is quite limited. And there is very little room to store construction materials and equipment. Hoist and construction elevator capacities are very small. The use of elevators is carefully scheduled by a computer, months in advance. See Fig. 57-8. Consider this example. The schedule calls for lifting electrical conduit. The conduit is to go to the twenty-sixth floor at 9:45 A.M. on June 6. The electrical contractor must have the conduit on the site at that time. Workers also must be there to load and unload it. If anyone is late, the whole project gets off schedule.

Summary

Building skyscrapers is different from building almost any other structure. Skyscrapers give the greatest amount of useful space on the least amount of land.

The development and widespread use of skyscrapers are mostly American. Skyscrapers have been built in other countries, too.

Skyscrapers are very complicated projects. They must be designed carefully. Schedules and procedures for building them must be carefully planned, organized, and controlled.

Terms to Know

beams
bedrock
columns
demolition
derrick
load-bearing walls
plumb
prefabricated
skyscrapers
topped out
tower crane
traveling crane
underpinnings

Think About It!

1. Why do you think the development and widespread use of skyscrapers is mostly American?
2. What is the tallest building in your community? How many floors does it have? Can you give some reasons why it was made so high?

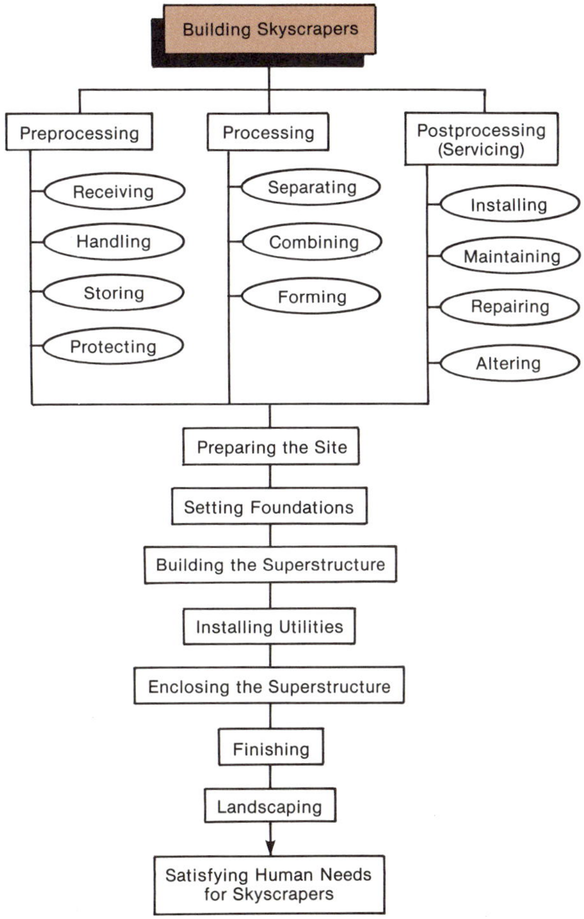

Reading 58

City and Regional Planning

Regulating Construction

Planning a dream home is important to many people. But they are also concerned about the area around their homes. They want to live in a safe, pleasant neighborhood. See Fig. 58-1. People want their homes and their neighbors' homes to be safe. For example, they want them to be safely wired. They do not want to risk damage to their homes by electrical fire. People also want their neighbors' homes to look nice. They do not want to live near buildings that look run-down.

Making sure a neighborhood is safe and pleasant requires joint planning. People must join together to *regulate* construction. They do this in several ways. First, they make *codes*. Codes tell what materials are acceptable for building. They also set the rules for how structures are to be built. Second, people create *zones*. Zoning regulations (rules) identify the kinds of structures that can be built in an area. A third way to regulate construction is to require *building permits* and *inspections*. This makes sure certain standards are met. These three types of regulating help produce good communities in which to build dream homes.

Fig. 58-1
Each home in a neighborhood is affected by the appearance of the homes next to it.

City and regional planning also helps produce safe, pleasant living conditions. This planning helps improve urban areas and communities near the city. City and regional planning affects all major kinds of construction. For example, the locations of streets, parks, and community buildings are the result of this planning. Efficient transportation is another result. The next six readings will help you understand city and regional planning.

As you read, remember three things. Cities develop in certain locations for special reasons. Construction leads to more construction. Certain conditions must be met before construction projects are built. In this reading, you will learn about how these things take place.

Primary Construction

A community begins to grow if its location meets a certain need. The need may be for a harbor or a missile launching site. A structure is built to serve that need. The construction creates a need for workers. These workers need goods and services. To meet these needs, more structures are built. These structures result in more and more jobs. This creates a need for more construction. Structures that produce other construction needs are called *primary construction*. Primary construction may cause the birth of a community. It always helps bring about community growth.

Primary construction is built to meet any of these six needs:

- Developing *natural resources* (dams and parks).
- Changing *modes of transportation* (seaports and airports).
- Changing the form of materials by *manufacturing* (factories).

- *Distributing* goods and services (shopping centers and warehouses).
- *Defending* the nation (missile silos and radar sites).
- *Governing* communities (city halls and capitols).

Developing Natural Resources

Hoover Dam is an example of primary construction. It was built because water, a natural resource, could be used to produce electrical power and to water crops. Hoover Dam was built on the Colorado River. It was first called Boulder Dam. Boulder City, Nevada, grew up around the construction.

Changing Modes of Transportation

New York City was settled at a place where land and river routes met at a great ocean harbor. There was need to transfer goods from one form of transportation to another. This meant that docks and railroad yards had to be built. Warehouses were needed to store goods until they could be distributed to people living inland. Workers were needed to handle the goods. As a result, there was a need for places for people to live and shop. Soon other construction, not directly related to transportation, was built.

Changing the Form of Materials

Many steel-producing cities are located on the shores of the Great Lakes. They are examples of cities that grew because of a need to change the form of materials. Iron ore could be brought to these cities by ship. Other resources were close at hand. For these reasons, many cities on the Great Lakes became manufacturing centers.

Distributing Goods and Services

People need access to goods and services. Businesses and shopping centers are built to meet this need. But goods must be shipped in. And materials are needed for providing services. A complex distribution system develops. A good distribution system causes businesses to grow. This occurs not only in cities but between cities. Much construction is needed to make distribution possible. Cities such as Denver, Chicago, and Atlanta are regional distribution centers. Much of their growth can be credited to that.

Defending the Nation

San Francisco, California, is an example of a community that grew from a need for defense. A Spanish fortress was built to guard the entrance to San Francisco Bay. The city grew from the construction of the fort. People at the fort had social and economic demands (wants). More construction was built to meet these demands.

Governing Communities

State capitals are examples of communities that grew from a need for government. One such city is Boston, Massachusetts.

More than one of the six needs may exist in the same place. Boston is an example. It has not only met the need for government, but also for defense and distribution. In addition, it has met needs for developing natural resources, changing modes of transportation, and changing the form of materials.

Building in the Community

Workers are needed for primary construction. Other workers are needed to maintain the structure. Still others may operate a business there. All these people need housing. They also need schools, roads, and stores. The community grows as its people have new needs that must be satisfied. See Fig. 58-2. The needs may be for new construction. People should express their needs. Only then will the needs be satisfied.

Construction Factors

A construction project should be *economical* (worth the cost). To be economical, the project must meet three conditions.

- There must be a demand for the construction.
- The construction must satisfy the demand by providing a service.
- The construction must be located on a suitable site.

Every construction project should meet these three factors. When it does, there is good reason to build it.

Demand

A *demand* is a reason for building. A person may want a project built. A community may want a project built. The project must fill a real need. The demand must be for the exact kind of project proposed. The number of projects to be provided must

Fig. 58-2

Kennedy Space Center, built at Cape Canaveral, Florida, is an example of primary construction. Suppose such a center were constructed near the area in which you live. How would your area be affected?

Kennedy Space Center occupies a large site. Would acquiring a site be difficult in your area? How many constructed products can you identify in this photo?

National Aeronautics and Space Administration

Construction at the Center continues. Launch pads such as this must be built.

National Aeronautics and Space Administration

Imagine the number of workers needed to build the parts of the Center. Some of these parts are —

Vehicle Assembly Building

National Aeronautics and Space Administration

Launch Control Complex (This is the firing room.)

National Aeronautics and Space Administration

Operations and Checkout Building

National Aeronautics and Space Administration

Churches

Schools

Offices

Utilities

Shopping facilities Keystone Group

Primary construction creates a need for more construction. Workers need goods and services. Perhaps your area would need more —

U.S. Steel Corporation

Health care facilities

Factories

Recreation areas Cocoa Chamber of Commerce

Streets and highways

Homes

match the demand. Also, people must be able to afford the project. It would not be economical to build a project that people could not afford.

Service

A *service* provides something that people want. The project must be useful. It must be located where it can best serve people. Also, the project must be located near services it needs, such as water supply and transportation. If it is near such services, it is economical.

Site

The *site* is the ground on which the project will be built. The cost of the site must be reasonable. Soils must be of types that cost little to develop. Also, for the project to be economical, the land slope must not increase construction costs.

Summary

Construction that causes the birth and growth of a community is called primary construction. Primary construction creates employment. This creates the need for more construction. Primary construction is built to meet needs. There may be a need to develop natural resources. Or, transportation or distribution systems may need to be changed. Sometimes materials must be changed. Primary construction may also be the result of a need for government or national defense.

Workers are needed to build and operate primary construction projects. Communities begin when these workers move to or near the site. Workers' families need housing, schools, and shops. A community grows as its people's needs grow and are met.

A construction project must be economical. This justifies its reason for being in the community. To be economical, the project must meet three conditions. First, there must be a demand for it. Second, it must satisfy the demand by providing a service. Finally, the project must be located on a suitable construction site.

Terms to Know

building permits
codes
defending
demand
distributing
economical
governing
inspections
manufacturing
modes of transportation
natural resources
primary construction
regulate
service
site
zones

Think About It!

1. Is your community growing in population? What plans do you know of that *manage* (plan, organize, and control) its growth?
2. What are some examples of primary construction in your community? When were they built?

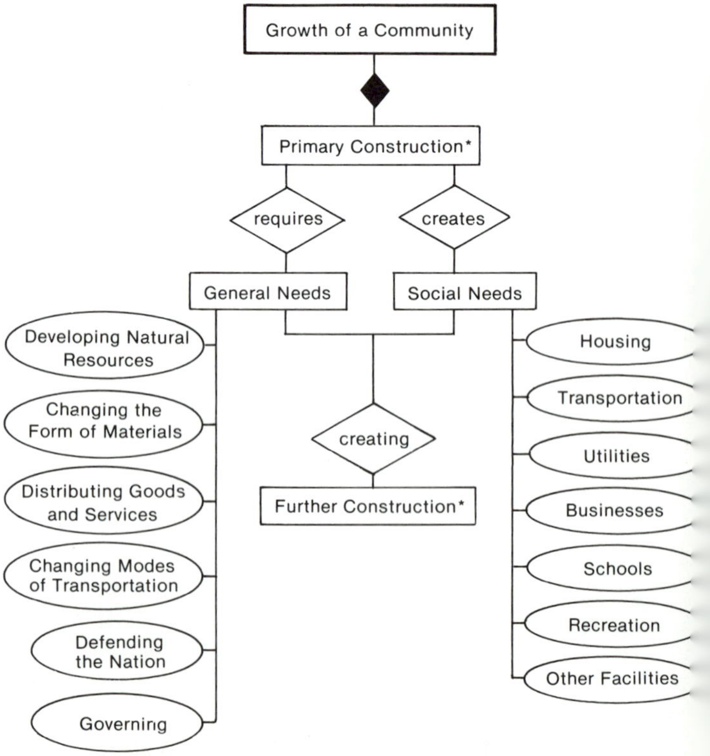

*Based on favorable conditions of demand, service, and site

Reading 59

Planning Community Services

Service Construction

Primary construction must have people to serve its needs. But it also must have *services.* There are two main types of services required. *Transportation services* move people and goods to the site. As a result, roads, railroads, docks, airstrips, and pipelines may be built. *Utility services* are also needed for primary construction. Water and sewer pipelines must be constructed. Electric and telephone lines must be built. All these bring utility services to primary construction projects. See Fig. 59-1. Without *service construction,* most primary construction could not serve its purposes.

Transportation Construction

There are five major kinds of *transportation construction.* These are roads (highways), railways, airstrips, waterways, and pipelines. All are needed to service primary construction projects.

Roads and Streets

Roads are car and truck routes outside an *urban area.* Before the development of modern roads, or *highways,* traffic had to slow down to travel in and around cities. Today, traffic from rural areas can run freely through urban areas. On modern roads, more than 2400 vehicles an hour can move in each direction.

Fig. 59-1
Utilities must be run to construction sites.

Loaded trucks have trouble climbing steep hills. Therefore, roads are generally built on flat or sloped land. Where necessary, tunneling, bridging, and earthmoving are done. These reduce or avoid steep grades (sharply sloping roads). But they also increase the cost of construction. Roads should be built on hard or firm soils. These soils hold up heavy truck loads. In addition, roads should be as straight as possible. This makes them safer and more efficient.

Streets are built to service urban areas. The three basic kinds of streets are *service streets, collector streets,* and *arterial streets.* See Fig. 59-2. They

Fig. 59-2
Street Capacities for the Three Basic Kinds of Streets

Type of Street	Number of Lanes Moving in One Direction	Street Capacity
Service Street	1	600 cars per hour
Collector Street	2	1200 cars per hour (600 cars per hour in each lane)
Arterial Street	3	1800 cars per hour (600 cars per hour in each lane)
Arterial Street (widened to 8 lanes at intersections)	4	2400 cars per hour (600 cars per hour in each lane)

Fig. 59-3
Three kinds of streets serve a community's needs.

A. *Service streets serve lightly traveled areas.*

B. *Collector streets serve traffic from many service streets.*

C. *Arterial streets serve traffic from many collector streets.*

have varying *street capacities*. This means that they can move only a certain number of cars per hour. Look again at Fig. 59-2.

Intersections with signals are the main cause of limited street capacity. Building *interchanges* can make signals unnecessary on arterial streets. Interchanges permit traffic to pass from one street to another without crossing traffic streams. See Fig. 59-3.

Many things must be considered in planning street locations.

- The kinds of streets needed.
- The shortest possible distance between origins and destinations.
- The use of as few streets as possible.
- Land slope and soil condition.
- Streets built at an earlier time.
- Structures in developed areas.
- The shapes of private properties. See Fig. 59-4.

Street building costs include the cost of the land for the street. They also include the cost of the street itself.

Fig. 59-4
Streets that are built around properties leave odd sized and shaped land that is difficult to develop.

Railways

Railway service is often the least expensive means of transportation. It is one of the best methods of moving large, heavy loads many miles over land. But heavy train loads are hard to pull up even very small hills. Therefore, railroad tracks are generally built on flat land. The land should be hard or firm to support heavy loads.

Airways

Some primary construction projects need air service. A manufacturing plant, for example, may want to ship its highly valuable products by air. Air service is also useful for fast transport of people and materials.

Airstrips (airplane landing strips) must be built on hard or firm soils. They require a large amount of flat, low-priced land. Also, a landing strip should be located near the primary construction that it serves. For example, it may be built to serve a factory or passenger terminal. See Fig. 59-5. Often, service construction leads to further primary construction. A manufacturing plant, for instance, may be built because air service is available nearby.

Waterways

Water transport is another way to move heavy loads. To be useful, *waterways* must be close to the primary construction. They also must be deep and wide enough for many types of boats, ships, and barges to pass through. Using waterways causes a need for more service construction. Bridges, dams, and locks may be needed. Roads and railroads must be close to docks. Docks for loading freight must be built at the water's edge. They must be built on flat or sloped land. Hard soils are needed near the docks. These soils must hold the weight of the freight being transported and loaded.

Pipelines

Some things can be carried more cheaply through *pipelines*. See Fig. 59-6. Oil and natural gas are often carried through them. Some solids, such as finely ground coal, are carried by water through the pipelines. Pipelines generally run underground. They follow the shortest distance between locations wanting pipeline service. Most pipelines are under pressure (pumped). Thus, their contents can follow the slope of the land. Pumping stations must be built along the path of the pipeline. Pumping keeps pressure in the pipeline system.

Fig. 59-5
The landing strip is only part of the construction needed for an airport. Facilities are built for loading and unloading passengers, for storing freight, and for repairing the airplanes.

Fig. 59-6
Many materials can be moved quickly and cheaply through pipelines.

U.S. Department of the Interior

Utility Construction

Primary construction needs many transportation services. But it also needs gas, electricity, communications, water, and sewage disposal services. The construction of these services is called *utility construction*.

Gas, electricity, and communications services can be brought to most areas. Usually extra lengths of pipelines or cable are all that is needed. These connect existing utility systems to the new sites.

Water Systems

Building water utilities requires more than adding lengths of pipe or cable. Much construction must be done. Water systems are made up of treatment plants, pumps and storage tanks, and piping. Water comes from a water supply. This may be a lake, river, or well. From there it is brought through pipes to the treatment plant. At the plant, the water is filtered. Chemicals are added. This treatment makes the water safe for human use.

After the water is treated, it is piped to the site at which it is needed. Sometimes the water source is on higher land than the highest site it serves. Then water is transported directly from a storage tank. The water does not have to be pumped. It will flow by *gravity*. However, the water source is not usually higher than all the sites it serves. Water must be moved uphill. A water pump is needed. A pressurized storage tank or water tower is needed, also. See Fig. 59-7.

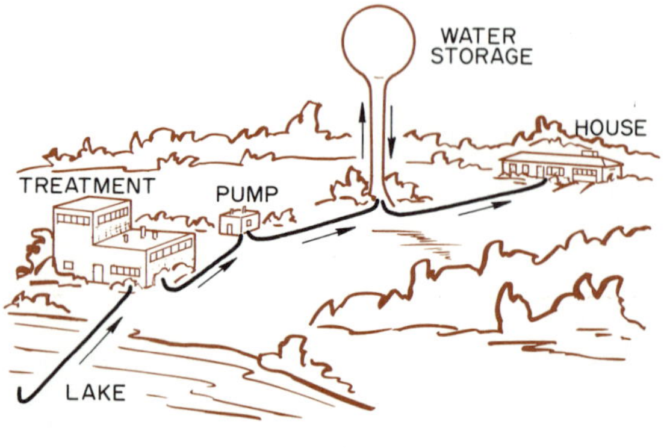

Fig. 59-7
A pump and storage tower are needed to move water to sites on ground higher than the water source.

Water leaves the storage tank in large, underground pipes. These pipes must be large. They carry all the water used by the service area. Smaller pipes branch from these large pipes. The smaller pipes take the water to the sites at which it will be used. A treelike network of pipelines is formed. The large pipes leaving the water tower are called *trunk lines*. They can be compared to tree trunks. Smaller pipes deliver water to the sites. These pipes can be compared to tree branches. They are called *branch lines*.

Site factors are less important for pipelines than they are for other construction. Trunk lines must be held by hard or firm soils. This is less important for smaller pipes. Since water can be pumped uphill, pipelines may be located in sloped or flat land. Water pipes are usually run along public roads. Land costs are lower.

Sewage Systems

A sewage system has two parts. The first is an underground collection system known as a sewerage. This is a network of sewer pipes, which carry *sewage* (waste). The second part is a sewage treatment plant.

A sewerage is a treelike piping system. The system looks much like the water piping system. In fact, the systems work in opposite ways. The sewage enters the ends of the small branches. It flows into the larger trunk lines. These carry the sewage to the treatment plant.

Sewerages differ from water systems in another way, too. They seldom operate under pressure. The flow in a sewerage is caused by gravity. Sewage is rarely pumped. If it is, it is usually only pumped to a higher lever, such as up from a basement. Then it flows down by gravity.

At the sewage plant, the solids are settled. This separates them from the liquid. Solid wastes are commonly manufactured into fertilizers. Next the liquid is treated (cleaned). Then, it is returned to a natural water channel. Or, it may be piped away and reused.

Like water pipes, sewer pipes are laid underground. They are commonly run along the roadbed, through publicly owned land. Therefore, land costs for sewers are not usually high.

The waste entering any pipe must flow downhill, toward the sewage plant. For this reason the plant must be placed at the lowest level in the system.

Summary

Work sites must have transportation services and utility services. Roads and streets, railways, airways, waterways, and pipelines provide transportation services. Gas, electrical, communications, water, and sewer lines are utility services. Both types of services must meet the construction factors of demand, service, and site. Primary construction requires many different kinds of service construction.

Terms to Know

arterial streets
branch lines
collector streets
gravity
highways
interchanges
intersections
pipelines
service construction
services
service streets
sewage
street capacities
transportation construction
transportation services
trunk lines
urban area
utility construction
utility services
waterways

Think About It!

1. Is automobile traffic slow in your community? If so, what do you think could be done to make it move faster?
2. What is the source of your drinking water? Is there ever a shortage of water? Why?

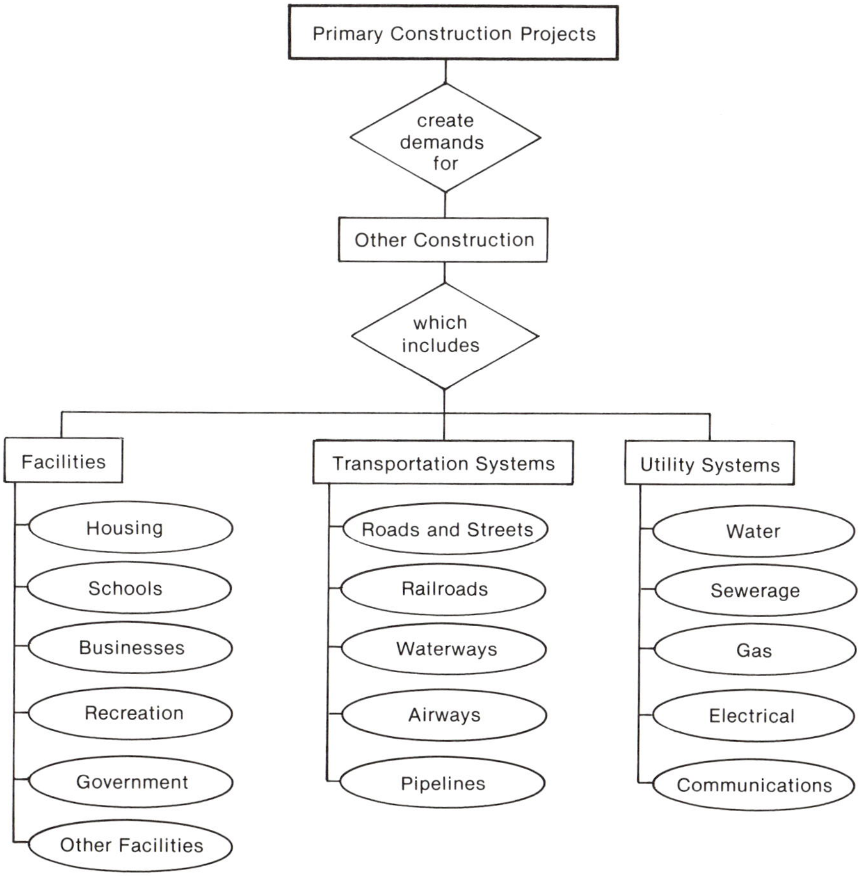

Reading 60

Housing People

The Need for Housing

Workers are employed to build primary construction. Others are employed to provide workers with the things they need. Both groups have families who need places to live. This reading will show how a demand for housing for workers adds to *community development*.

The Population Multiplier

The average number of people in an American family is about $3\frac{1}{2}$. Because of this, each job site will usually have a predictable *population* around it. The population will be about $3\frac{1}{2}$ times the number of workers. See Fig. 60-1. Suppose, for example, that a production plant hires 100 workers. About 350 people will be directly supported by the plant.

The production workers are called *primary workers*. Their families need goods and services. This means that bankers, doctors, dentists, and shopkeepers are needed. The people who provide these goods and services are called *secondary workers*.

A city or town can support about $1\frac{1}{2}$ secondary workers for each primary worker. Thus, the plant with 100 workers will require about 150 secondary workers. They will provide services to the workers and their families. Housing is needed for all the workers. Each job in primary construction creates a demand for about $2\frac{1}{2}$ *housing units*. One of these units is needed by each primary worker's family. And $1\frac{1}{2}$ are needed by the secondary workers and their families. See Fig. 60-2.

Fig. 60-1
On the average, a primary worker supports $3\frac{1}{2}$ people.

Housing Density and Income Levels

Houses must be built to fit the incomes and wants of workers. Housing may be more expensive than most workers can afford. If so, it will not be sold. Or houses may not be the right size. Again, the houses will not be sold.

One way to find out about the housing needed is to study family incomes. These will suggest what general *housing density* workers can afford. Housing density means the number of houses or apartments per acre.* The density of an area's housing often determines its price.

Low-Density Housing

Single family housing units on large lots in new areas are generally expensive. This is partly

*In metric measure, land is divided into hectares. However, using metrics to show land size or to figure measurements will not be common until some time in the future.

Fig. 60-2
One primary worker creates a demand for 1 house directly and 1½ houses indirectly.

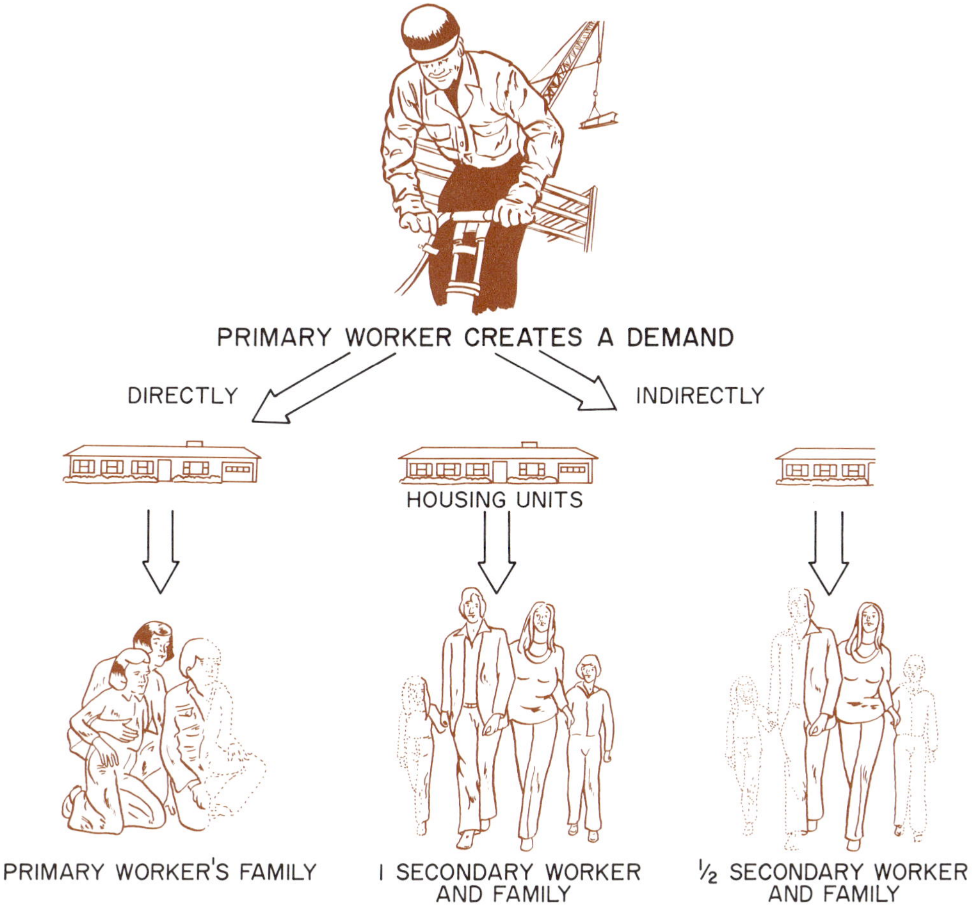

Fig. 60-3
Low-density housing requires large lots. This increases the cost per unit.

Fig. 60-4
Medium-density housing is usually medium-priced. More houses are built per acre than in low-density housing areas.

Fig. 60-5
Several high-density units such as these can be built on an acre of ground. These can house many families.

because they are constructed at a low density. That is, there are only one to three houses on an acre of land. See Fig. 60-3. This makes the land cost per unit high. *Low-density housing* like this is referred to in one of three ways. One single family unit per acre may be written as SF (1). Two single family units per acre is shown by SF (2). Three single family units per acre is written as SF (3). Houses in low-density areas are usually larger than houses in other areas. This adds to their cost.

Medium-Density Housing

Single family housing units for middle-income families are constructed at a higher density. There are four or five houses on each acre. This is called *medium-density housing*. See Fig. 60-4. The densities are written as SF (4) and SF (5).

High-Density Housing

Lower-income housing units are generally constructed at a density of six or more units per acre. Such densities are written SF (6), SF (7), and so on. *Multiple* (or multi-) *family units* (apartments) are often used at these densities. This is called *high-density housing*. Land, street, and utility costs per housing unit are lower. See Fig. 60-5. But there are also types of high-density housing planned for higher-income families. Condominiums and large apartment buildings called high rises may house many families per acre. Usually the cost per unit is higher than for other high-density housing. This is because these units have more luxuries.

Income Factors

Housing units are generally planned, built, and sold according to the number of workers and their income levels. In planning for housing density, it helps to know the income levels of the people who need housing.

Service and Site Factors in Housing Construction

The price of a house is made up of three costs.

- The developer's cost to develop the sites and build the housing units.
- The cost of the land on which the housing is built.
- The developer's business costs (such as maintaining an office and paying taxes) and profit (money left over after all expenses are paid).

Housing should be close to all utilities. It should also be close to places of work. This reduces travel cost and time. But housing should not be right next to a worker's job site. There is often traffic or noise near work sites. Such things are not desired in areas where people live.

All housing units need firm or hard soils to support them. Soft soils cause structures to settle unevenly. The slope of housing sites is not very important. It is true that steep slopes raise construction costs. But the higher costs usually can be made up in higher sale prices. Owners of low-density housing may pay more for a site with a good view or for an unusual site. Site costs for high-density housing can be divided among a large number of units. Each unit owner pays a smaller amount of the total land cost.

The Effect of Land Price on Housing Density

Community growth causes a rise in the price of sites near places of employment. Sometimes land prices are very high. If so, the developer cannot make a profit by building single family housing. The developer must build a greater number of housing units on a site in order to make a profit. Each person pays less but there are more people. See Fig. 60-6.

Land may be high-priced near employment centers. There also may be a high demand for low-cost housing there. If so, high-density housing may be what is needed. Incomes of workers and demand for housing help determine what kind of housing should be built.

Summary

To check what you know about population and housing density, follow this example. An industrial plant has 450 primary workers. The area can also support 675 secondary workers ($1\frac{1}{2}$ for each primary worker). The total number of people employed in the area will be about 1125 (450 + 675). The total population of the community is about 4000 people (number of families × $3\frac{1}{2}$ people per family). This number of families will require 1125 housing units.

Fig. 60-6

In this high-density housing, SF (6+), the land cost is spread across a large number of housing units.

In general, housing must be built to fit the income levels and needs of the people. For most high-income families, one to three houses are built on an acre. For most low-income families, six or more houses are built on an acre.

Near work centers, there may be a great demand for housing. This makes the land more costly. High-density housing is commonly built in such areas.

Terms to Know

- community development
- high-density housing
- housing density
- housing units
- low-density housing
- medium-density housing
- multiple family units
- population
- primary workers
- secondary workers

Think About It!

1. Do you live in low-density, medium-density, or high-density housing? How do you know?
2. Why is high-density housing usually found near places of major employment?

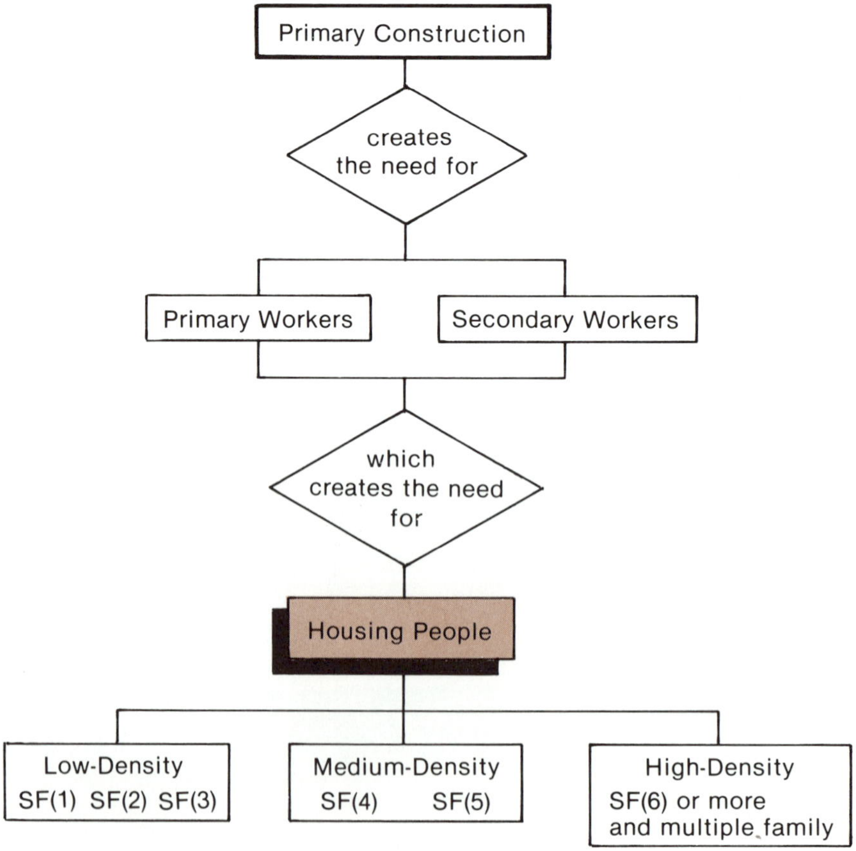

Reading 61

Planning Other Facilities

Other Types of Facilities

Housing makes up more than half of all construction. But other facilities are needed, too. Business places, schools, and recreational areas also serve major area needs.

This reading describes basic knowledge used in planning business facilities. It also tells how planners provide for schools and recreation.

Business Facilities

Businesses include, for example, banks, markets, and cafeterias. In addition, they may include offices for people such as doctors, insurance agents, and realtors.

Businesses serve housing areas called *trade areas*. Trade areas vary for different types of businesses. For example, grocery stores and drugstores generally serve people living within 1.6 kilometers. Department stores and speciality shops serve larger areas. Large cities have many overlapping trade areas. These include neighborhood shops, shopping centers, and the central business district. See Fig. 61-1.

Central Business Construction

Central business districts are built to service (meet the needs of) a whole area. To serve the needs of a large area, much construction is required. Offices, theaters, stores, and parking lots, for example, must be constructed.

In the past, all *central business construction* was at the heart of a city. This area is often called

Fig. 61-1
Trade areas often overlap.

downtown. Major roads and streets all lead there. As cities have grown, downtown areas have become tightly packed. Originally, most buildings did not fill their lots. Today, most lots are nearly covered by buildings. See Fig. 61-2A. This has made it hard to widen streets. The extra traffic clogs them.

Car usage also increases as cities grow. This often causes *traffic jams.* To allow for this, people need more time to travel to and from the downtown area. As cities have grown, new housing has been built farther from downtown. This has further increased the time it takes to make a trip downtown.

Central businesses are still built and maintained as long as there is a need for them. In general, each group of 350 housing units in a community will support no more than one acre of central business construction. Much less space is needed if central businesses are housed in tall buildings.

Location of Central Business Sites

Sites at intersections are best for central business construction. Structures should be built where the most important streets cross. This permits quick access to the structures from the entire trade area.

Sometimes, *inner belts* are constructed to speed inner city travel. These are streets that collect and move traffic quickly and safely. In very large cities, they may be linked to *outer belts.* These are "rings" of highways. Traffic on outer belts does not need to travel on long, narrow streets. This reduces driving time.

Cost of Central Business Sites

Central business sites generally are the most costly sites in an area. However, profits that result from their development are also high. As with high-density housing, the higher the land price, the greater the number of units that must be built on the site to make a profit. Therefore, taller buildings are built on costly sites. This distributes the land cost among the many businesses that occupy a structure.

Shopping Center Construction

In large communities, there is much trade. Traveling downtown is not always convenient. The community may be divided into smaller areas, also called communities. These trade areas are served by *shopping centers.*

Fig. 61-2

Business facilities may have various site requirements.

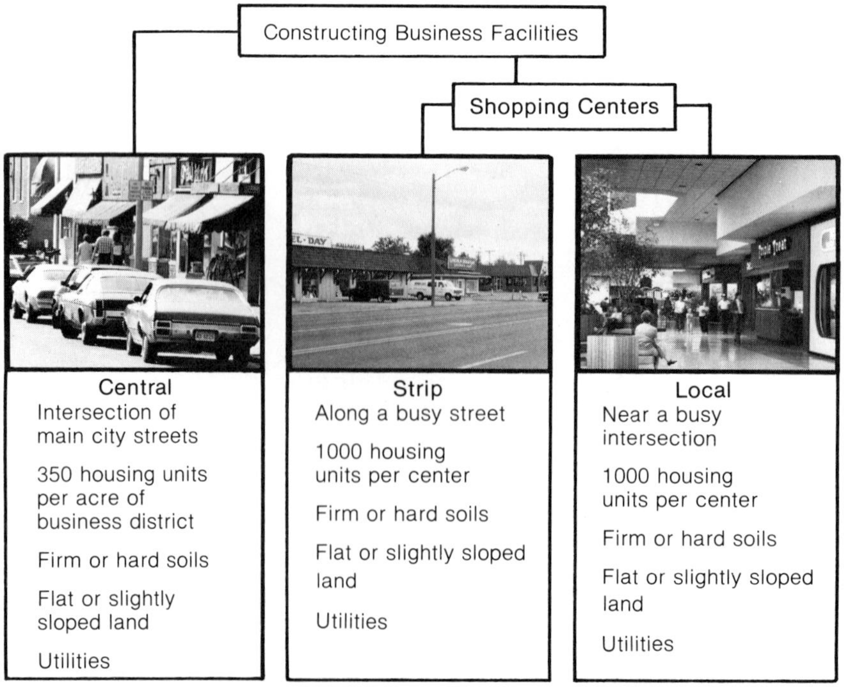

The needs for local business services vary. These needs are created by the construction and sale of housing units. The location, kind, and number of shopping centers are determined by these needs.

Kinds of Shopping Centers

There are two kinds of shopping centers. They are called strip shopping centers and local shopping centers.

A *strip shopping center* is made up of business places spread along a length of busy street. See Fig. 61-2B. Each business is separate from the others. There are also many small, separate parking areas. Signs are needed along the street. Drivers must park on the streets. Or, they must enter parking areas from heavily traveled streets. This can slow traffic. It may cause accidents.

Local shopping centers differ from strip shopping centers. They are often more attractive. Also, they are generally constructed on large sites. The sites are set back from busy streets. A local shopping center is made up of a group of businesses. Each business is a separate store or office in a larger structure. Look at Fig. 61-2C. Sometimes only one large sign marks the shopping center. Often the businesses are not even named on the sign.

Local centers are generally planned much better than strip centers. There is most often one parking lot for all the businesses. Service roads connect the main streets to the parking lot. This helps to keep traffic moving smoothly.

Supporting Shopping Centers

All housing units need business services. However, not all communities can support shopping centers. There must be at least 1000 housing units in a trade area to support even the smallest shopping center. If there are fewer than 1000 housing units, a shopping center will probably not be built. Separate stores will satisfy business needs.

Site Requirements for Shopping Centers

The best location for a shopping center site is where important streets cross. The cost of such land

Fig. 61-3

Many factors are kept in mind when planning school construction.

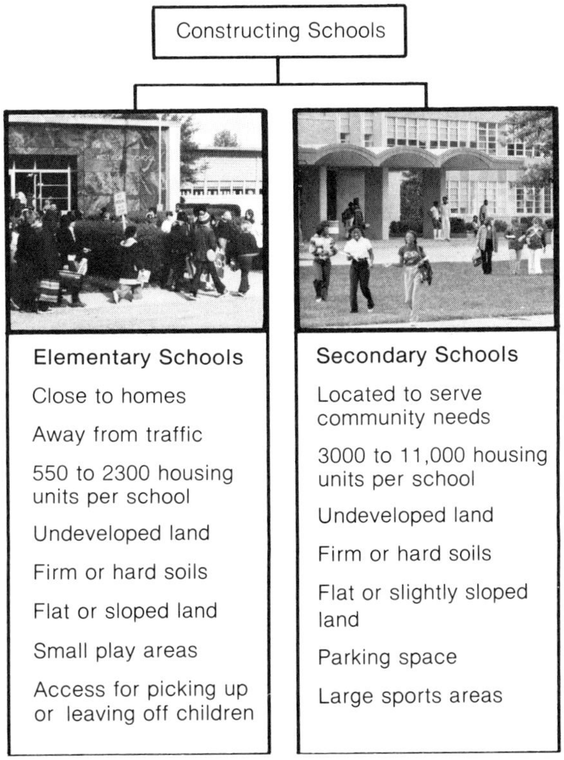

is high. It is even more expensive if structures have already been built on the land. The cost of developing such a site may be too high for a profit to be made. If so, a site is chosen along another important street.

Firm or hard soils are needed to support the weight of shopping center buildings. The land also should be flat or only slightly sloped. Steep land raises construction costs. And parking areas on steep land are unsafe and inefficient.

Local shopping centers are generally between one and ten acres in area. A parking lot takes up at least two-thirds of the site.

School Construction

Families moving into an area create a need for school construction. Two main types of schools are needed. These are *elementary schools* and *secondary* (junior-senior high) *schools.* An elementary school may include kindergarten or grade one through grade six or eight. A secondary school may include grades nine through twelve or seven through twelve.

Supporting Schools

A community that has 550 or more housing units can generally support the construction of an elementary school. If there are fewer than 550 housing units, there are usually too few children of elementary-school age to support a school. In such cases, communities often go together to construct a school.

Still more homes are needed to support construction of a secondary school. A secondary school is usually built when there are more than 3000 housing units in the community.

There is also a maximum (highest) number of housing units that a school can service. If there are too many students in a school, it cannot operate efficiently. An elementary school generally can service no more than 2300 housing units efficiently. Secondary schools cannot efficiently service more than 11,000 housing units. If there are more housing units, additional schools should be built.

Location of Schools

School locations must be planned carefully. See Fig. 61-3. Schools should be close to the housing units they service. Ideally, a child should have to walk no more than 0.8 kilometers to an elementary school. Also, the walk must be safe. Children should not have to cross heavily traveled streets.

Students attending secondary schools are able to walk farther. They can walk up to 2.4 kilometers to school. Secondary schools must be on or near major streets. Many activities take place in these schools. This results in much traffic to and from the schools.

Service Needs of Schools

Schools have their own service needs. All schools need utility services. They also need parking lots and streets.

Schools need large amounts of land. An elementary school generally requires a site at least five to ten acres in size. A secondary school needs at least twenty acres. All schools need sites with firm or hard soils. They also need flat land for sports playing fields.

Recreational Facility Construction

There are many kinds of *recreational facilities.* Examples are playgrounds, playfields, and neighborhood and community parks. See Fig. 61-4. Each meets a different need.

Locating Recreational Facilities

Playgrounds are built for children of elementary school age. They generally have equipment such as swings, slides, and sandboxes. Playgrounds generally need between one and ten acres of land. This need will vary with the number of children to be served.

Playfields are constructed for youths and adults. Often, these fields contain recreational equipment such as for basketball or tennis. But construction for playfields consists mainly of facilities for games and sports. Seats are built for people who watch these events. Parking areas are also constructed nearby. Indoor areas for dancing, table tennis, and arts and crafts may be built on playfields. Usually these fields cover between ten and twenty acres of land. Their sizes should be determined by the number of people they serve.

Neighborhood parks are built near housing areas. They generally have shady places and large lawns. The sizes of neighborhood parks depend on how much land is available. The number of people needing the parks is also a factor.

Community parks have a different purpose. They provide a place for outings for a whole community. Community parks are usually twenty or more acres in area.

Fig. 61-4
Recreational facilities serve the needs of people in the community.

Supporting Recreational Facilities

The total amount of land used for recreation must match the demands of the community. Generally, at least ten acres of space are needed for every 300 housing units. The following formula can be used for figuring the amount of land needed for recreation:

$$\frac{\text{number of housing units}}{300} \times 10 \text{ acres} = \text{total acres of land needed}$$

For example, a community of 3000 housing units would need at least 100 acres of land for recreation.

Site Requirements for Recreational Facilities

All types of land may be used for recreational areas. Playfields require large, flat areas. Parks may be developed in areas where little other construction is likely to be done.

Summary

Central business construction fills the needs of the total area. A community must reach a certain size before this construction can begin. At least 350 housing units are needed to support one acre of central business construction.

Trade areas in large communities are often divided up. Small communities within the area are serviced by shopping centers.

Local shopping centers generally serve communities better than strip shopping centers. They also cut down on traffic problems. Shopping centers must be near at least 1000 housing units.

Schools should be built on sites close to the homes they serve. They need playground spaces. They also require services such as utility services.

Communities need recreational areas. These include playgrounds, playfields, neighborhood parks, and community parks. Usually, there should be at least ten acres of recreational space for every 300 housing units.

Terms to Know

- central business construction
- downtown
- elementary schools
- inner belts
- local shopping centers
- outer belts
- recreational facilities
- secondary schools
- shopping centers
- strip shopping center
- trade areas
- traffic jams

Think About It!

1. Does your community have a central business area? Have new shopping centers been developed in the past few years? Why or why not?
2. How would a rapid transit system affect the central business district of a large city? Would central business people be likely to support the construction of such a system? Why or why not?
3. How can you determine whether your community has adequate recreational facilities?

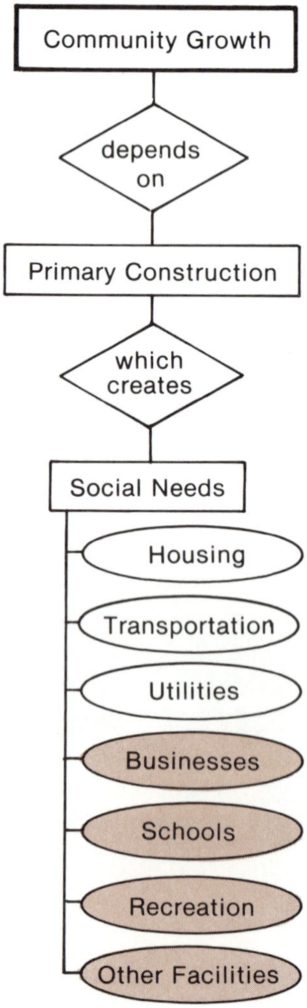

Reading 62

Economics of Community Development

Growing Communities

When a community is "born," primary construction makes employment available. The community grows as more construction is needed for the workers and their families. In a *prosperous* community, this cycle continues.

Deterioration of Construction

In a healthy community, all structures are used. Also, all are in good physical condition. They are *maintained*. This helps the structures to serve their intended purposes. And it helps them to hold their *value*.

Most communities, however, have at least a few structures that are poorly maintained. Such structures are in poor condition. They are said to have *deteriorated* (decayed). See Fig. 62-1. At first, they may need only a coat of paint. If time passes and no repairs are made, wooden parts of a structure rot. Also, metal parts rust, bricks crack, roofs leak, and the structure sags. Finally, it collapses. All structures deteriorate if they are not cared for properly.

Many communities have areas in which structures have not been maintained. As time passes, repairs become too costly. Then the areas "die." This generally happens when primary construction has been abandoned (given up). When there is no work in an area, people leave. Those who remain cannot provide businesses with the support they need to keep operating. With no money for repairs, businesses deteriorate. Construction stops.

Fig. 62-1

Regular maintenance should be done. Then, housing will not begin to deteriorate.

Fig. 62-2
This community in Nevada was a thriving business center in the middle 1800s. It is now a "ghost town," a decaying reminder of what used to be.

Fig. 62-3
This home's location brings its value down.

When a community dies, it becomes a "ghost town." See Fig. 62-2. No one lives there. The structures decay until they become piles of rubble (broken stones or bricks). This has happened over the centuries, even to many great cities. For example, a total of seven cities once existed near what is now New Delhi, India. Only rubble remains of five of the cities. Old Delhi, the sixth city, still exists. But it is too crowded and decayed to serve modern needs. New Delhi has replaced it.

The Reason for Deterioration of Structures

There is a reason why structures are not always maintained. Owners must be able to get back the money they spend for repairs. There are two main ways owners can make up this cost. First, the owner can make a bigger *profit*. This profit comes from rental or sale of the structure. People will pay more for a structure that is maintained. The second way is by preventing *loss* in the structure's value. Making repairs prevents this loss.

Reasons for Deterioration of an Area

An area may deteriorate for many reasons. For example, a major employer may move to another city or state. This leaves many people without work. They may leave the community to get work elsewhere. This reduces the *market* (need) for quality construction. In turn, property owners have less reason to spend money on *improvements*.

Sometimes an area deteriorates because a structure loses its *locational value*. This usually occurs when a community changes. The "old" location is no longer desirable. For instance, a new highway may cause business locations along the old road to lose value.

Areas also may decay because of what is around the structures. Structures may lose their *environmental value*. For example, the value of housing units generally goes down if the units are close to a railroad. See Fig. 62-3. Heavy industry, sewage treatment plants, or very heavily traveled roads also reduce the value of housing units. The noise, smells, and danger to children make the houses less wanted. When these conditions occur, it becomes difficult to sell the property. *Sales prices* fall rapidly. Then few repairs are made. The structures decay. And the area decays.

Economic Loss from Poor Planning

Losses from decay are not the only ones a community may have. Much value is also lost through poor planning. This occurs mainly when good facilities must be replaced. For example, suppose a costly four-inch water pipe is laid. At first, it serves the needs of the area. But the area grows. The small pipe cannot provide all the water service needed. The small, expensive pipe must be replaced by a larger pipe. The small pipe is still usable. But it is too small. People did not plan ahead.

Such waste could have been prevented. If the growth of the area had been foreseen, a larger pipe could have been laid at the start. Then taxpayers would not have had to pay for pipe twice.

Another example is in street interchange construction. Often, communities must buy expensive, developed property for such construction. Businesses or families have to be relocated. Usable structures must be torn down. The same ground could have been bought for much less before it was developed. Again, this would have saved taxpayers' money.

Preventing Economic Illness

Community decay and poor planning can lead to *economic illness* or *depression*. Such things take away the *economic strength* of the community. They can be prevented by long-range planning. A community must look years ahead. It must make plans for any expected growth. This is efficient community development. See Fig. 62-4. It results in lower taxes because money is spent wisely. It is spent according to carefully made plans.

Fig. 62-4

Here are two different "faces" of a major U.S. city, Pittsburgh, Pennsylvania.

A. *This is the "Golden Triangle" downtown area, as it looked in 1947.* Pittsburgh Chamber of Commerce

B. *Here is the same area almost twenty years later, following vast redevelopment. Notice how traffic facilities, buildings, and park and recreation areas were improved.*

Pittsburgh Chamber of Commerce

Summary

The decay of construction in a community usually results from a reduced market or from a loss of locational or environmental value. Run-down property can cause great economic loss for a community. Losses also may result from a lack of planning.

Communities should be economically strong. This helps them to be satisfying places in which to work and live. Planning community construction will help ensure that communities remain healthy. They remain efficient, and their economies grow.

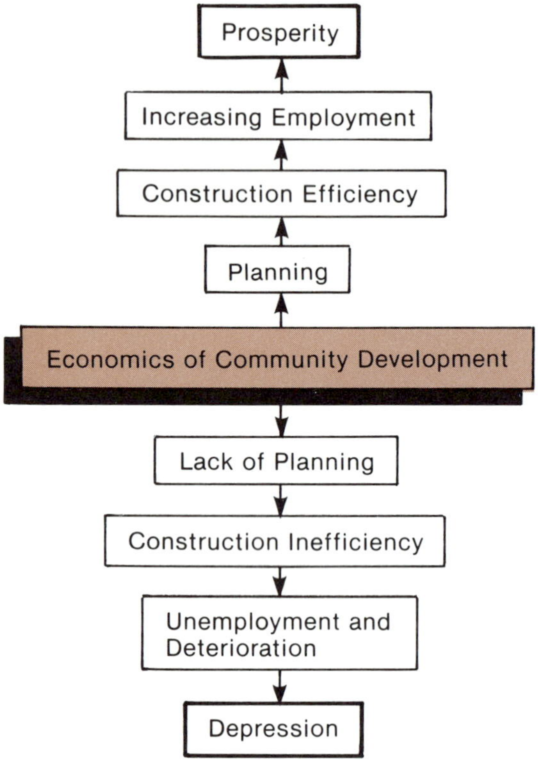

Terms to Know

depression
deteriorated
economic illness
economic strength
environmental value
improvements
locational value
loss
maintained
market
profit
prosperous
sales prices
value

Think About It!

1. Is any area of your community decaying? If so, what caused the deterioration? Was it a reduced market, loss of locational value, or loss of environmental value (or some combination of these)?
2. List several ways to return prosperity to an area.

Reading 63

Managing Community Development

Community Development

Most people are concerned about their community. They care about its appearance, convenience, and efficiency. They want pleasant, low-cost living. They also want their community to continue to meet their needs in the future. They may want recreation areas. They may need better roads and more housing. Meeting such needs is community development. Most large communities manage this development. Managing helps make development possible. It involves planning, organizing, and controlling. Managing is done by *city planners*. They work for the improvement of communities. See Fig. 63-1.

Fig. 63-1
The city planner studies plans of the whole community.

The General Plan

City planners estimate the expected growth of a community. They base this on predictions of the number of jobs to become available. Planners must think of ways to manage the expected growth. They state their plans as community *objectives* (goals). Then the community decides whether to accept the stated goals. If it does, these statements guide the area development.

All the statements for development are combined. Together, they form a *general plan*. A general plan may include such goals as these:

- Providing high-speed inner and outer traffic belts.
- Improving conditions in the inner city area.
- Providing elementary schools within 0.8 kilometers of most homes.
- Providing a park for every township.

All of these goals help to improve community life.

The Development Plan

Many details must be added to a general plan before the goals can be reached. The statements of details become a *development plan*. This plan sets the standards for street widths and for sizes of school sites. It sets the standards for the sizes and locations of neighborhood parks and for sizes of housing sites. The development plan affects all area construction.

The development plan shows what facilities will be built in the following few years. City planners must estimate the total cost of this construction.

The development plan includes a schedule for building the projects. This schedule is called a *capital improvements program*. Scheduling ensures that each facility will be built where it is wanted and when it is needed. The plan for paying for the projects is called a *capital budget*.

The Planning Board

A city planner does much to organize community development. But many other people are involved, too. Early in the community planning process, a *planning board* or *commission* is chosen. This board is made up of citizens who represent the people of the area. They tell the city planner what people want. They help the planner make a plan that the citizens will support. See Fig. 63-2.

Community Development Laws

The general plan is approved by the planning board. Then the city planner and the board suggest regulations that will control community development. Elected public officials review these suggestions. They pass the laws that will regulate community construction.

These laws set standards for construction methods. They also control the location and size of buildings. Even the use of the property around buildings is regulated. Laws state how much of it may be used for parking or for landscaping. Laws also set many other standards for land development.

Control of Community Development

Planning and organizing community development is important. But it means little without control. Laws serve no purpose unless they are followed. For this reason, the *building inspector* and the planning board are given the power to enforce laws.

The building inspector's job is to see that every building in the community is safe for human use. The inspector checks the plans for new construction. This is done to make sure that all legal requirements are met. If they are, a *building permit* is issued. It states that the structure is approved for construction.

The planning board reviews proposed plans. The board must determine whether the plans violate land development laws. If a plan does not meet the legal requirements, it is not approved. The site cannot be developed as planned. The plans must be changed before they can be approved.

Fig. 63-2

People like to live in well-planned areas. This carefully developed suburb helps make the whole community a good place to live and work.

By enforcing laws, a community can control its development. This is done to fulfill the goals of the general plan. Yet these goals are never completely fulfilled. The community always grows or changes. Then the general plan must be changed to fit the new conditions. Also, as the first goals are reached, the development plan is pushed farther into the future. Managing community development is a process that never ends. See Fig. 63-3.

The Community and Its Region

The community is made of many individual construction projects. Each is related to the others. And each is related to the community. Every community is a part of a larger area called a *region*. Regions may be made up of a few communities. Or they may include an entire state, or even several states.

Communities in a region have some common concerns and interests. For example, they may all be involved in farming or in industry. Or, they may be located in a common area. For example, they may be in a valley or around a lake.

Many community problems affect the whole region. They must be solved by the whole region. For example, suppose a highway must be built to connect communities in a region. Each community alone cannot build the highway. The highway must be built for the whole region. The same would be true for a highway between regions. It would be built by all the regions. It would meet regional needs.

Fig. 63-3

Communities continue to change as community needs change. Here old buildings have been removed to make room for many new facilities.

Fig. 63-4

A system of channels regulates the flow of the Sacramento River on its way to a pumping plant, and later to a canal. A department of the federal government was involved in this project to help meet needs of communities in nearby areas.

J.C. Dahilig — U.S. Department of the Interior

Fig. 63-5

This area model shows the planned development of a region near Niagara Falls, New York.

Port Authority of the State of New York

There are many other regional problems. Generating electricity is an example. Can you think of others? See Fig. 63-4.

Managing Regional Development

Like community development, *regional development* must be managed. See Fig. 63-5. The well-being of the whole region depends on this. It is the most efficient way to solve problems of the whole region. Lack of planning results in great losses. This is true for communities. It is also true for regions.

Many states have state planning divisions. Regional planners are hired to plan, organize, and control each region's development. Through good regional managing, the best possible construction projects are built in each community of the region. Regional managing also helps solve regional problems. These include regional transportation, flood protection, recreation, and utility services.

Summary

Community development programs must be managed. Successful management depends on several things. First, people must want a better community. They must support the city planner, the planning board, and the elected officials who plan the community.

Second, laws to control community development must be passed. This is done by elected officials. They also see that the laws are strictly enforced.

Finally, planners must be able to plan for the future. They must effectively guide community development.

Each constructed project is related to all other construction in the community. In the same way, each community is part of a larger region.

Every housing unit in a community relates to all other construction in the area. It is part of a larger constructed system. This system is the world of construction.

Terms to Know

building inspector
building permit
capital budget
capital improvements program
city planners
commission
development plan
general plan
objectives
planning board
region
regional development

Think About It!

1. What plans do your city and region have for growth and development? If you do not know, how could you find out?
2. How might well-planned construction improve your community and region?

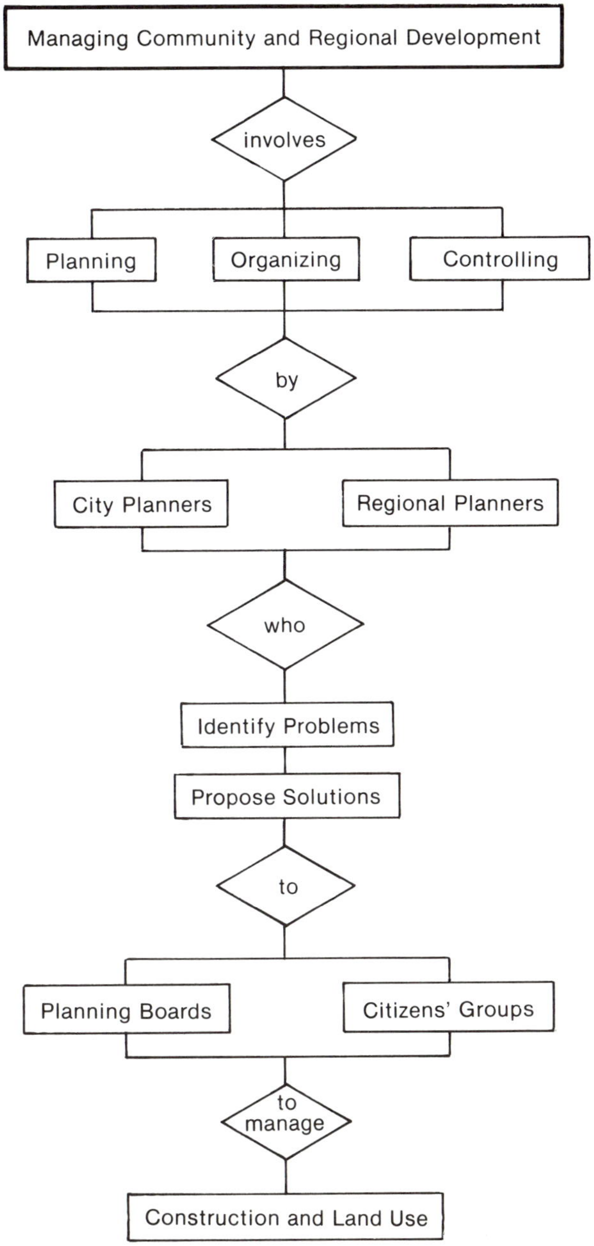

Reading 64

Construction in the Future

The Changes

Construction technology must change as the world changes. The changes can take many forms. New designs are created. Construction techniques are improved, and better materials are developed. Equipment and workers become more *productive*.

Fig. 64-1
Construction must be planned to make the best use of available land and energy sources. This model suggests a high-rise building with a tapered, reinforced-concrete frame and glass walls. Such a design would make good use of heat and light from the sun.
Libbey-Owens-Ford Glass Co.

The Challenges

In the years ahead, the construction industry faces major *challenges*. It must provide for the needs of an ever-growing population. And it must prepare for the future. It should make changes that will improve the lives of people in years to come. See Fig. 64-1.

A great building job lies ahead for the construction industry. Think about the number of structures already standing in the United States. Then, imagine. This number will probably be doubled within your lifetime! It could happen within the next thirty years! And it is not just the United States that is growing. Other countries are changing, too. Nations in Latin America, Africa, and Asia are examples. They will continue to develop. They will have great needs for construction.

Rebuilding is also very important. Poor housing is a major concern. Better homes must be built at a lower cost. Large parts of cities will need to be rebuilt.

Building must be done to make the best use of our resources. For example, the United States generally has enough rainfall. Some areas, however, are too dry. Major construction projects, such as irrigation systems, are needed to provide fresh water. These will aid agricultural production. Just one project may cost several billion dollars. But both industry and the public will benefit.

Another source of water is needed. For many years people have been searching for a cheap method of removing the salt from sea water. If such a method is developed, construction will have to be done. Production plants and distribution systems will be needed.

Many other challenges exist for the construction industry.

- Transportation facilities must be made more *efficient*. Changes must be made to make airports, highways, railroads, and waterways meet needs. New facilities must be built to handle the increasing traffic.
- Manufacturing plants must be expanded to meet the needs of a larger population.
- More efficient facilities are needed for education, government, and recreation.
- Facilities for reducing air and water pollution are needed.
- We must build to obtain fuels. We must build to conserve them.

All these needs must be met at an affordable cost. See Fig. 64-2.

Obstacles to Be Overcome

Construction technology faces many *obstacles*. An obstacle is something that stands in the way. It takes steady effort to overcome obstacles.

One obstacle to construction is slow development of technology. In some fields, new technology develops quickly. Electronics is an example. In construction, technology is changing more slowly. Improvements have been made. Many more have been started. But these are not yet ready for everyday use. See Fig. 64-3.

Another obstacle is resistance to change. People tend to use the materials and methods that they know will work. Because of this, many buildings today are much like those of 100 years ago. The methods used to build them may also be much the same. For example, brick buildings have been built for centuries. Age-old materials and methods are still used to build them.

It usually takes years to introduce a new item or material. And it takes a long time for people to accept it. Laws and building codes are based mostly

Fig. 64-2

A. *Improvements are continually being made on transportation systems and in the construction technology required to build them.*

B. *During your lifetime, many new structures will be built to serve a growing population. You will notice changes in structure design. This will happen as designers learn new techniques and as new materials and construction methods are developed. This modern structure is the Sydney Opera House in Australia.*

Australian Information Service

Fig. 64-3

New construction processes are being developed. In this laboratory, experiments are being done using the laser for welding, metal cutting, and drilling. New processes will make many types of construction work faster and more efficient.

Aluminum Company of America

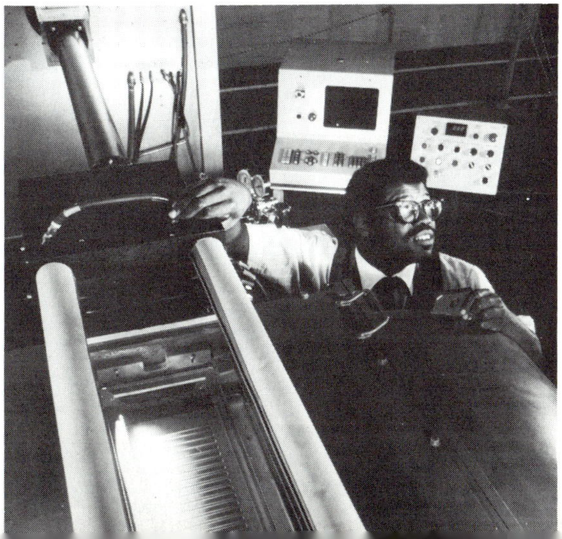

on today's products. These are not revised often. And labor practices and testing laboratory requirements are slow to change. Some areas still have not accepted many aluminum and plastic products.

Hidden costs of building are another obstacle. Hidden costs are legal, financing, insurance, and administrative charges. They are in addition to labor, materials, and equipment expenses. But they are a large part of total construction costs. Ways need to be found to lower these costs.

Meeting the Challenge

New developments are underway in design, methods, materials, and labor. These will help meet the construction needs of the future.

Design

Advanced and imaginative design techniques will cause changes. For example, the *modular system* is likely to become more important. In this system, buildings are designed in standard stock sizes. This makes designing and building easier. It also involves less labor and reduces waste. See Fig. 64-4. Soon *metric measurements* will be used. Sizes of standard stock will change.

Prefabricated units are being used more and more. These are sections of buildings that are built at a factory. They are assembled at the site. See Fig. 64-5.

Entire structures have been built of prefabricated units. For a hotel in San Antonio, for example, each room unit was built in a factory or yard. The units came complete with furniture. The beds were even made! The room units were transported by truck to the site. Then they were hoisted into place with a crane. The twenty-one story structure was completed in less than ten months.

Designs for *remodeling* older structures are being developed. Structures that are substandard may be

Fig. 64-4
Modular units made in standard sizes may become more and more common in the future.

Portland Cement Association

United Aircraft Corporation

Fig. 64-5
Buildings made of prefabricated modular units can be constructed quickly.

remodeled. However, they must be structurally sound. Many old homes are being modernized. They are converted (changed) into modern buildings. In Philadelphia, for example, many buildings 100 to 200 years old have been modernized inside. Their original outward appearance has been kept. See Fig. 64-6. Prefabricated units have even been used in remodeling older housing.

Methods

Other technological methods show promise. For example, controlled atomic blasting could be used in excavating a new Panama Canal. Excavation costs would be less. It has been estimated that using this method could save a billion dollars.

Improved management techniques lower costs. They also increase construction efficiency. Contractors may use one of two methods to schedule construction. One is the Critical Path Method (CPM). The other is the Program Evaluation and Review Technique (PERT). They enable the contractor to get work done in sequence. For example, construction of forms would be scheduled before delivery of concrete.

Other methods also can be used for managing construction. Computers are used for data processing (sorting facts) and for *systems analysis*. This is finding the best or most efficient way to get a job done. Other methods are *input-output* and *simulation models*. See Fig. 64-7. These provide needed information or aid in design work. Their use improves management techniques.

U.S. Department of Housing and Urban Development

U.S. Department of Housing and Urban Development

Fig. 64-6
Remodeling is done to give sound but worn-looking buildings a fresh look.

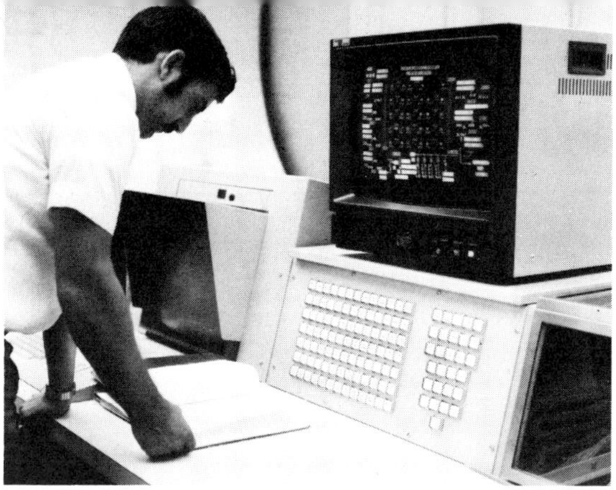

Pacific Gas and Electric Company

Fig. 64-7
Computer systems are installed for monitoring purposes. The computer shown here is used to monitor a power system.

Materials

Products used in building are changing. They will change even more. See Fig. 64-8. Wood will be molded into many new shapes. There will be greater use of *laminates* such as plywood. Better *preservatives* and finishes will greatly lower the maintenance needed.

Plastics will be used much more frequently. Wall panels and prefabricated kitchens and bathrooms will be plastic. Plastic forms for concrete will replace wood or metal forms. Fiberglass will be used more in construction. Fiberglass structural shapes are in use now. An example of this is an I-beam.

The durability and strength of metals will be improved. And new materials are being sought. For example, materials are needed that are as strong as steel but lighter in weight.

Plasterboard and clay products will be greatly improved. They will be stronger and more fire-resistant. They will need less maintenance.

The use of *prestressed* and *precast* concrete will increase. Better roofing materials will be used.

Adhesives will continue to replace welds, nails, bolts, and mortar. *Epoxy resins* are quite strong.

Fig. 64-8
Special waterproof, fireproof fabrics may be used in construction.

Birdair Structures

Fig. 64-9
Someday, capsules much like this one may be common.

Labor

A major trend toward more skilled labor will emerge. Fewer common laborers will be used. Many changes in labor practices will be needed. For example, prefinished, low-maintenance materials may be used more. This would reduce the need for painters. Workers will need new skills to work with new materials. And workers will have to learn new methods of performing tasks. Retraining workers will be more and more important.

Other Improvements

A compact utility package may be developed for houses. It would contain heating, ventilating, and air-conditioning systems. Water supply and waste-disposal units may also be included.

In general, construction equipment will be more powerful. It will be able to do more. And it will become more specialized. It will also operate at greater speeds than today's models. One new development will be a more flexible crane. Use of this crane will make some scaffolding and falsework unnecessary.

The Distant Future

Future cities may look much different than those of today. In the next century, a city may be a completely enclosed unit. The trend toward this is seen frequently today. Covered shopping malls are common. And there are covered sports arenas, such as the New Orleans Superdome.

A city of the future could be completely roofed over. It would be *encapsulated*. This means "to be placed within a covering or capsule." See Fig. 64-9. Such a city would be heated and air-conditioned with purified air. It could have its own nuclear power plant. The city would be free from all pollution. All wastes would be recycled.

Future transportation systems could be powered by engines that do not pollute. A high standard of housing may be provided for everyone. It could happen. Many changes are already possible with today's technology. See Fig. 64-10.

Fig. 64-10
Construction today and tomorrow.

A. ***Today.*** *Using a guy derrick to place a geodesic dome.*

American Hoist and Derrick

B. ***Tomorrow.*** *Using automated beam builders to construct the frame of a satellite.*

National Aeronautics and Space Administration

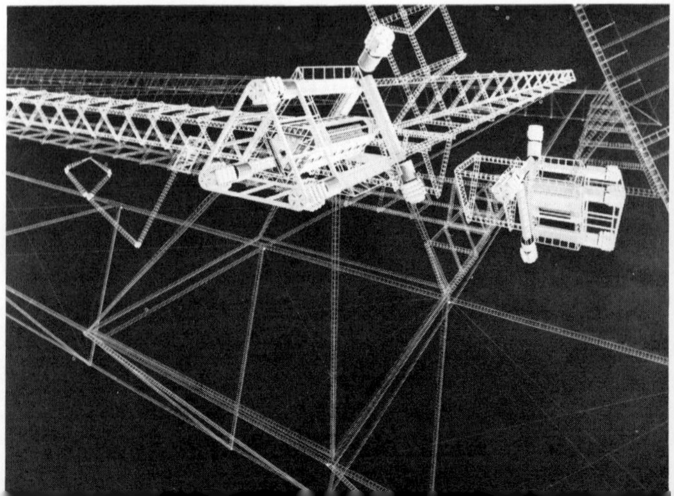

Summary

Unfulfilled needs of the present and future are a challenge to construction technology. Housing, transportation, fresh water, and power are some of these needs. Industries must be developed. Pollution must be eliminated.

To meet these challenges, new designs must be created. New methods, materials, and additional labor skills must be developed. Then, construction technology can meet the challenges.

Terms to Know

challenges	obstacles
efficient	precast
encapsulated	prefabricated units
epoxy resins	preservatives
hidden costs	prestressed
input-output	productive
laminates	remodeling
metric measurements	simulation models
modular system	systems analysis

Think About It!

1. Why has the development of construction technology been slow in the United States?
2. Within the next twenty years, what changes will have to be made if we are to meet the demands for structures of all kinds?

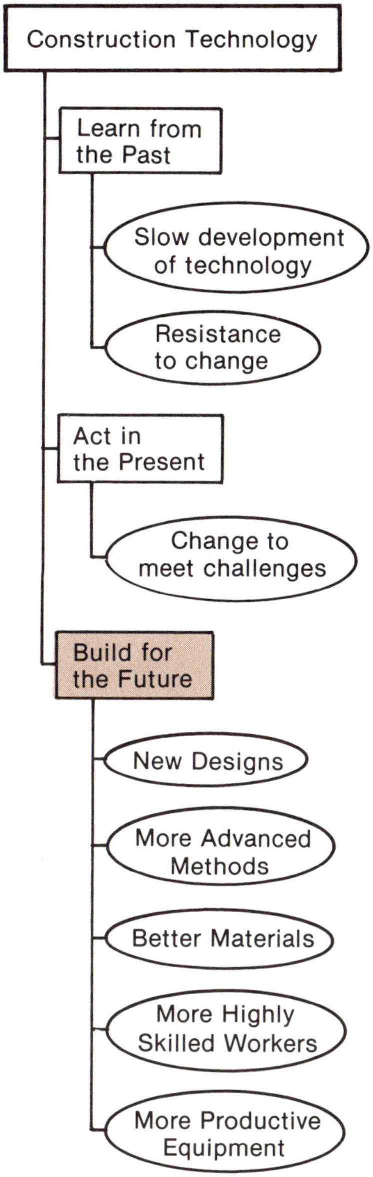

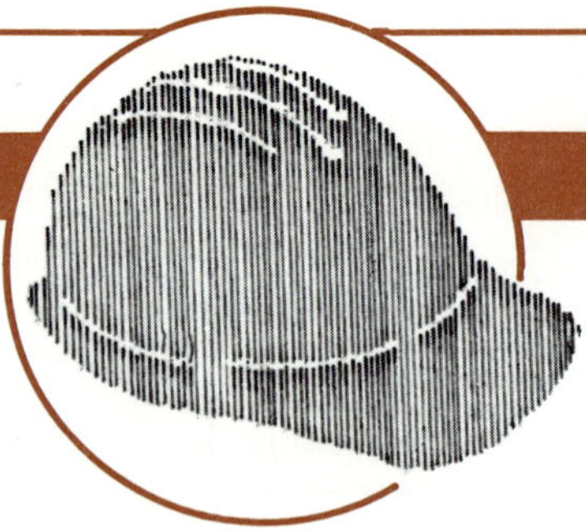

Reading 65

Safety and Health in Construction

Safety and Health Concerns

Safety and *health* are concerns in all walks of life. In this reading, you will learn about safety and health concerns in the construction industry. As you read, think of ways that construction safety relates to you in school and at home.

The problem of preventing accidents and protecting the health of workers is a complex one. Accidents may be minor. For example, a worker may simply bruise an arm. But some accidents are very serious. A mistruck nail could fly up and hit a worker's unprotected eye. Some accidents can even be fatal. But using knowledge, care, proper safety equipment, and common sense can prevent most accidents. See Fig. 65-1.

Accidents are one concern. Occupational diseases are the other. These are sicknesses that workers may develop because of job conditions. Some are not very serious. For example, cement rash is easily cured. But some other diseases are serious. Examples are lung or eye diseases caused by too much dust in the air. Great care should be taken to make the site a safe place to work.

Everyone in construction must care about safety. Workers must learn to be responsible for their own safety. And they must look out for other people on or near the site.

Fig. 65-1

Use common sense! Wear the proper safety gear for the job you are doing.

A. *Wear safety glasses when nailing.*

Paslode Company

B. *Welding goggles, long sleeves, and gloves are musts when welding.*

E.I. DuPont DeNemours and Company

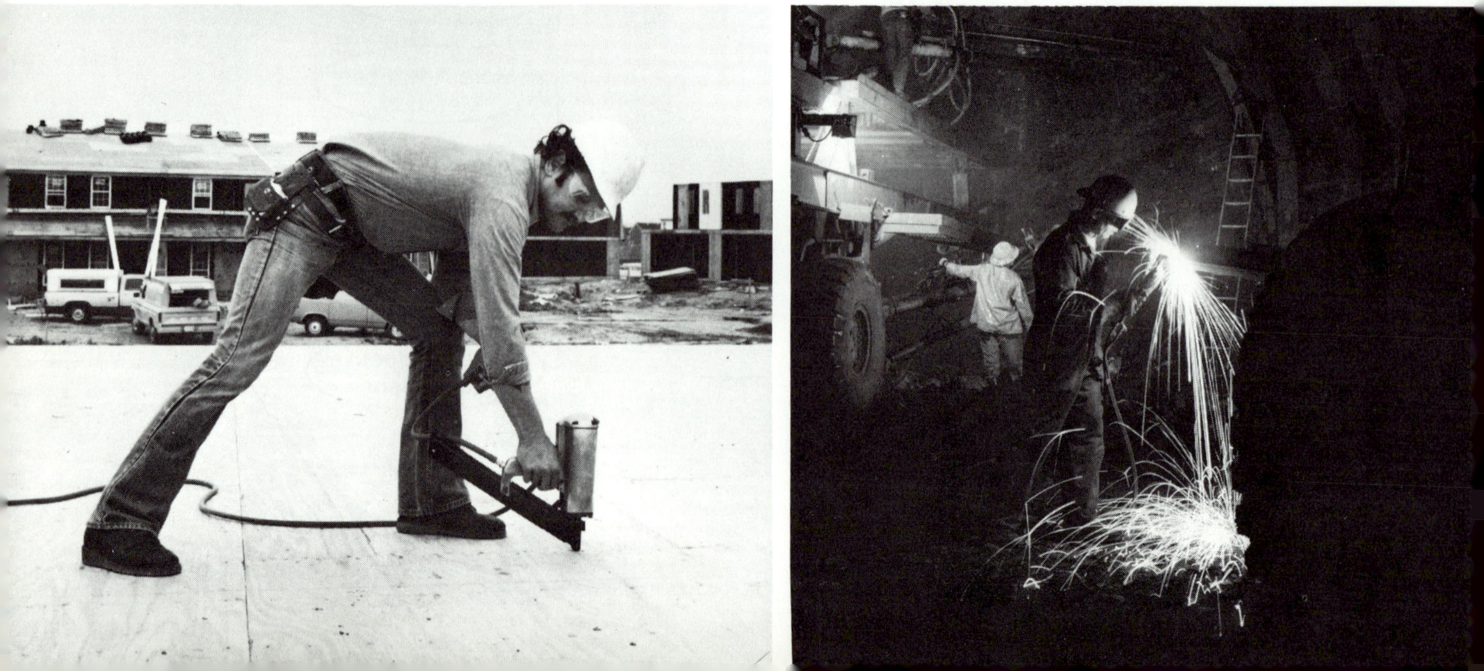

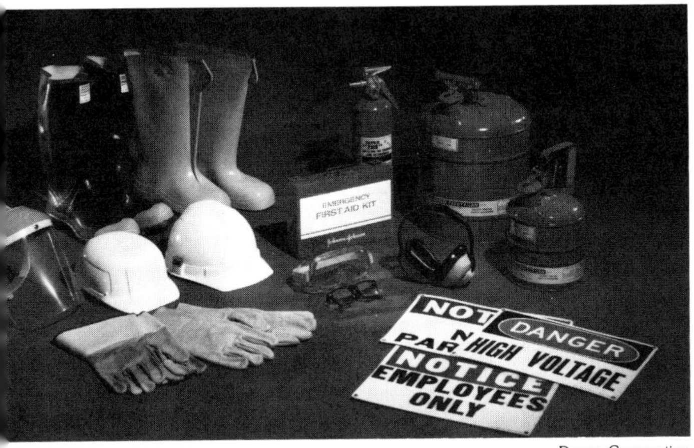

Dayco Corporation

Fig. 65-2
Contractors care about safety. They supply much of the safety equipment for workers.

Contractors have a great responsibility for construction safety and health. See Fig. 65-2. They develop company safety procedures. They try to make sure all safety practices are followed.

The government also has an important part in safety and health in construction. It sets guidelines that must be followed at work sites.

Safety on the Site

The construction site can be a dangerous place. People on the site must find out what dangers exist there. They must know how to avoid them.

Several conditions on the site could be dangerous.

- Working with tools.
- Using equipment.
- Handling materials.
- Working underground or high above the ground.
- Weather and air conditions.
- *Hazards* caused by worker carelessness.

Construction workers use tools. If used improperly, tools can strike or cut workers. Beginners must be taught to use tools safely. See Fig. 65-3.

Large, heavy equipment is used in construction. Such equipment must be operated correctly. If it is not, workers may be injured or killed. Equipment operators must learn how to handle equipment carefully. Safe work habits help prevent accidents. See Fig. 65-4.

Fig. 65-3
Used properly, tools make work much easier. Used improperly, they can be dangerous. Learn the safe way to use tools.

Fig. 65-4
Know-how and an alert, watchful attitude make this equipment operator a safe worker.

Heavy materials must be properly fastened or supported. Otherwise they may fall or collapse on workers. Great care must be taken to protect workers on the job. See Fig. 65-5.

The location of work may be dangerous. Some construction employees work in trenches or deep underground. Trenches and tunnels must be braced and shored to prevent cave-ins. See Fig. 65-6. Other workers perform tasks hundreds of meters above the ground. If safety precautions are not taken, they could easily fall from the structures. See Fig. 65-7.

Weather and air conditions, such as sun, rain, or wind, can cause accidents. Weather may affect the general health of a worker. And dirty air may affect the lungs and breathing. Weather and air quality can affect health and safety on the site. See Fig. 65-8.

Portland Cement Association

Fig. 65-5
Safely positioning heavy precast slabs of concrete requires care and teamwork.

Fig. 65-6
The walls of excavations must be properly supported or they may cave in on workers.

Aluminum Company of America

Fig. 65-7
Inspecting cables can be dangerous. But, with a safety belt attached to a cable, this inspector carefully performs the task.

Fig. 65-8
Construction work often creates dusty conditions. Breathing protection such as this mask may be needed.

Celanese Corporation

Fig. 65-9
The construction site is no place for carelessness. This ironworker attaches his safety belt when working up high.

Last, and very important, worker carelessness presents dangers. For example, workers may leave materials in the wrong place. Other people may fall over them. There are good reasons for work rules. They should be followed. See Fig. 65-9.

Worker Responsibilities

Safety begins with the worker. Each person must work with great care and use tools and equipment properly. All safety *precautions* must be followed.

Following safety rules, regulations, and practices may seem like a waste of time. For example, cleaning safety glasses may seem a bother. But workers must be able to see what they are doing. Otherwise they may be injured. Cleaning and wearing safety glasses is important. Safety glasses protect eyes. Remember, no one is safe from accidents. It is always best to be prepared in case something goes wrong.

Many types of safety equipment should be used on the construction site. See Fig. 65-10. For example, a hard hat should be *standard equipment*. And safety devices may be needed for the eyes, nose, and ears. Workers often need to wear special protective clothing and shoes.

Each worker should think of other workers' safety. Many accidents are caused because of carelessness. Every person must take great care not to injure others on the job.

Fig. 65-10
A hard hat, safety shoes, and eye and ear protectors are standard equipment for this construction worker.

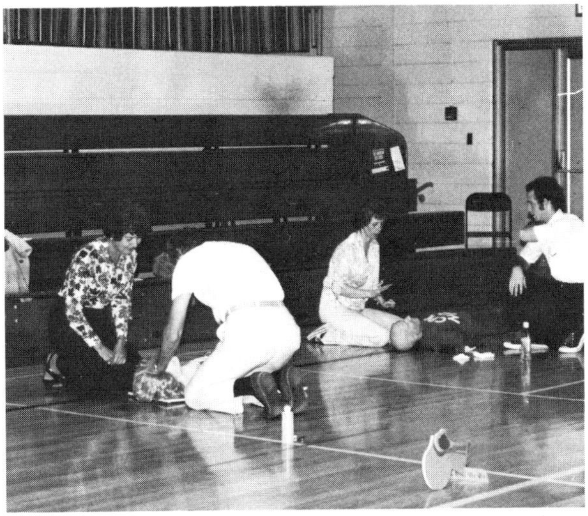

Fig. 65-11 American Red Cross, McLean County Chapter
Contractors often provide special safety classes for workers. Shown here is a CPR (cardiopulmonary resuscitation) instruction session.

Company Responsibilities

Construction contractors are very interested in safety and health practices. They encourage safety and health *awareness*.

Planned safety programs are very important to employers and employees. See Fig. 65-11. Contractors often hold meetings for workers on safe

Fig. 65-12
Construction areas must be clearly marked or fenced in. Onlookers must not get too close. They could be injured.

Fig. 65-13
Think safety, act safely, and always observe the safety rules of construction.

work practices. Contractors and employees work together on safety and health problems. They try to make work on the site healthful and safe.

Contractors are responsible for providing safety equipment for workers. Safety boots and other such items may be purchased by the workers.

Construction jobs interest the public. Spectators may want to stand close to the work to watch. Contractors need to provide protection for the public. See Fig. 65-12. No one should be allowed to get hurt.

Government Responsibilities

Government at all levels has established safety and health *regulations* (rules). The Occupational Safety and Health Act of 1970 was one federal law. Since 1970, states have passed similar *legislation*. Federal and state laws set down rules that must be observed.

The Occupational Safety and Health Administration (*OSHA*) is part of the U.S. Department of Labor. OSHA helps keep work sites safe. It sets standards for safety and good health. OSHA also sees that the standards are met.

OSHA inspects construction sites. It can give an employer a *citation of non-compliance* or even charge a fine. The citation is an official note that the company has poor safety and health regulations. It means that the company has not complied with (met) standards.

Most employers try on their own to meet OSHA standards. They understand that everyone benefits from following safety and health regulations. See Fig. 65-13.

You and Safety

When you are constructing, whether in school or at home, you must be responsible for your own safety. Injuries you get now may hurt for the rest of your life. But you can protect yourself and others. Following are a few precautions you can take now. There are many more for each activity you do.

- Handle materials carefully. Watch what you are doing at all times.
- Use tools with care. Sharp tools cut materials. They can also cut you.
- Wear eye protection when you are working in the shop.
- Do not breathe fumes or too much dust. Make sure the work area is well-ventilated or wear a safety mask.

Professional construction workers must follow many rules to be safe. This is also true for you. Think safety, act safely, and always observe safety rules.

Summary

Work in the construction industry can be dangerous. Precautions must be taken. These protect health and help prevent accidents. Everyone in construction must take special care to make work safe. And everyone will benefit.

You must take special care, too, both in school and at home. Following good safety practices will help keep you safe and healthy.

Terms to Know

awareness	OSHA
citation of non-compliance	precautions
hazards	regulations
health	safety
legislation	standard equipment

Think About It!

1. Why has government set up safety and health regulations?
2. Who loses when someone gets hurt on the job? Explain.
3. Listed below are five work situations. Select any two and tell
   - what safety equipment you think the worker would need.
   - what precautions you think must be taken.
   - what hazards a worker would need to be aware of.

   a. Nailing paneling on an interior wall.
   b. Finishing an exterior wall while roofers are working above.
   c. Laying pipe in a trench.
   d. Installing exterior panels to enclose the eighth floor of a building.
   e. Applying hot asphalt to a roof.
4. Name at least two safety rules that you should follow at school or home.

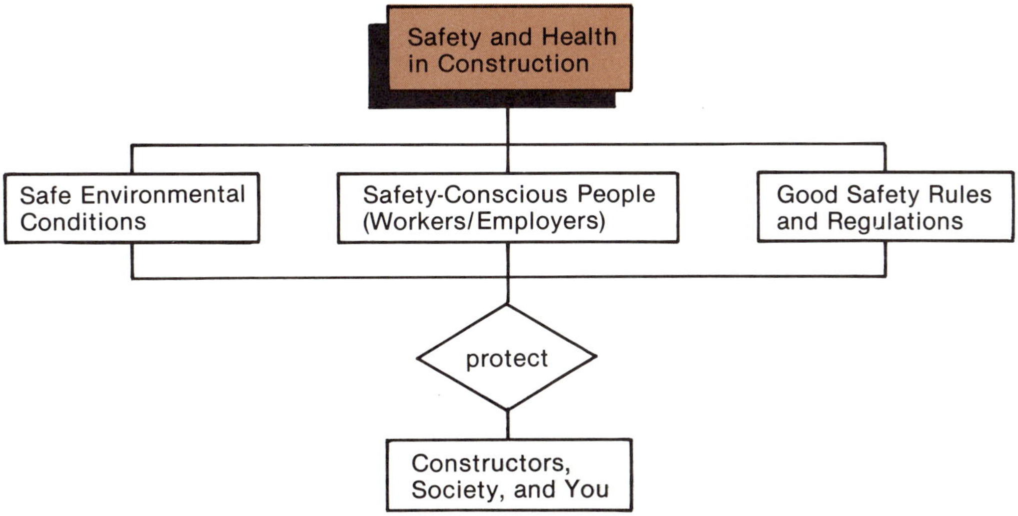

Index